# MODERN PHYSICS

## for Scientists and Engineers

Lawrence S. Lerner

*California State University*
*Long Beach*

**Jones and Bartlett Publishers**
*Sudbury, Massachusetts*

BOSTON ▪ LONDON ▪ SINGAPORE

*Editorial, Sales, and Customer Service Offices*
Jones and Bartlett Publishers
40 Tall Pine Drive
Sudbury, MA 01776
1-800-832-0034
508-443-5000

Jones and Bartlett Publishers International
7 Melrose Terrace
London W6 7RL
England

**Library of Congress Cataloging-in-Publication Data**
Lerner, Lawrence S., 1934–
    Modern physics for scientists and engineers / Lawrence
S. Lerner
        p.     cm.
    Includes index.
    ISBN 0-86720-487-7 (pbk.)
    1. Relativity (Physics)   2. Quantum theory.   I. Title.
QC173.55.L47   1995
539—dc20                                95-25596
                                             CIP

Photo and illustration credits appear on page A12, which constitutes a continuation of the copyright page.

Printed in the United States of America
99 98 97 96 95     10 9 8 7 6 5 4 3 2 1

*Chief Executive Officer*: Clayton E. Jones
*Chief Operating Officer*: Donald W. Jones, Jr.
*Vice President, Production and Manufacturing*: Paula Carroll
*Vice President and Editor-in-Chief*: David P. Geggis
*Director of Sales and Marketing*: Rob McCarry
*Marketing Manager:* Anne T. King
*Acquisitions Editor*: David E. Phanco
*Associate Editor:* Laura Maier
*Editorial Assistant*: Deborah L. Haffner
*Manuscript Editor*: Patricia Zimmerman
*Production Administrator*: Mary Sanger
*Senior Manufacturing Buyer*: Dana L. Cerrito
*Design/Layout*: Deborah Schneck
*Editorial Production Service*: Lifland et al., Bookmakers
*Illustrations*: Network Graphics, JAK Graphics Ltd.,
    Tech-Graphics
*Photo Research*: Kristi Heffron, Mary Sanger
*Cartoons:* Sidney Harris
*Typesetting/Separations*: University Graphics, Inc.
*Cover Design*: Mimi Ahmed
*Printing and Binding*: R. R. Donnelley & Sons Company
*Cover Printing*: Henry N. Sawyer Co., Inc.
*Cover Photograph*: © Comstock Inc., 1995.

To the memory of Isidor Lerner

*physicist, teacher, father*

# About the Author

Lawrence S. Lerner has been a member of the Department of Physics and Astronomy at California State University, Long Beach since 1969. There he has taught physics, history of science, and many interdisciplinary courses. Earlier, he held research positions at Hughes Aircraft, Hewlett-Packard, and Lockheed Aircraft Research Laboratories, working mainly in condensed-matter physics. A native of New York City, he attended Stuyvesant High School. After the eleventh grade, he entered the University of Chicago, where he earned his bachelor's, master's, and doctoral degrees.

Lerner has been honored with his university's Distinguished Teaching Award and three other awards of merit. He has been active for many years in efforts to encourage more young women to enter the sciences and engineering. He has served as Director of Cal State's General Honors Program and as Founding President of the campus Phi Beta Kappa chapter and is a member of the National Faculty for the Humanities, Arts, and Sciences.

Lerner's current research centers on history of science and science education. He is the author of more than a hundred scholarly papers, two textbooks, and a translation of a work by Giordano Bruno. Lerner and his wife, Dr. Narcinda R. Lerner, a research chemist, live in Woodside, California with their two Newfoundland dogs, and they attend the opera whenever they can.

# Contents

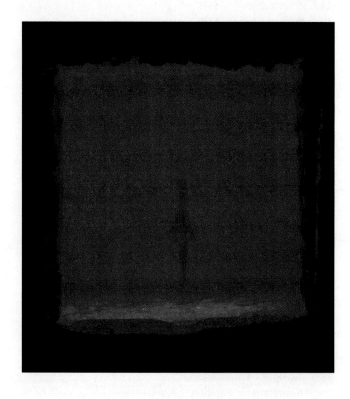

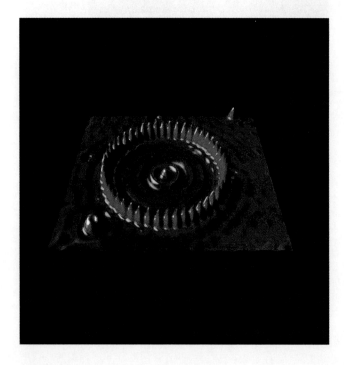

# Preface

*Modern Physics for Scientists and Engineers* continues my introductory text, *Physics for Scientists and Engineers*, for those students who require a solid grounding in the areas beyond classical physics.

I am gratified that the publisher has found it possible to offer both texts in a combined package. To accommodate the wide variation in the placement of relativity in physics courses, the chapters on relativity are included in both texts.

My motivation for writing *Modern Physics* was to extend to this vital subject the principles of friendly, clear explanation that guided the writing of *Physics*. The excitement and fascination of the worlds of the very small, the very fast, and the very large are known to every physicist; the challenge is to introduce them to the student in such a way as to mitigate the difficulties that accompany counterintuitive ideas.

As in *Physics*, I consider it paramount to provide students with clear, accessible explanations that they can read and study on their own. In my experience, students are not engaged by the abstract, formal style of traditional texts derived from scientific papers and advanced texts. I have found that a literate but colloquial, friendly style provides students with a much more accessible approach to difficult concepts.

A second objective is to help students build a solid conceptual mastery and a maturing physical intuition. Because students spend so much time working problems and preparing for exams that require solving similar problems, they can easily fall into the serious misconception that physics is nothing more than the set of techniques required to solve the end-of-chapter problems. Although deep insight into the nature of physics takes a long time to develop, both the classroom instructor and the textbook author can help students to step back once in a while to see what they have been doing and why it is worth doing. The ''Thinking Like a Physicist'' pieces distributed throughout the text are one mechanism for doing this.

My third aim is to provide problems—even the simpler ''confidence builder'' ones—that require the student to think before reaching for the nearest equation. I do not claim that my problems are completely original, but they do offer a fresh alternative to those found in other texts.

A few comments on how the problem sets are organized:

- Each problem set begins with Queries that are qualitative or semiquantitative in nature and serve both to enhance the students' awareness of their gain in understanding and to challenge them to attain still higher levels.

- The Queries are followed by quantitative Problems arranged in three groups. Group A comprises single-step problems designed to help build the student's confidence. The middle-level Group B problems require and help to build a broader understanding of the subject matter. Group C problems are more challenging.

- Most Queries and Problems are keyed to the relevant text section by an Arabic number given parenthetically immediately after the title. More general exercises are denoted by (*G*).

Because problems are the tool by which students reinforce their skills, a *Student's Study Guide and Solutions Manual* with a problem-solving emphasis has been developed to accompany the text. Full solutions with detailed explanations are provided for 25% of the end-of-chapter problems, and Hints and Strategies are offered for solving both Queries and Problems.

Teachers of physics know that physics is stimulating, challenging, beautiful, and satisfying, as well as broadly useful and professionally necessary. A major goal of every instructor is to help students in their efforts to achieve this perspective. My ultimate purpose in writing this book is to join in the efforts of both instructor and student, by making the most of the wondrous opportunities to combine the science of physics with the art of teaching.

## Supplements to the Text

*Instructor's Solutions Manual*, compiled by Dean Lee of Harvard University and a team of graduate students from Harvard, MIT, Boston College, and Boston University, contains complete solutions to all end-of-chapter problems. All solutions have been thoroughly checked by the author and a group of additional reviewers to ensure accuracy and consistency with text notation. The *Instructor's Solutions Manual* is also available on disk.

*Student's Study Guide and Solutions Manual*, prepared by Paul French of the State University of New York at Oneonta, summarizes important physics concepts, addresses common student misconceptions, identifies strategies for problem solving, and includes numerous worked examples as well as detailed step-by-step solutions to approximately half of the odd-numbered end-of-chapter problems from the text.

*Overhead Transparencies* and a *Test Bank* are available to qualified adopters. The set of full-color transparencies consists of key illustrations from the text. The complete test bank includes fill-in-the-blank, true/false, matching, multiple choice, and conceptual essay questions.

*Internet Access to Physics Problems*, provided by The University of Texas at Austin, is an interactive homework system that utilizes the World Wide Web as its platform. Adopters of the textbook may establish access to this system by submitting a class roster and making problem selections. More than 2000 algorithm-based problems are available, and the problem parameters vary so that each student must do independent work. All grading is done by computer, with results automatically posted on the World Wide Web. The system has been class tested at the University of Texas, with more than 100,000 questions answered electronically each month.

A demo using the World Wide Web interface is available at the **URL http://hw.ph.utexas.edu** by clicking on the **demo** link. Further information for instructors interested in using the system in their institutions is available from **see @ physics.utexas.edu.**

*f(g) Scholar* is a software package for science, engineering, and mathematics. This tools program, developed by Future Graph, Inc., has sophisticated spreadsheet, calculator, and graphing capabilities that may be used independently or in conjunction with one another. The spreadsheet provides access to more than 500 mathematical, statistical, scientific, engineering, and financial functions devised to make mathematical manipulations and calculations much easier. The calculator can define functions, and results can be easily graphed. *f(g) Scholar* is also an ideal tool for writing laboratory reports because it can analyze and graph data, allowing for the insertion of descriptive text and illustrations. This independent package is available at a special low student rate to adopters of the textbook.

*Experimental Research Notebook*, made of recycled carbonless paper, is ideal for compiling laboratory data; students can submit the original and keep a copy. The labs use a quadrille graph design, and the three-hole punched pages are numbered consecutively, making the manual easy to use. The notebook is available in 100 duplicate sets of labs, to meet individual course needs.

# Acknowledgments

I take seriously a maxim that I often quote to my students: Anyone can write; the real task of the serious writer is in the rewriting. Many skilled physicists with long teaching experience played an essential role in the transformation of a first-draft manuscript into this book through careful and creative reviews of all or part of the work at various stages of its development. Their comments have contributed much to the clarity and accuracy of the text, and I thank them sincerely:

Donald Abernathy, *DeVry Institute of Technology*
Harold D. Bale, professor emeritus, *University of North Dakota*
John Paul Barach, *Vanderbilt University*
Robert Bauman, professor emeritus, *University of Alabama*
Colston Chandler, *University of New Mexico*
Roger W. Clapp, *University of South Florida*
Sumner P. Davis, professor emeritus, *University of California, Berkeley*
Mildred S. Dresselhaus, *Massachusetts Institute of Technology*
Lowell Eliason, *California State University, Long Beach*
Raymond Enzweiler, *Northern Kentucky University*
Donald R. Franceschetti, *Memphis State University*
Anthony P. French, *Massachusetts Institute of Technology*
Paul French, *State University of New York at Oneonta*
Edward F. Gibson, *California State University, Sacramento*
Thomas P. Greenslade, *Kenyon College*
Patrick Hammill, *San Jose State University*
Edwin Kashy, *Michigan State University*
Isidor Lerner (deceased), *University of Illinois, Chicago Circle*
Robert H. Lieberman, *Cornell University*
Ralph Llewellyn, *Central Florida University*
Bo Lou, *Ferris State University*
Oscar Lumpkin, *University of California, San Diego*
David Markowitz, *University of Connecticut*
Charles E. McFarland, *University of Missouri, Rolla*
Jack Munsee, *California State University, Long Beach*
Austin Napier, *Tufts University*
Lawrence Pinsky, *University of Houston*
Wendell H. Potter, *University of California, Davis*
C. W. Price, *Millersville University*
John P. Ralston, *University of Kansas*
Sema'an I. Salem, *California State University, Long Beach*
Scott Shepard, *Baylor University*
Christopher M. Sorensen, *Kansas State University*
Julien C. Sprott, *University of Wisconsin–Madison*
Edwin F. Taylor, *Massachusetts Institute of Technology*
George Williams, *University of Utah*
Arthur M. Yelon, *Ecole Polytechnique, Montréal*

Professors Eliason, Munsee, and Salem were particularly helpful in testing parts of the manuscript in their classes.

A special note of thanks to Anthony P. French of The Massachusetts Institute of Technology, who offered encouragement as well as many constructive comments in the course of his review of the manuscript and who later scrutinized the entire text for accuracy, reading every page of proof. He truly merits the epithet ''lynx eyed'' that Galileo took so to heart.

Grateful as I am for all of these contributions to accuracy and clarity, I wish to emphasize that any remaining errors or obscurities are solely my responsibility. If you find errors, please let me know; they will be corrected. I warmly welcome any responses to the text from instructors and students alike, as well as any suggestions for future editions.

My very special thanks are due to Sidney Harris, who contributed cartoons that offer his unique combination of a whimsical world view and insight into the nature of the physical principles that his cartoons highlight.

I also thank Dean Lee of Harvard University for the careful preparation of solutions and answers to problems. He and a team of graduate students and faculty meticulously worked all of the problems in the text.

Writing a book begins as a solitary activity but expands rapidly into a team effort as the book approaches production and publication. It is impossible to overestimate the dedication, talents, and effort applied to this work by the experts who turn a draft manuscript and scribbly sketches into a finished book. Among the many members of this team, I am particularly grateful to those who interacted with me most directly and intensely: William Bennetta, Paula Carroll, Dave Geggis, Deborah Haffner, Kristi Heffron, Anne King, Sally Lifland, Gail Magin, Laura Maier, Irene Nunes, David Phanco, Mary Sanger, Deborah Schneck, Madge Schworer, and Patricia Zimmerman. Finally, I owe much to the physics publishing experience of Don Jones, Sr., and to Art Bartlett and Paul Prindle for their faith in this work.

# MODERN
# PHYSICS

# Part XI

# Relativity

Up to this point, our study of mechanics has been limited to the classical Newtonian realm—the realm in which objects that are neither too big nor too small move at speeds much less than the speed of light. In this context, "neither too big nor too small" means that the objects are no smaller than molecules (preferably a good deal bigger) and no bigger than the sun (preferably a good deal smaller).

But the world stretches beyond the boundaries of the classical realm. We will see explicitly that, far from being universal, the writ of classical physics does not run beyond its proper boundaries. You have already seen a few examples of how applying the laws of classical physics beyond these boundaries leads to unsatisfactory agreement between theoretical predictions and experimental results. Remember just two of many possible examples. Electrons in solids do not behave like charged billiard balls (Section 27.5), and the electrical resistance of solids reflects that fact. Electromagnetic waves cannot be consistently described in terms of the mechanical oscillation of an ether (Section 34.4), and attempts to describe the properties of the ether lead to contradictions.

In the remainder of this book, we will move decisively beyond the slow, middle-sized world where our senses operate directly and where we acquire our "common sense" into the realms of the very fast-moving, the very massive, and the very small, where classical physics fails to explain what is observed. Satisfactory explanation requires that we develop and apply one or more of several physical theories developed in the twentieth century. We will by no means abandon classical physics. Rather, we will set classical physics in the context of a much broader physics. In doing so, we will deepen our understanding of classical physics at the same time that we broaden our overall perspectives on the physical world.

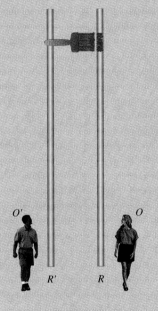

The realm where objects do not necessarily move slowly is the subject of *special relativity*, the main topic of Chapters 38 and 39. The realm of the very small is the subject of *quantum mechanics*, the topic of Part XII.

There is a price to be paid when we move away from the realm of our everyday experience. Our common sense fails because we have no familiar common-sense experience to use as a basis for new insights. We continue to rely on experimental evidence, as we have committed ourselves always to do in science. But the relevant experiments can no longer be simple extensions of familiar experience, as they often are in the classical realm. We will by no means abandon homely analogies, but we will become warier of taking them too seriously.

This price paid, we will have a reward: We will become familiar with a world far broader and far richer than that within the confines of everyday experience. We will doubt-

less retain a fondness for that "homeland"; people usually do. But we will gain confidence in a wider role as citizens of a vaster physical world.

# Relativistic Kinematics

———  Maxwell's equations predict that the speed of light is $c = 1/\sqrt{\mu_0 \epsilon_0}$, where the product $\mu_0 \epsilon_0$ can be measured in a laboratory. How, then, does a light wave appear to an observer who is moving with respect to the lab?

———  Careful investigation of this and related issues leads to the conclusion that the concepts of absolute time and absolute simultaneity, which are central to Newtonian mechanics, are untenable.

———  The relativistic world view reestablishes the foundations of physics on two postulates called the principles of relativity.

———  From these principles can be deduced a set of rules, called the Lorentz transformations, that enable each of two observers moving at constant velocity with respect to one another to use his own position and time measurements of an event to predict the position and time measurements for the same event made by the other observer.

*Left:* One of the unprecedentedly precise gyroscopes that will go into space on the Gravity Probe B satellite. The probe will provide a stringent test of general relativity theory. The quartz rotor is spherical within one wavelength of light and is coated with an equally precise thin niobium coating to render it superconducting. Part of the housing, which serves to spin the rotor up and contains suspension and orientation readout devices, is shown.

## SECTION 38.1

# Introduction

In the first twenty-two chapters of this book, we considered Newtonian (classical) mechanics and its application to a wide variety of subject areas. Some of these applications were straightforward and direct. Examples are the study of the motion of blocks sliding down inclined planes and of spinning wheels. Some applications, equally direct, required a mathematically more sophisticated approach. An example is the study of waves propagating through mechanical media. Some applications had sensational consequences. The study of celestial mechanics, for example, made it possible to understand the interactions and motions of heavenly bodies with exquisite precision centuries before it was physically possible to visit them. And some applications were in areas where it was far from obvious in advance that the science of mechanics was even relevant. An example is the study of heat and thermal interactions.

Given these far-flung triumphs—and there have been many others as well—it is small wonder that by the end of the nineteenth century scientists and others had tremendous confidence in the soundness of Newtonian mechanics. In many ways, this confidence is no less justified today. Physicists have not found, and do not expect to find, anything ''wrong'' with the way in which the subjects in this book up to this point have been treated.

Nevertheless, there is a glaring gap in the fundamental assumptions on which Newtonian mechanics is based. This gap remained hidden for a long time because it shows up directly and obviously only in the study of phenomena not accessible to experiment until after 1900. Even before 1900, indirect experimental evidence piled up gradually that, taken as a whole, suggested puzzling inconsistencies in the Newtonian world view.

As is usually true of fundamental inconsistencies, the inconsistencies underlying Newtonian mechanics crop up in several apparently independent places. One of these places lies at the boundary between mechanics and electromagnetism. It is tempting to unify electromagnetism with mechanics by ''explaining'' electromagnetic phenomena—notably electromagnetic wave propagation—in mechanical terms. But, as you saw in the final subsection of Section 34.4, the study of electromagnetism has led us beyond the bounds of mechanics. In particular, Maxwell's equations, which describe the propagation of light, do not rest on a mechanical model. Attempts to put Maxwell's equations on a mechanical basis result in a variety of paradoxes. Among them are the contradictory properties of the luminiferous ether, discussed in Section 34.4.

A number of first-rate physicists wrestled with these paradoxes for many years. It remained for Albert Einstein to cut the Gordian knot in 1905. His solution and its consequences are the subject of this chapter and the next one.

Einstein said years later that his intensive decade-long inquiry into the foundations of mechanics, leading to the development of the special theory of relativity, began when he was sixteen years old. At that time, he asked himself what a light wave, traveling

*Our everyday expectations and actions are built on common-sense kinematic rules. These rules work fine as long as $v \ll c$. But remember Einstein's aphorism: ''Common sense is nothing more than a deposit of prejudices laid down by the mind before you reach eighteen.''*

with speed $c$ with respect to one observer, would look like to a second observer who was riding along with it—that is, moving with speed $c$ with respect to the first observer. The question interested him because of the paradox it appeared to pose.

Here is one of several possible ways to look at the paradox. As you learned in Chapter 34, an electromagnetic wave is described by wave equations having the form of Equations 34.22 and 34.23:

$$\frac{\partial^2 \mathscr{E}_y}{\partial t^2} = \frac{1}{\mu_0 \epsilon_0} \frac{\partial^2 \mathscr{E}_y}{\partial x^2} \quad \text{and} \quad \frac{\partial^2 \mathscr{B}_z}{\partial t^2} = \frac{1}{\mu_0 \epsilon_0} \frac{\partial^2 \mathscr{B}_z}{\partial x^2}.$$

As with all wave equations, the constant on the right side of the equation is the square of the wave speed; according to Equation 34.24, we have

$$\frac{1}{\mu_0 \epsilon_0} = c^2.$$

But $\mu_0 \epsilon_0$ is a product of constants that can be determined by laboratory measurements made by an observer at rest with respect to the apparatus. An entirely analogous statement can be made about a stretched string: The quantity corresponding to $\mu_0 \epsilon_0$ is [tension]/[mass per unit length] $= v^2$ (Equation 21.7), and this quantity can be determined by laboratory measurements on the string—the *wave medium*. But, for the stretched string (or any other mechanical medium), $v$ is the speed of the wave *with respect to the medium, not the observer*. An observer who is moving with respect to the medium measures a wave speed different from $v$.

If we insist on keeping the analogy—that is, on considering the electromagnetic wave a form of mechanical wave—we must seek the medium in which electromagnetic waves propagate. Nineteenth-century physicists called this medium the *luminiferous ether*. We have already discussed some of the difficulties posed by the hypothetical properties of such an ether. But we now consider a deeper difficulty, one that does not depend on the peculiar properties of the ether. Suppose that $c$ is the speed of light with respect to the ether. An observer moving with speed $V$ in the same direction as a light pulse ought to observe its speed as $c - V$. Similarly, an observer moving with speed $V$ in the opposite direction ought to observe the speed of the pulse as $c + V$. Other directions of motion should result in other variations. But no such variations have ever been observed. In particular, the famous **Michelson-Morley experiment** of 1887 detected no effect due to the motion of an observer relative to the ether. (The details of the experiment are discussed in Problem 38.32. For a sketch of Michelson, see Section 36.2.)

Einstein's solution to this paradox is at once radical and conservative. It is radical because its fundamental postulates completely overturn many notions of what is "obvious" in the physical universe. It is also radical because it opens up to physical study horizons inconceivable in Newtonian terms. But it is conservative in a vital sense: It is compatible with Newtonian mechanics when it is applied to those areas where Newtonian mechanics is undeniably successful. That is why there is nothing "wrong" in the approach we have taken in the preceding chapters of this book.

---

**The Galilean Transformation**

To see what is logically wrong with Newtonian mechanics, we must ask, "What is basic to Newtonian mechanics?" In large measure, the answer is the set of equations to which Einstein gave the name **Galilean transformation** (in recognition of Galileo's pioneering contribution to the study of relative motion at constant velocity, later incorporated into Newton's laws as the law of inertia).

### *The Galilean Position Transformation*

Figure 38.1 shows observers $O$ and $O'$ located, respectively, at the origins of coordinate frames of the same names. The $x$ and $x'$ axes are collinear, and $O'$ moves along the $x$ axis at constant velocity $\mathbf{V} = V\hat{\mathbf{x}}$ with respect to $O$. (It is equally valid to say that $O$ moves along the $x'$ axis at constant velocity $-\mathbf{V}$ with respect to $O'$.) The corresponding axes $y$, $y'$ and $z$, $z'$ are parallel. As you saw in Section 3.1, we call such reference frames **inertial frames**. Observers $O$ and $O'$ make measurements of the motion of the same particle $P$. For convenience, the two observers synchronize their clocks, and they call the instant when the origins of their two frames coincide $t = t' = 0$.

**FIGURE 38.1** The coordinate frames and clocks of observers $O$ and $O'$. Both make repeated simultaneous observations on the position of $P$.

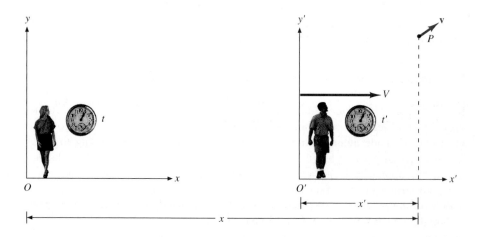

The two observers make simultaneous measurements on particle $P$. From the Newtonian point of view, the meaning of the word "simultaneous" is self-evident. Observers $O$ and $O'$ may agree in advance, for example, that each of them will measure the position of $P$ at $t = t' = 0$ s, 0.1 s, 0.2 s, and every 0.1 second thereafter.

Each observation results, for $O$, in a set of position coordinates $(x, y, z)$ and, for $O'$, in a set $(x', y', z')$. From these observations, $O$ deduces the velocity of $P$ as seen from her coordinate frame, which she calls $\mathbf{v}$. She also deduces the acceleration of $P$, which she calls $\mathbf{a}$. Similarly, $O'$ uses his data to deduce the velocity $\mathbf{v}'$ and the acceleration $\mathbf{a}'$, as seen from his coordinate frame.

It is evident from the start that the numerical values of the position $(x, y, z)$ measured by $O$ will not be the same as the values $(x', y', z')$ measured by $O'$. But that is not important. What *does* count is that $O$ and $O'$ be able to communicate about the position of $P$ in mutually intelligible language. That is, each must be able to check his or her own results against the other's, and the two observers must ultimately be able to agree that their data and their conclusions are mutually consistent. Or, what amounts to the same thing, $O$ must be able to *predict*, on the basis of her own measurements $(x, y, z)$, the values of the simultaneous measurements $(x', y', z')$ of $O'$—and vice versa. When $O$ and $O'$ come together again after the series of measurements is complete, they can check to see if they have been successful in their predictions.

All this explanation is prelude to a result that you may consider obvious and that all physicists prior to the twentieth century considered obvious. The result,

$$x' = x - Vt \quad \text{or} \quad x = x' + Vt, \qquad \text{(38.1a,b)}$$

$$y' = y \quad \text{and} \quad z' = z, \qquad \text{(38.1c,d)}$$

is called the **Galilean position transformation**. It is simply the set of rules that $O$ and $O'$ need to predict the results of each other's measurements on the basis of their own. Because we have chosen to consider the special case in which the motion of $O'$ relative to $O$ is along the $x$ axis, the time $t$ appears only in the equations relating $x$ and $x'$ and not in the corresponding equations for $y$ and $z$. But we have implicitly built another "obvious" assumption into Equations 38.1a and 38.1b. We should really write Equation 38.1b in the form $x = x' + Vt'$, because it is his own time measurement $t'$ that $O'$ uses in predicting the coordinate $x$ measured by $O$. But we have taken for granted the statement

$$t' = t \qquad \text{(38.1e)}$$

and have saved ourselves the trouble of writing the prime on $t$ in Equation 38.1b.

It would never occur to anyone to "save the trouble" of omitting the prime from the quantity $x'$ in the same equation; this would result in nonsense. It would seem that we have not thought twice about giving time a primacy over position in our logical framework. Although we are content to have transformation rules of position, we have asserted that time is *absolute*. This assertion is embodied in Equation 38.1e. Because this is one of the Galilean transformation equations, we say that time is **invariant** under the Galilean transformation. (Why do we not say the same thing about $y$ and $z$?)

## The Galilean Velocity Transformation

Example 38.1 suggests how you can begin with the Galilean position transformation (Equation 38.1a–e) and deduce the Galilean transformation equation for velocity.

---

**EXAMPLE 38.1**

Find the rules connecting the velocities $\mathbf{v}$ and $\mathbf{v}'$ of particle $P$, as measured by observers $O$ and $O'$ in Figure 38.1.

**SOLUTION:** You begin by writing the definition of velocity, $\mathbf{v}' = (v'_x, v'_y, v'_z)$, in terms of the fundamental measured quantities $x'$, $y'$, $z'$, and $t'$ (or $t$):

$$v'_x \equiv \frac{dx'}{dt'} = \frac{dx'}{dt}.$$

Because you want to predict the value of $v'_x$ in terms of the measurements made by $O$, you use Equation 38.1a to obtain

$$v'_x = \frac{d}{dt}(x - Vt) = \frac{dx}{dt} - \frac{d}{dt}(Vt).$$

We have stipulated that $\mathbf{V}$, the velocity of $O'$ with respect to

$O$, be constant, and so this equation simplifies to the form

$$v'_x = \frac{dx}{dt} - V.$$

The quantity $dx/dt$ is simply $v_x$, the $x$ component of the velocity of $P$ as measured by $O$. So you have

$$v'_x = v_x - V. \qquad \text{(38.2a)}$$

You can work backward in identical fashion to obtain the equivalent transformation in the opposite direction:

$$v_x = v'_x + V. \qquad \text{(38.2b)}$$

These are the results we discussed informally at the end of Section 38.1.

The other two components of the velocity are found even more directly. Using Equation 38.1c, you have $v'_y \equiv dy'/dt' = dy'/dt = dy/dt$, or

$$v'_y = v_y, \quad \text{and likewise} \quad v'_z = v_z. \quad (38.2\text{c,d})$$

Combining the results for the three components of the ve-

locity of particle $P$, you have

$$\mathbf{v}' = \mathbf{v} - V\hat{\mathbf{x}} \quad (38.3)$$

for the **Galilean velocity transformation**. How would you express $\mathbf{v}$ in terms of $\mathbf{v}'$?

The Galilean velocity transformation is a precise statement of something you have probably long taken for granted—that velocities add and subtract like vectors. Your experience outside as well as within the formal study of physics supports this view. For example, suppose you are driving on a straight road at 120 km/h. How fast are you overtaking a car going 100 km/h?

## Absolute Time and the Simultaneity Paradox

The problem with the Galilean transformations, as Einstein saw, was that the Newtonian view they underlie leads logically to a tangle of inconsistencies, one of which—the light-wave paradox—we mentioned in Section 38.1. As will soon become evident, the most glaring inconsistency has to do with the "obvious" assumption of Equation 38.1e, $t' = t$. But we cannot simply abandon the Galilean transformations without substituting something else. Einstein therefore proposed to substitute a set of transformation equations based on two fundamental assumptions, called the **principles of relativity**:

1. *The laws of physics are the same for all observers situated in inertial frames.*
2. *The speed of light in vacuum is the same for all observers and in all directions.*

The first principle may seem to be nothing new, but it is violated the moment we assume the presence of a luminiferous ether. To see this, let us rephrase the first principle in negative terms: No conceivable physical observation can be used to distinguish between one inertial frame and any other. But, if the ether exists, it must be at least conceivably detectable. (Otherwise there is absolutely no point in saying it exists.) If the ether is detectable, we can distinguish between a reference frame that is at rest with respect to the ether and another frame that is not. It follows logically that all inertial frames are not equivalent in every way, in contradiction to the first principle of relativity.

The restriction of the first principle of relativity to inertial frames limits us to a special case of relativity for which the mathematical treatment is fairly simple. The resulting limited theory is called the **special theory of relativity**. Removing the restriction leads to the **general theory of relativity**, originally proposed by Einstein in 1915. The general theory can be used to treat the case of two observers who accelerate with respect to one another. Although the general theory leads to profound insights into the relation between matter and space, it is mathematically and conceptually complex and we will not consider it in this book.

So confident have physicists become of the soundness of the second principle of relativity that it has been taken to underlie the definition of one of the SI base units.* As you learned in Section 35.2, the speed of light is *defined* to be $c \equiv 299\ 792\ 458$ m/s. The meter is a derived unit whose value is based on the value of $c$; similarly, the value of the constant $\epsilon_0$, the permittivity of free space, is based on the value of $c$.

---

*You should not take this confidence in the constancy of $c$ on faith but should reserve judgment until you have completed your study of this chapter.

## The Fallacy of Absolute Simultaneity

"Everyone knows" that either two events occur at the same time or they do not. We will now show that this common-sense view is incompatible with the second principle of relativity. Consider the following thought experiment, first suggested by Einstein in a slightly different form. In Figure 38.2, observer $O$ stands beside a railroad track. Next to her is a flashbulb. Also next to the track are two clocks. Clock 1 is located a distance $L$ up the track from $O$, and clock 2 is located an equal distance down the track. On the track is a boxcar with glass sides. Its length is $2L$, and at its ends are clocks $1'$ and $2'$. In the middle of the car stands observer $O'$. Next to him is a flashbulb. The two observers carefully synchronize all the clocks. After nightfall, the boxcar is moved up the track and accelerated to a speed $V$, which it maintains thereafter throughout the experiment. (You will probably find it helpful to imagine that $V$ is an appreciable fraction of the speed of light $c$, but the argument does not depend on this.)

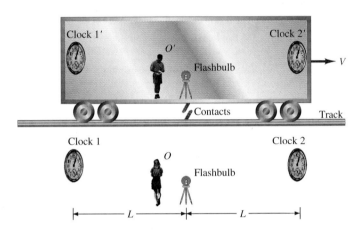

**FIGURE 38.2** A thought experiment to demonstrate that simultaneity cannot be absolute.

The car rushes past observer $O$. Because it is dark, neither observer can see any clock. However, there is an electrical contact mounted on the underside of the boxcar just beneath $O'$, and another one is mounted on the track next to $O$. Just as $O'$ passes $O$, the contacts touch and set off the flashbulbs, which are adjacent to each other at that instant. By the brief flash of light, each observer can read all four clocks (Figure 38.3).

What does $O$ see? For clocks 1 and 2 by the side of the track, the answer is simple. The light takes a time $\Delta t_{1,2} = L/c$ to reach clock faces 1 and 2. Although $O$ will not see the light reflected from the clock faces until a later instant, it is the time $\Delta t_{1,2} = L/c$ after the flash that she will read and record for each clock. We can say that $O$ perceives the two *events*—light reaches each of the two clock faces—as simultaneous events; that is what we commonly mean by simultaneity.

Observer $O$ also reads the faces of clocks $1'$ and $2'$ on the boxcar, which are moving with respect to her. But, while the light pulse is traveling from the flashbulbs toward

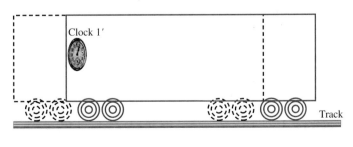

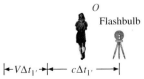

**FIGURE 38.3** Illustration for Example 35.3, showing the boxcar in black at the instant when the flashbulb flashes and in red at the instant when the light pulse reaches clock $1'$ and illuminates it.

clock $1'$, the clock is approaching $O$ and meets the light pulse part way. The light therefore illuminates clock face $1'$ at a time earlier than $L/c$ after the flash.

Next, $O$ turns her attention to the light flash heading toward clock $2'$. She argues that this flash has a distance greater than $L$ to travel, because clock $2'$ is receding from her. An argument similar to that in the preceding paragraphs shows that $O$ reads the interval of time from the flash of the bulb until the clock face is illuminated as *longer* than $\Delta t_{1,2} = L/c$.

Observer $O$ therefore makes the following record of the sequence of events: *First*, clock $1'$ was illuminated at $\Delta t_{1'}$; *then* clocks 1 and 2 were illuminated simultaneously at $\Delta t_{1,2}$; and *then* clock $2'$ was illuminated at $\Delta t_{2'}$.

So far so good. But now let us find what observer $O'$ sees. From his point of view, clocks $1'$ and $2'$ are motionless and equidistant. Because, according to the second principle of relativity, the speed of light is the same for him as it is for $O$, he will read clocks $1'$ and $2'$ simultaneously at a time after the flash given by $\Delta t'_{1',2'} = L/c$. Because he sees clock 2 to be approaching him, he sees it meeting the light pulse part way and so reads an earlier time $\Delta t'_2$ on it than on clocks $1'$ and $2'$. And, because $O'$ sees clock 1 to be receding from him, he reads it later than clocks $1'$ and $2'$, the time interval being $\Delta t'_1$.

Thus observer $O'$ makes the following record of events: *First*, clock 2 was illuminated at time $\Delta t'_2$; *then* clocks $1'$ and $2'$ were illuminated simultaneously at $\Delta t'_{1',2'}$; and *then* clock 1 was illuminated at $\Delta t'_1$.

When $O$ and $O'$ compare their records, they have a surprise waiting. Each one claims that the events involving his or her own clocks were simultaneous, and those involving the other's clocks were not. They cannot even agree as to the *sequence* of nonsimultaneous events. Observer $O$ claims that the order was $1' \rightarrow (1, 2) \rightarrow 2'$, but $O'$ claims it was $2 \rightarrow (1', 2') \rightarrow 1$. As you have just seen, this disagreement is a direct and necessary logical consequence of the second principle of relativity: The speed of light (in vacuum) is the same for all observers.

We are thus obliged to abandon the notion of absolute simultaneity and, with it, the idea of absolute time. The two are intimately connected, as Einstein pointed out in 1905:

> We have to take into account that all our judgments in which time plays a part are always judgments of simultaneous events. If, for instance, I say, "The train arrives here at seven o'clock," I mean something like this: "The pointing of the small hand of my watch to 7 and the arrival of the train are simultaneous events."

## THINKING LIKE A PHYSICIST

Well, you may argue, the boxcar paradox is just a special case depending on the special way in which the thought experiment was cooked up. After all, Newton made a distinction between "absolute, true, and mathematical time," which "of itself, and from its own nature, flows equably without relation to anything external," and "relative, apparent, and common time," which "is some sensible and external . . . measure of duration by the means of motion [for example of a pendulum], which is commonly used instead of true time." Perhaps all we have here is an example of what can happen if we use common time instead of absolute time.

But, if you take this point of view, you are stuck with a terrible difficulty. The Absolute Clock—if it indeed exists at all—is forever inaccessible, at least for purposes of physical measurement. We have to make do with "common time." We had therefore better formulate the laws of physics in terms of common time and banish absolute time from our

physics books.

Can we do so satisfactorily? The answer is unequivocally Yes! For, though $O$ and $O'$ will have to abandon the idea that simultaneity and the order of events are identical for both of them, each one can predict perfectly well what the other sees, in terms of what his or her own observations have been. This point of view is an extension of the one we took in formulating the Galilean transformations. In Figure 38.1, we do not insist that $O$ and $O'$ measure the same position coordinates for particle $P$ but only that each can predict what the other measures. It is this very breadth of vision—the ability to put ourselves in someone else's shoes—that we lose when we insist on the notion that our own point of view is somehow privileged. For then, the other person will inevitably take the same position about his own point of view, and the only result (besides perhaps a sense of smug satisfaction) will be the inability to communicate.

# The Time Dilation

Now that we have disposed of the idea of absolute simultaneity, we must find the rules that two observers such as $O$ and $O'$ must use to transform each other's time measurements into the terms proper to their own coordinate frames. In other words, we would like to know exactly how the rate at which a clock indicates the passage of time depends on its speed $V$ with respect to the observer who reads it.

The argument that follows is applicable to any clock at all, regardless of the details of its construction. But it is particularly simple to understand what happens if we use a **light clock**, described in Figure 38.4.

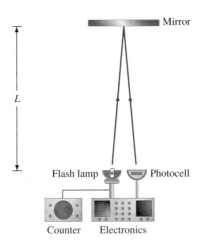

**FIGURE 38.4** A light clock. When the flash lamp flashes, the time required for the light pulse to reach the photocell via the mirror is $\Delta t_0 = 2L/c$. The arriving light pulse triggers the photocell, which activates the very fast electronic circuit to flash the lamp again. Thus the lamp flashes (the clock "ticks") at intervals $\Delta t_0$. The counter tallies the flashes and thus measures the time elapsed since the clock was started.

Suppose that observer $O$ has a light clock at rest in her reference frame. As she sees it, the time required for a light pulse to travel from the lamp to the mirror and back to the photocell depends only on the distance $L$ and the speed of light $c$. Specifically, the time elapsed is $\Delta t_0 = 2L/c$. (The subscript zero is a reminder that the speed of the clock is zero with respect to $O$.)

Now, suppose that observer $O'$, moving with respect to $O$ at velocity $\mathbf{V} = V\hat{\mathbf{x}}$, has an identical clock at rest in *his* coordinate frame. He will see his clock ticking at intervals $2L/c$, just as $O$ sees her own clock. What we want to know is *how each observer sees the other's clock*. Figure 38.5 shows how $O$ sees the clock of $O'$ as it passes her. Suppose that it flashes shortly before it passes her, when it is at $x = A$, and that the next flash takes place when the clock is at $x = C$. Suppose also that $A$ and $C$ are equidistant from $O$.

The first light pulse travels outward in all directions at a speed that must be $c$ for both observers, as required by the second principle of relativity. From the point of view of $O$, however, the entire clock has moved by the time the reflected light hits the photocell and triggers a new flash. Because the speed of light is the same in all directions, the light must have struck the mirror when the clock was at $B$, directly opposite $O$.

Let us give the name $\Delta t_V$ to the time interval between flashes, as $O$ sees the clock of $O'$. (The subscript $V$ is a reminder of the speed of the clock with respect to $O$.) Because of the way the clock moves with respect to $O$, the light from her point of view must make a round trip not of length $2L$ but of the greater length $2D$. She therefore sees the time interval between flashes to be

$$\Delta t_V = 2\frac{D}{c}. \tag{38.4}$$

We use Pythagoras's theorem to find the distance $D$ and thus to determine the value of $\Delta t_V$. The distance the clock travels between flashes is $V \Delta t_V$. Referring to Figure

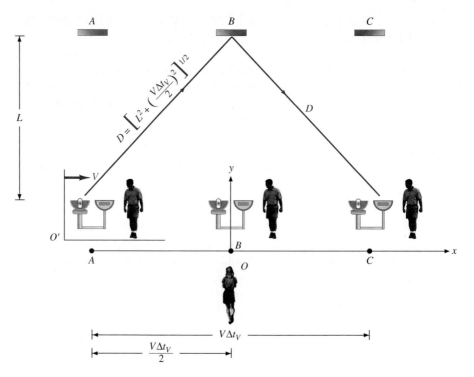

**FIGURE 38.5** The light clock moves at velocity $\mathbf{V} = V\hat{\mathbf{x}}$ with respect to $O$. $O'$ rides along with the clock.

38.5, you will see that the distance $D$ is the hypotenuse of a right triangle whose sides are $L$ and $V \Delta t_V/2$. Thus we have

$$D^2 = L^2 + \left(\frac{V \Delta t_V}{2}\right)^2.$$     **(38.5)**

We now square Equation 38.4 to obtain $\Delta t_V{}^2 = 4D^2/c^2$. Substituting the value of $D^2$ given by Equation 38.5 into this expression gives us

$$\Delta t_V{}^2 = \frac{4}{c^2}\left(L^2 + \frac{V^2 \Delta t_V{}^2}{4}\right),$$

or      $$\Delta t_V{}^2 = \frac{4L^2}{c^2} + \frac{V^2 \Delta t_V{}^2}{c^2}.$$

We bring all the terms in $\Delta t_V{}^2$ to one side of the equation and factor to obtain

$$\Delta t_V{}^2\left(1 - \frac{V^2}{c^2}\right) = \frac{4L^2}{c^2}.$$

Taking the square root of both sides of this equation, we have

$$\Delta t_V\sqrt{1 - \frac{V^2}{c^2}} = \frac{2L}{c}.$$     **(38.6)**

But the quantity $2L/c$ is just $\Delta t_0$, the time interval between flashes that $O$ measures for her own clock. We substitute this value into Equation 38.6 and solve for $\Delta t_V$, obtaining

$$\Delta t_V = \frac{\Delta t_0}{\sqrt{1 - \dfrac{V^2}{c^2}}}.$$     **(38.7)**

The interval $\Delta t_0$, read by an observer with respect to whom the clock is at rest, is called the **proper time**.* The interval $\Delta t_V$ is called the **dilated time**.

The denominator of Equation 38.7 is always equal to or less than 1, so $\Delta t_V \geq \Delta t_0$. That is, a clock always runs slower as seen by an observer with respect to whom it is moving than as seen by an observer with respect to whom it is at rest. Consequently, the relation expressed by Equation 38.7 is called the **relativistic time dilation**.

Be careful not to think that the time dilation is a result of the *clock*'s being "in motion" in some absolute sense. Motion has meaning only in terms of the clock *and* the observer. Observer $O'$ will see his clock as running at the rate $\Delta t_0$ and the clock of $O$ as running at the slower rate $\Delta t_V$, which is the opposite of what $O$ observes.

## Applications of the Time Dilation

As always in physics, the argument leading to the time dilation must ultimately be verified by experiment. Example 38.2 shows why time dilation is not observed in everyday events—that is, why we so easily take for granted the Newtonian assumption that time is absolute. We then consider experimental evidence that supports the validity of the time dilation and exemplifies the circumstances under which it must be taken into account.

---

**EXAMPLE 38.2**

A motorist sets his wristwatch by using the phone company time service. He then drives from a point in Reno to a point in Salt Lake City, a distance of 850.000 km, at a constant speed of 110.000 km/h, or 30.5556 m/s. Finally, he checks his watch again with the phone company time service. Will the time dilation be significant? That is, will he have to reset his watch if it can be read to the nearest second? Use $c = 2.997\,92 \times 10^8$ m/s.

**SOLUTION:** Imagine a pair of observers, one located in Reno and the other in Salt Lake City. Each has a wristwatch synchronized with the phone service. They can respectively observe the start and finish times of the trip and thus determine a proper time $\Delta t_0$ from their point of view:

$$\Delta t_0 = \frac{850.000 \times 10^3 \text{ m}}{30.5556 \text{ m/s}} = 27\,818.1 \text{ s}.$$

On the basis of this result, they can calculate the dilated time. This is the time interval they would find directly if the first observer could read the driver's watch as he passed the starting line and the other could read it again as he passed the finish line. According to Equation 38.7, they calculate

$$\Delta t_V = 27\,818.1 \text{ s} \times \frac{1}{\sqrt{1 - \dfrac{(30.5556 \text{ m/s})^2}{(2.997\,92 \times 10^8 \text{ m/s})^2}}}.$$

Because $[30.5556/(2.997\,92 \times 10^8)]^2 = 1.038\,85 \times 10^{-14}$, the quantity after the multiplication sign has the approximate value $(1 - 1 \times 10^{-14})^{-1/2}$. Use the approximation $(1 - x)^{-1/2} \simeq 1 + \frac{1}{2}x$, which is accurate for $x \ll 1$. This gives you the value $1 + 5 \times 10^{-15}$ for the factor that relates the dilated time to the proper time. That is, the time dilation amounts to only about 2 parts in $10^{14}$ of the total time. But the total time of the journey is about $3 \times 10^4$ s, and wristwatches can be read only within 1 s. Expressed as a fraction of the total time, this is about 3 parts in $10^5$. Thus the time dilation effect is entirely negligible in this case. Indeed, it would have to be a billion times as great to be barely detectable by using the wristwatches.

---

We now consider experimental evidence, first reported by Rossi and Hall in 1941, that supports the validity of the time dilation relation (Equation 38.7).

*Cosmic radiation*, consisting of very energetic nuclear particles, is present everywhere in the universe and constantly bombards the earth. A cosmic ray particle, de-

---

*Note that "proper" in this sense does *not* mean "correct," because neither time is correct or incorrect. Rather the meaning "intrinsic" is intended, because it is the time read by an observer at rest in the same coordinate system as the clock.

scending toward the earth through the upper atmosphere, collides with the nucleus of an atom of air. Such collisions occur for the most part in the region between 10 and 30 km above the earth's surface. A collision results in a shower of energetic particles of various types, the most common being the type called the *pion*. The pion, however, is unstable; before it has moved very far from its point of origin, it decays into another particle called a *muon*. Subsequently the muon also decays into still other particles. As with all unstable particles, the decay of the muon is random, and there is no way of telling when a particular muon will decay. Nevertheless, the lifetime can be characterized statistically. Given a large number of muons, the *half-life* is the time it takes for just half of them to decay. This half-life can be measured by using muons produced in the laboratory, which are moving with negligible speeds with respect to the observer; its value is $\tau_0 = 1.52 \times 10^{-6}$ s.

The muons produced by pion decay at the top of the atmosphere are moving with a speed close to the speed of light $c$. Let us assume for the moment that the speed is $V = c$. Then we can calculate the time it takes, from the point of view of an observer on the earth, for a muon born at altitude $h = 10$ km to reach the earth's surface if it happens to be heading straight down. We have for this descent time

$$t = \frac{h}{V} \simeq \frac{h}{c} = \frac{1.0 \times 10^4 \text{ m}}{3.0 \times 10^8 \text{ m/s}} = 3.3 \times 10^{-5} \text{ s.}$$

This is equal to about 22 times the half-life $\tau_0$. Because each half-life "sees" a reduction of the number of remaining muons by a factor 2, we would expect the number of muons reaching the earth's surface to be about 1 in $2^{22}$, or about 240 in each billion of those created at $h = 10$ km (and even fewer of those created at higher altitudes).

The muon production rate is known from high-altitude measurements. If the actual attrition rate were the one we have just calculated, the detection of a muon at the earth's surface would be a rare event. In fact, cosmic-ray muons are quite commonly observed at the earth's surface. Example 38.3 explains the reason why.

## EXAMPLE 38.3

Calculate the half-life $\tau_V$ of cosmic-ray muons traveling straight downward toward the earth from the point of view of an observer on the earth. From energy measurements, the average speed of cosmic-ray muons is known to be $V = 0.994c$.

**SOLUTION:** The muons constitute a clock, which is read by counting the fraction of the original muons that survive. But it is observed as a moving clock, and you therefore read the dilated time $\tau_V$. So you use Equation 38.7 to write

$$\tau_V = \frac{\tau_0}{\sqrt{1 - \left(\dfrac{0.994c}{c}\right)^2}} = \frac{\tau_0}{\sqrt{1 - (0.994)^2}} = 9.1\tau_0.$$

In contrast with the case considered in Example 38.2, the value of $V^2/c^2$ here is close to 1 rather than close to 0. As a result, the dilated half-life $\tau_V$ of the muons is 9.2 times the proper half-life $\tau_0$. Inserting the value $\tau_0 = 1.52 \times 10^{-6}$ s obtained from laboratory measurements, you have

$$\tau_V = 1.4 \times 10^{-5} \text{ s.}$$

You have seen that the descent time for the muon, as seen by the observer on the earth, is $t = 3.3 \times 10^{-5}$ s, or about $2.4\tau_V$. Consequently, between one-quarter and one-eighth of the muons produced at $h = 10$ km will survive to reach the earth's surface, and the fairly common penetration of cosmic-ray muons to the surface is explained.

The cosmic-ray experiment just described is far from being the only evidence in support of time dilation. As is often the case with such fundamental phenomena, evidence in its support arises directly or indirectly from a very wide variety of experimental results. Moreover, because we have obtained the time dilation by assuming the validity of the principles of relativity, such experimental results support those principles as well.

# The Fitzgerald-Lorentz Contraction

Once we have granted that the observed rate of a clock depends on its speed with respect to the observer, we are impelled to consider that other "old faithful," the length of an object. In Newtonian terms, measuring the length of an object implies measuring the positions of its two ends simultaneously. But we must now be suspicious of any measurement process involving the uncritical use of the concept "simultaneous."

We begin by reviewing a tacit assumption we made in deriving the time dilation equation. Specifically, we assumed that the distance $L$ between the photocell and the mirror in the light clock of Figure 38.5 was independent of the speed of the light clock with respect to the observer.

## *Length of an Object Perpendicular to Its Observed Velocity*

We now prove that it was correct to assume $L$ to be independent of $V$ when, as in the light clock, the length is perpendicular to the direction of motion. Suppose that observers $O$ and $O'$ have rods $R$ and $R'$ of equal length (Figure 38.6). A paintbrush is tied to $R'$ at a certain distance from its lower end. If $O$ and $O'$ bring the rods slowly together so that they are parallel with their lower ends coinciding, the brush will leave a paint mark on $R$ at a distance from its lower end equal to the distance of the paintbrush from the lower end of $R'$. Now let observer $O'$ with his rod move past $O$ and her rod with speed $V$ in the $x$ direction. As $R'$ goes by her, $O$ must see $R'$ either as having its length unaffected by its motion past her, or as being contracted, or as being expanded. Suppose that she sees it to be contracted. That is, suppose she observes its length to be less than what she measured it to be when it was at rest with respect to her. If this is so, the paintbrush will leave a mark on her rod $R$ at a point below the first mark.

FIGURE 38.6 Thought experiment to prove that the observed length of an object perpendicular to the direction of its velocity **V** with respect to the observer is independent of $V$.

Observer $O'$ must agree that the second mark is below the first mark. But from his point of view, rod $R'$ is still at rest, and its length must therefore be unaffected. He must conclude that rod $R$, which he sees to be moving past him with speed $V$, is *expanded*. But the first principle of relativity requires that the laws of physics be the same for both observers. It is not possible that relative motion results in an observed contraction for $O$ and an observed expansion for $O'$. The only remaining possibility is the conclusion that the length of an object is unaffected in a direction perpendicular to its motion relative to the observer. That is,

$$y' = y \quad \text{and} \quad z' = z \qquad \text{(38.8a,b)}$$

for motion in the $x$ direction, just as for the Galilean transformation.

## *Length of an Object Parallel to Its Observed Velocity*

When an object is observed in motion parallel to its length, the result is dramatically different from the perpendicular case. As shown in Figure 38.7, $O$ and $O'$ measure the length of a rod and find it to be $L_0$. Next, $O'$ moves away from $O$, taking the rod with him. He then comes past her with speed $V$, the rod being parallel to his motion in the $x$ (or $x'$) direction. Observer $O$ has a light clock. When the front end of the rod comes past her, she starts counting time. She continues to do so until the rear end of the rod passes her. If the passage of the rod requires a time $\Delta t_0$, she can say that its length $L_V$ is

$$L_V = V \, \Delta t_0. \qquad \text{(38.9)}$$

(Here again, the subscripts $V$ and 0 remind us of what is moving and what is at rest with respect to the observer.)

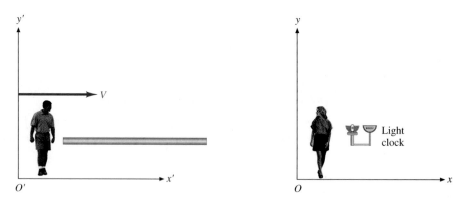

**FIGURE 38.7** Thought experiment to determine the observed length of a rod as a function of its speed $V$ with respect to the observer, when the rod is oriented parallel to the direction of motion.

Observer $O'$, who is at rest with respect to the rod but is moving with respect to the light clock, makes the corresponding measurement. But his task is different in detail from that of $O$. While she measures a moving rod with a stationary clock, he measures a stationary rod with a moving clock. His result, corresponding to Equation 38.9, is

$$L_0 = V \, \Delta t_V. \qquad \text{(38.10)}$$

Observer $O'$ is aware that the time interval $\Delta t_V$ is dilated with respect to the proper time $\Delta t_0$. He uses Equation 38.7 to reexpress Equation 38.10 in the form

$$L_0 = V \frac{\Delta t_0}{\sqrt{1 - \dfrac{V^2}{c^2}}}.$$

According to Equation 38.9, the quantity $V \, \Delta t_0$ in this equation is $L_V$, the length of the rod as seen by $O$. Making this substitution and solving for $L_V$ yields

$$L_V = L_0 \sqrt{1 - \frac{V^2}{c^2}}. \qquad \text{(38.11)}$$

The length $L_0$ measured by an observer with respect to whom the rod is at rest is called the **proper length**. The length $L_V$ is called the **contracted length**; the quantity $\sqrt{1 - V^2/c^2}$ is always equal to or less than 1, so $L_V \leq L_0$. Equation 38.11 expresses the **relativistic length contraction**, also called the **Fitzgerald-Lorentz contraction** or, most commonly, the **Lorentz contraction**. Working independently in the early 1890s, the Irish physicist George Fitzgerald (1851–1901) and the Dutch physicist Hendrik A. Lorentz (1853–1928) proposed the contraction as an artificial hypothesis to account for contradictory observations, some of which are considered in Section 38.1. The details of Fitzgerald's approach are the subject of Problem 38.33.

There is a nice complementarity between the points of view of $O$ and $O'$ in the thought experiment leading to Equation 38.11. For $O$ the length of the rod is contracted; for $O'$ the time is dilated. The complementarity is explored further in Example 38.4.

---

**EXAMPLE 38.4**

Consider the muon experiment of Example 38.3 from the point of view of a hypothetical observer who moves with a group of muons as they descend through the atmosphere with speed $V = 0.994c$. From the point of view of this observer, the muons are at rest. Their half-life therefore has the proper-time value $\tau_0 = 1.52 \times 10^{-6}$ s. How can any substantial proportion of them survive to reach the earth's surface?

**SOLUTION:** The key to the answer lies in the fact that, from the hypothetical observer's point of view, it is the earth (including its atmosphere) that is moving with speed $V$. The muon's path from the point of its creation to the earth's surface is contracted from its proper length $h = 10$ km; according to Equation 38.11, its contracted length is

$$L_V = L_0 \sqrt{1 - \frac{(0.994c)^2}{c^2}}$$

$$= 1.0 \times 10^4 \text{ m} \times 0.109 = 1100 \text{ m.}$$

This is 1/9.2 times the proper length, 10 km. Compare this result with that of Example 38.3, where you found that the dilated half-life of the muon was 9.2 times the proper half-life. The time required to traverse the distance $L_V$, measured by the observer's clock, is the proper time $\Delta t_0$. It is given by

$$\Delta t_0 = \frac{L_V}{V} \simeq \frac{L_V}{c}$$

$$= \frac{1100 \text{ m}}{3.0 \times 10^8 \text{ m/s}} = 3.7 \times 10^{-6} \text{ s.}$$

This time is $2.4\tau_0$. Compare with the result of Example 38.3, where the dilated time was found to be $2.4\tau_V$. Thus, no matter which coordinate frame you choose to calculate in, you will find the same proportion of the original muons reaching the earth's surface.

---

# The Lorentz Transformation

We are now prepared to derive from the principles of relativity the set of equations that constitute the relativistic equivalent of the Galilean transformations and thus to fulfill the task proposed at the beginning of Section 38.3.

## *The Position Transformation*

Consider again the situation depicted in Figure 38.1. Again the observer $O'$, fixed in the $O'$ coordinate frame, moves past observer $O$ at a velocity that she observes to be $\mathbf{V} = V\hat{\mathbf{x}}$. Again we choose the time $t = 0$ to be that when the two frames coincide. For the moment, let particle $P$ be at rest in the $O'$ frame (that is, set $\mathbf{v} = 0$), so the particle, too, moves past $O$ at velocity $\mathbf{V}$.

At time $t = 0$, $O$ measures the $x$ coordinate of the position of $P$. She knows that this is a contracted length $x_{t=0} = x_V$, and she knows that, if $O'$ measures the $x'$ coordinate of $P$ in his frame, he will obtain a proper length $x' = x_0$. The lengths measured by the two observers are related by Equation 38.11, and so we have

$$x_{t=0} = x' \sqrt{1 - \frac{V^2}{c^2}}.$$

The value of $x'$ does not change with time because particle $P$ is at rest with respect to $O'$. But the primed frame, and everything at rest with respect to it, moves away from $O$ with speed $V$. At an arbitrary time $t$, the distance from the origin of $O$ to the origin of $O'$ is $Vt$. Thus, for $O$, the $x$ coordinate of the particle $P$ at any time is

$$x = x_{t=0} + Vt,$$

or

$$x = x'\sqrt{1 - \frac{V^2}{c^2}} + Vt.$$

We solve this equation for $x'$ and obtain

$$x' = \frac{x - Vt}{\sqrt{1 - \dfrac{V^2}{c^2}}}. \qquad (38.12a)$$

This relativistic equation corresponds to the Galilean Equation 38.1a. Note that there is really no inconsistency between the two if we restrict attention to everyday observations in which all speeds are very much smaller than the speed of light. If $V \ll c$, the value of the denominator in Equation 38.12a is approximately equal to 1 and we have $x' = x - Vt$, which is Equation 38.1a.

The transformation equation inverse to Equation 38.12a is

$$x = \frac{x' + Vt'}{\sqrt{1 - \dfrac{V^2}{c^2}}}. \qquad (38.12b)$$

The derivation is the subject of Problem 38.29. Equation 38.12b reduces to Equation 38.1b for $V \ll c$.

For relative motion along the $x$ and $x'$ axes only, the transformations for the $y$ and $z$ coordinates follow directly from the argument at the beginning of Section 38.5. They are

$$y' = y \quad \text{and} \quad z' = z. \qquad (38.12c,d)$$

These are identical with the corresponding Galilean transformation equations (Equations 38.1c,d).

Finally, we need the transformation equation for time. We begin by noting that Equations 38.12a and 38.12b contain $t$ and $t'$, respectively. We substitute the values of $x'$ given by Equation 38.12a into Equation 38.12b. This gives

$$x = \frac{x - Vt}{1 - \dfrac{V^2}{c^2}} + \frac{Vt'}{\sqrt{1 - \dfrac{V^2}{c^2}}}.$$

We next multiply through by the denominator of the first term on the right side to obtain

$$x\left(1 - \frac{V^2}{c^2}\right) = x - Vt + Vt'\sqrt{1 - \frac{V^2}{c^2}}.$$

Collecting terms in $x$ and dividing through by $V$, we have

$$-\frac{xV}{c^2} = -t + t'\sqrt{1 - \frac{V^2}{c^2}}.$$

We now solve for $t'$, obtaining the desired result

$$t' = \frac{t - \dfrac{xV}{c^2}}{\sqrt{1 - \dfrac{V^2}{c^2}}}. \qquad (38.12e)$$

Here again, we obtain $t' = t$ for $V \ll c$. This says that the concept of absolute time that underlies the Galilean transformations and all of Newtonian mechanics *is* valid if all speeds of interest are so small that we can just as well regard the speed of light to be infinite. Can you see why this is so?

The transformation equation inverse to Equation 38.12e is

$$t = \frac{t' + \dfrac{x'V}{c^2}}{\sqrt{1 - \dfrac{V^2}{c^2}}}. \tag{38.12f}$$

The derivation is the subject of Problem 38.30.

## *The Velocity Transformation*

Suppose now that the particle $P$ in Figure 38.1 is no longer at rest in the $O'$ frame but is moving parallel to the $x'$ axis (and the $x$ axis as well) in such a way that observer $O'$ measures its velocity to be $v'_x$. Observer $O$ also measures the velocity of the particle and finds it to be $v_x$. What is the relation between $v_x$ and $v'_x$? (By now you should be wary enough not to respond too quickly that $v'_x = v_x - V$.)

In the $O$ frame, the velocity is defined to be

$$v_x \equiv \frac{dx}{dt}.$$

Similarly, in the $O'$ frame, the velocity is defined to be

$$v'_x \equiv \frac{dx'}{dt'}.$$

Equation 38.12a gives an expression for $x'$ in terms of $x$ and $t$, and Equation 38.12e gives an expression for $t'$ in terms of $x$ and $t$. By differentiating these two equations, we will be able to express both $dx'$ and $dt'$ in terms of $dx$ and $dt$. Having done this, we will be able to express $v'_x$ in terms of $v_x$, as desired.

Beginning with Equation 38.12a, we have

$$dx' = \frac{dx - V\,dt}{\sqrt{1 - \dfrac{V^2}{c^2}}}. \tag{38.13}$$

And differentiating Equation 38.12e gives

$$dt' = \frac{dt - \dfrac{V}{c^2}\,dx}{\sqrt{1 - \dfrac{V^2}{c^2}}}. \tag{38.14}$$

We divide the first of these equations by the second to obtain

$$v'_x = \frac{dx'}{dt'} = \frac{dx - V\,dt}{dt - \dfrac{V}{c^2}\,dx}.$$

We now use the definition $dx/dt \equiv v_x$ and find

$$v'_x = \frac{v_x - V}{1 - \dfrac{Vv_x}{c^2}}. \tag{38.15a}$$

An argument in the same spirit leads to the transformation equations for the perpendicular velocity components. The result is

$$v_y' = \frac{\sqrt{1 - V^2/c^2}}{1 - \dfrac{Vv_x}{c^2}} v_y \quad \text{and} \quad v_z' = \frac{\sqrt{1 - V^2/c^2}}{1 - \dfrac{Vv_x}{c^2}} v_z. \qquad \textbf{(38.15b,c)}$$

The derivation of Equations 38.15b and 38.15c is the subject of Problem 38.40. Equations 38.15a,b,c constitute the **Lorentz velocity transformation**. There is good reason for the name. In the 1890s, Lorentz wrestled with the same paradox that puzzled the younger Einstein—how could several observers moving with respect to one another apply Maxwell's equations to describe the propagation of an electromagnetic wave? Or, to put it in precise terms, how could Maxwell's equations be made *invariant* under coordinate transformations? They were certainly *not* invariant under the Galilean transformation. Lorentz showed that the transformation equations named after him would satisfy the invariance requirement but was unable to find a satisfactory physical explanation that would justify their use in preference to the Galilean transformation equations.

Equation 38.15a reduces to the familiar Galilean result $v_x' = v_x - V$ (Equation 38.2a) in the case $v_x \ll c$ or $V \ll c$. Equations 38.15b and 38.15c reduce in similar manner to Equations 38.2c and 38.2d, $v_y' = v_y$ and $v_z' = v_z$.

Equation 38.15 implies that velocities comparable to the speed of light do not simply "add," as small velocities do in the Galilean limit. This statement runs counter to everyday experience and is therefore difficult for many persons to accept at first. But everyday experience does not include observation of objects moving at speeds comparable to $c$!

## EXAMPLE 38.5

Luke Skywalker is on an urgent interstellar mission when he suddenly sees an Empire battlecruiser headed straight for him from due galactic north at speed $0.9990c$. Before he can take evasive action, he sees a second Empire battlecruiser coming straight at him from due galactic south, also at speed $0.9990c$. He deftly executes a sharp turn and watches with understandable if not excusable satisfaction as the two less maneuverable cruisers collide with one another. Just before the collision, what is the speed of the first cruiser as seen by the unhappy captain of the second?

**SOLUTION:** You begin by drawing a sketch of the situation like that in Figure 38.8. Let Luke be observer $O$, as shown. Then cruiser 1 takes the role of particle $P$ and has velocity $v_x = -0.9990c$, because it is moving toward Luke in the negative $x$ direction. Observer $O'$ must then be the captain of cruiser 2. Because $V$ is the velocity of the primed frame seen from the unprimed frame, you have $V = 0.9990c$. You can now employ Equation 38.15:

$$v_x' = \frac{-0.9990c - 0.9990c}{1 - \dfrac{-0.9990c \times 0.9990c}{c^2}}$$

$$= \frac{-2 \times 0.9990c}{1 + (0.9990)^2} = -0.9999c.$$

**FIGURE 38.8**

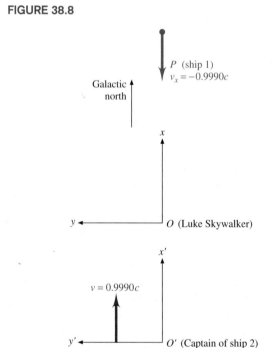

The result of the fanciful space story in Example 38.5 is instructive. It is indeed true that the captain of cruiser 2 sees cruiser 1 approaching him at a speed greater than the speed Luke observes. But the speed, though it is very close to the speed of light $c$, is still less than $c$. It is certainly not $2c$. Even if the two given speeds were $v_x = -c$ and $V = c$, the result would be $v'_x = c$, as you can quickly verify by calculation. *The speed of light constitutes an absolute upper limit on the speed of physical objects in the universe*, regardless of the observer. You will see in Chapter 39 that there are dynamical as well as kinematic reasons for this.

In hindsight, Einstein's solution to the electromagnetic wave paradox seems simple: As Lorentz and others already knew, the Galilean transformations cannot be applied to light waves. Instead of trying to ''fix things up,'' Einstein elevated this property of light to a fundamental law of nature. According to the second principle of relativity, *all* observers see the speed of light as $c$. We *begin* with this principle and derive the laws of kinematics—that is, the Lorentz transformation equations—in a straightforward way, as we have done in this section and Section 38.5. One *cannot* ''ride along with a light wave''!

---

## Symbols Used in Chapter 38

| | | | |
|---|---|---|---|
| $c$ | speed of light | $\mathbf{v}, \mathbf{v}'$ | velocity of an object as observed by $O$, $O'$ |
| $D$ | half-length of light path of light clock, as seen by an observer with respect to whom the clock is in motion | $v_x, v_y, v_z$ | components of $\mathbf{v}$ |
| | | $v'_x, v'_y, v'_z$ | components of $\mathbf{v}'$ |
| $L$ | length of an object | $x, y, z$ | position coordinates measured by $O$ |
| $t$ | time as measured by observer $O$, at rest in frame $O$ | $x', y', z'$ | position coordinates measured by $O'$ |
| | | $_0$ (subscript) | labels quantities measured by observer at rest with respect to object observed |
| $t'$ | time as measured by observer $O'$, at rest in frame $O'$ | $_V$ (subscript) | labels quantities measured by observer in motion at speed $V$ with respect to object observed |
| $\Delta t_j, \Delta t_{j'}$ | time interval indicated by clock $j$ or $j'$, as observed by $O$ | | |
| $\Delta t'_j, \Delta t'_{j'}$ | time interval indicated by clock $j$ or $j'$, as observed by $O'$ | $\tau_0, \tau_V$ | proper half-life, dilated half-life of a muon |
| $V, \mathbf{V} = V\hat{\mathbf{x}}$ | constant velocity of frame $O'$, as observed by $O$ | | |

---

## Summing Up

When we attempt to interpret a number of crucial experiments within the framework of Newtonian mechanics, the result is a series of paradoxes. Resolution of these paradoxes begins with a careful examination of the concepts of absolute time and its twin, absolute simultaneity, on which the Galilean transformations are based.

A set of thought experiments leads to the conclusion that Equation 38.1e, $t' = t$, is untenable and that the Galilean transformations are therefore not universally valid. This conclusion in turn mandates a reformulation of the transformation equations for the kinematic quantities time, displacement, and velocity. Such a set of equations is necessary if observers who are moving with respect to one another are to be able to interpret each other's observations. The reformulation is based on two fundamental assumptions, called the **principles of relativity**, which replace the assumptions underlying Newtonian mechanics.

The resulting theoretical structure is called the **theory of relativity**. The **special theory of relativity** is limited to the case in which each observer sees the other to be in an inertial frame.

A thought experiment leads to a quantitative expression for the time dilation, Equation 38.7:

$$\Delta t_V = \frac{\Delta t_0}{\sqrt{1 - \dfrac{V^2}{c^2}}}.$$

A second thought experiment leads to the conclusion that measurements of the $y$ and $z$ dimensions of an object are independent of its velocity with respect to the observer. A third and final thought experiment leads to Equation 38.11, the length contraction in the $x$ direction:

$$L_V = L_0\sqrt{1 - \frac{V^2}{c^2}}.$$

The predictions embodied in Equations 38.7 and 38.11 are amply confirmed by direct and indirect experimental evidence.

The **Lorentz transformations** fulfill the mandate for re-formulation of kinematics on the basis of the principles of relativity. These equations, together with the corresponding Galilean transformation equations for purposes of compari-son, are given in Table 38.1. A particularly dramatic and important implication of Equation 38.15a is that no observer can ever see a physical object moving faster than the speed of light.

**TABLE 38.1 The Galilean and Lorentz Transformations**

| Galilean Transformations | | Lorentz Transformations | |
|---|---|---|---|
| $x' = x - Vt$ | (38.1a) | $x' = \dfrac{x - Vt}{\sqrt{1 - V^2/c^2}}$ | (38.12a) |
| $y' = y$ | (38.1c) | $y' = y$ | (38.12c) |
| $z' = z$ | (38.1d) | $z' = z$ | (38.13d) |
| $t' = t$ | (38.1e) | $t' = \dfrac{t - xV/c^2}{\sqrt{1 - V^2/c^2}}$ | (38.12e) |
| $v'_x = v_x - V$ | (38.2a) | $v'_x = \dfrac{v_x - V}{1 - Vv_x/c^2}$ | (38.15a) |
| $v'_y = v_y$ | (38.2c) | $v'_y = \dfrac{\sqrt{1 - V^2/c^2}}{1 - Vv_x/c^2}\, v_y$ | (38.15b) |
| $v'_z = v_z$ | (38.2d) | $v'_z = \dfrac{\sqrt{1 - V^2/c^2}}{1 - Vv_x/c^2}\, v_z$ | (38.15c) |

## KEY TERMS

### Section 38.1  Introduction
Michelson-Morley experiment

### Section 38.2  The Galilean Transformation
inertial frame ▪ Galilean position transformation, velocity transformation ▪ invariant

### Section 38.3  Absolute Time and the Simultaneity Paradox
principles of relativity ▪ special theory of relativity, general theory of relativity ▪ absolute time, absolute simultaneity

### Sections 38.4 and 38.5  The Time Dilation and the Fitzgerald-Lorentz Contraction
light clock ▪ proper time, dilated time ▪ proper length, contracted length ▪ relativistic time dilation ▪ relativistic length contraction

### Section 38.6  The Lorentz Transformation
Lorentz position and velocity transformations

## Queries and Problems for Chapter 38

## QUERIES

**38.1** *(3) Relatedness of principles.* Show how the second principle of relativity can be derived logically from the first. Specifically, if the second principle were not true, how would this affect the first principle?

**38.2** *(3) Special circumstances.* Two light clocks are located at opposite ends of a spaceship. Observer $O'$, on the ship, sees that they flash simultaneously. Observer $O$, on the ground, also observes the flashes as the spaceship passes by at high speed. Are there any circumstances under which she also sees the two flashes simultaneously?

**38.3** (3) *Fast Narcissus.* As a young man, Einstein imagined a runner who could run at a speed only slightly less than $c$. The runner holds a hand mirror in front of himself as he runs. Explain what the runner sees in the mirror.

**38.4** (4) *Stout fellows!* A spaceship is prepared to break the record for the running-start time from Venus to Mars. A clock record is to be presented to the Guinness Book of World Records for verification. Is it better to use a clock on the spaceship or two synchronized clocks, one on Venus and the other on Mars?

**38.5** (4) *Track star.* Comment on this limerick:

> There was a young lady named White,
> Who could travel much faster than light.
> She started one day
> In a relative way,
> And arrived on the previous night.

**38.6** (5) *You can't get there from here!* A friend offers the following argument to show the limited utility of manned interstellar travel over large distances: A one-way trip that takes more than about 25 years of spaceship time is impractical because the crew would likely die of old age before they could return, and thus they could not tell what they saw; one might as well just send cameras and other recording devices. Be-

cause the spaceship cannot travel faster than $c$, that limits all practical manned voyages to a radius of about 25 light-years. But there are very few stars within 25 light-years. So why bother? Criticize this argument.

**38.7** (5) *Einstein was wrong!* Here is a standard "disproof" of relativity offered by various persons hoping to win a Nobel Prize by showing that Einstein was wrong. A very rich man buys a very fast, very big car whose length is 12 m. His garage is only 10 m deep, but he wishes to put the car into the garage, however briefly. He installs fast-acting electric doors on both ends of the garage. While he stands next to the garage, he has his chauffeur approach the garage at such a high speed that the car is contracted to a little less than 10 m. He opens the front door, leaving the back door closed. Just as the car comes fully into the garage, he shuts the door. Thus both doors are closed at the same time, confining the car. Then he quickly opens the back door in time for the car to emerge. But the chauffeur argues that it was the garage, not the car, that was contracted. Thus, he claims, the owner could not have closed the doors, and Einstein was wrong!

Use the Lorentz transformation to reconcile what the owner sees with what the chauffeur sees, and show that there is no inconsistency.

**38.8** (G) *Act of Congress?* Suppose that the speed of light were reduced to 30 m/s. Consider a number of commonplace experiences, and describe how they would be changed.

# PROBLEMS

## GROUP A

**38.1** (4) *Speedometer check zone?* Observer $O'$ has a light clock that flashes once per second. He moves along a special track at speed $V$ m/s. Set up along the track are photocells that can detect and record the flashes of the clock. Checking later, $O'$ finds that the distance between photocells that have detected flashes is $2V$ m. Calculate $V$.

**38.2** (4) *Another point of view.* By working out the derivation leading to Equation 38.7 from the point of view of $O$, show explicitly that $O'$ sees the clock of $O$ running slower than his own, contrary to what $O$ sees.

**38.3** (4) *Longevity on the fast track.* The half-life of a stream of particles in an accelerator beam is measured and found to be 7.5 times that of the same particles observed at rest. How fast are the particles moving? Express your answer as a fraction of the speed of light and in m/s.

**38.4** (4) *Real commitment.* Astronomers detect a civilization on the other side of our galaxy, a distance of $10^5$ light-years away. In an attempt to make person-to-person contact, a spaceship is sent off with a crew of teenagers. If the crew are to have a reasonable chance of reaching their goal before they die of old age, approximately how fast must the ship go?

**38.5** (4) *Stretching things out.* When observed at a speed of $2.90 \times 10^8$ m/s, a beam of particles has half-life 28.2 μs. What is their half-life when observed at rest?

**38.6** (4) *Easy as* $\pi$. Pions have a proper half-life $\tau = 1.8 \times 10^{-8}$ s. A beam of pions passing through a set of detectors is observed to have a half-life of $3.1 \times 10^{-8}$ s. **(a)** What is the speed of the pions? **(b)** What is the length of the track of a particular pion whose life happens to be just equal to the half-life?

**38.7** (4) *The twin paradox, I.* Two precise, identical atomic clocks are synchronized. One of them is loaded on a commercial airplane, which flies in one direction, turns around, and returns to the starting point. The total flight time is 10 hours, and the speed of the plane is 300 m/s. **(a)** What is the difference in the readings of the two clocks? **(b)** Which clock reads the greater elapsed time? Neglect the small time during which the plane is accelerating. [Hint: To carry out the calculation, you will need to take advantage of the (very good) approximation $V \ll c$.]

**38.8** (4) *The twin paradox, II.* Olivia and Oliver are twins. Olivia becomes an astronaut and, on their twenty-first birthday, bids her brother a fond farewell and makes a voyage to a nearby star. The journey takes 7 years by the clock on the spaceship. During most of the voyage, the ship travels at the constant speed $0.96c$ with respect to the earth. (Neglect the relatively brief periods of acceleration and deceleration.) After a brief period of data gathering at the star, Olivia reverses her course and returns home. When she arrives, her age is close to thirty-five. She hastens to visit her twin brother, Oliver. How old is he?

**38.9** *(4) Fast, faster, fastest . . .* **(a)** Show that, for speed $V$ close to $c$, you can use the approximation

$$\frac{1}{\sqrt{1 - V^2/c^2}} = \sqrt{\frac{c}{2\delta}},$$

where $\delta \equiv c - V$. What is the half-life of muons traveling past the observer at **(b)** $0.99c$, **(c)** $0.999c$, and **(d)** $0.9999c$?

**38.10** *(4) . . . and still faster!* How fast is a beam of muons moving with respect to an observer who measures their half-life to be 1 s? Express your result in terms of the quantity $\delta$ defined in Problem 38.9.

**38.11** *(4) It's all a plot! I.* Make a plot of $1/\sqrt{1 - V^2/c^2}$ as a function of $V/c$. (Hint: What is the range of interest for $V/c$?)

**38.12** *(4) It's all a plot! II.* In doing Problem 38.11, plotting becomes awkward for values of $V/c$ approaching 1. To get around this difficulty, repeat Problem 38.11, using semilog paper. Alternatively, plot $\log 1/\sqrt{1 - V^2/c^2}$ as a function of $V/c$.

**38.13** *(5) It's all a plot! III.* Make a plot of $\sqrt{1 - V^2/c^2}$ as a function of $V/c$.

**38.14** *(5) Mama's little baby loves . . .* An angry baker hurls a *baguette* of French bread 1 m long at your head. It misses narrowly. As it flies past you, you observe that its length is only 0.5 m. How fast is it going?

**38.15** *(5) Deflation?* Observer $O'$ moves past $O$ with speed $0.62c$, holding a rod parallel to the direction of his motion. **(a)** If $O$ sees the rod to be 1.3 m long, what is its proper length as seen by $O'$? **(b)** At the same time, $O$ holds an identical rod perpendicular to the direction in which $O'$ is moving. How long is this rod as $O'$ sees it?

**38.16** *(5) Half-measures.* If you find the length of a moving object to be one-half its proper length, how fast is it moving?

**38.17** *(5) Pushing it.* You have a fast sports car whose length is 3.6 m. You drive it at full speed to impress your date, who expresses a preference for standing on the roadside and watching you go by. If you push the car to 230 km/h, how much shorter does the car appear to your date? Compare the contraction to the wavelength of blue light, $\lambda \approx 400$ nm.

**38.18** *(5) Depends on your point of view.* $O$ and $O'$ have identical stopwatches. $O'$ moves past $O$ with speed $0.9c$. Just as he passes her, both start their watches. $O'$ waits until his watch reads 100 s and flashes a light on its face. Using a telescope, $O$ can observe the reading of the watch at the instant of the flash. **(a)** What did the watch of $O$ read at the instant

of the flash? **(b)** According to $O$, how long did the flash take to reach her? What does her watch read at the instant when she sees the flash? **(c)** According to $O'$, how long did the flash take to reach $O$? What does his watch read when the flash reaches her?

**38.19** *(6) Catching up.* A spaceship leaves the earth for Antares, traveling at $0.8c$ with respect to the earth. One year later, an improved model leaves for Antares, traveling at $0.9c$ with respect to the earth. **(a)** At what speed does the improved ship overtake the older one? **(b)** From the point of view of observers on the earth, how far away do the two ships meet? **(c)** At that instant, how far away do the crew members of the older ship see the earth? How long have they been en route? **(d)** How far away do the crew members of the improved ship see the earth? How long have they been en route?

**38.20** *(6) Pair production.* Under proper conditions, an electron-positron pair can be created in the laboratory. (A positron has the same mass as an electron, but opposite charge.) The two particles move away from each other, each having a speed $0.95c$. What is the speed of the positron as seen by an imaginary observer located on the electron?

**38.21** *(6) That fast!* **(a)** Show that, if the speed of light were infinite, all the Lorentz transformation equations would boil down to corresponding Galilean transformation equations. **(b)** Explain the implications of the conclusion of part **a** for the validity of the Galilean transformation.

**38.22** *(6) Help from Hendrik.* Observer $O$ sees event 1 occur at time $t_1$ at a point $x$. A little later, she sees event 2 occur at the same point at time $t_2$. The time interval between the events is $\Delta t \equiv t_2 - t_1$. Observer $O'$ sees the same events and measures a time interval $\Delta t'$ between them. Express $\Delta t'$ in terms of $\Delta t$.

**38.23** *(6) High-speed sum.* Show that if $v_x = c$ in the reference frame of $O$, the Lorentz velocity transformation yields $c$ as the velocity measured by an observer $O'$ moving at *any* velocity $V$.

**38.24** *(G) Low-speed approximation.* **(a)** Show that, if $V \ll c$, the approximation

$$\sqrt{1 - \frac{V^2}{c^2}} \simeq 1 - \frac{1}{2}\frac{V^2}{c^2}$$

is valid. **(b)** Write a similar approximation for the factor $1/\sqrt{1 - V^2/c^2}$. **(c)** At what speed does use of the approximations of parts **a** and **b** introduce an error of 1% in the result? Express your result in m/s and in km/h, and thus show that the approximation is useful over a quite wide range of speeds $V$.

---

## GROUP B

**38.25** *(2) Relaxing the rules.* Equation 38.3 applies to the special case in which the coordinate frames $O$ and $O'$ have collinear $x$ and $x'$ axes and in which $\mathbf{V} = V\hat{\mathbf{x}}$. Generalize this result to any pair of coordinate systems with parallel corresponding axes and with $\mathbf{V}$ constant but having arbitrary direction.

**38.26** *(5) No SLACker!* The Stanford Linear Accelerator is

3000 m long. It accelerates electrons to a speed that differs from the speed of light by only 8 parts in $10^{11}$. [That is, $c - V = (8 \times 10^{-11})c$.] How long does the accelerator appear to be from the point of view of an observer riding on one of these electrons? (Hint: Your calculator will probably balk at handling the necessary numbers. Use the approximation suggested in Problem 38.9.)

**38.27** *(5) Far and fast.* A spaceship on a mission to a distant star travels with constant speed $V$ relative to the earth. When the crew members note that 1 year has passed, they flash a signal to the earth. **(a)** How long after the departure of the ship do observers on the earth receive the signal? **(b)** The observers on the earth determine that the signal was flashed when the spaceship was just 1 light-year from earth. How fast is the spaceship going?

**38.28** *(6) Funny coincidence.* Observer $O$ sees event 1 and then 3 μs later sees event 2 occurring at the same place. Observer $O'$ sees the same two events separated in time by 4 μs. In his reference frame, what is the distance between the points at which the two events occur?

**38.29** *(6) In Lorentz's footsteps, I.* Following a path similar to the one that leads to Equation 38.12a, derive Equation 38.12b.

**38.30** *(6) In Lorentz's footsteps, II.* Following a path similar to the one that leads to Equation 38.12e, derive Equation 38.12f.

**38.31** *(6) Speed of light in flowing water.* In 1851 and again in 1859, Fizeau measured the speed $v$ of light in water that was flowing through a pipe with velocity $V$. If you approached the question classically, you might predict the result $v = c/n + V$, where $n$ is the index of refraction of water. But Fizeau showed—and Michelson and Morley confirmed with a more precise measurement made in 1885–86—that the measured value of $v$ was close to

$$v = \frac{c}{n} + V\left(1 - \frac{1}{n^2}\right).$$

Fresnel had predicted this value in 1818, on the basis of artificial assumptions concerning the way in which the water drags the ether along as it flows. (For this reason, the term in parentheses is called the *Fresnel drag coefficient*.) But show that the equation can be derived without special assumptions, by using the Lorentz velocity transformation.

**38.32** *(G) The Michelson-Morley experiment.* This renowned 1887 experiment provides a road to the special theory of relativity, though it is one that Einstein seems not to have taken. Take the Newtonian view that light travels at a speed $c$ with respect to the ether. Riding on the earth, you move at some speed $V$ with respect to the ether. Now, set up a Michelson interferometer as shown in the next column. The arms have equal length $L$. The $M_2$ arm faces "upstream," in the direction in which you are moving; the $M_1$ arm, of equal length, is perpendicular to your direction of motion. **(a)** Show that, according to the Newtonian view, the time required for light to travel from the beam splitter $H$ to $M_2$ and back is

$$t_\parallel = \frac{2L}{c}\left(1 + \frac{V^2}{c^2}\right).$$

**(b)** Show that the Newtonian speed of light "across the stream" from $H$ to $M_1$ and back is $\sqrt{c^2 - V^2}$. **(c)** Find the Newtonian time required for the round trip from $H$ to $M_1$ and back. **(d)** Make the assumption that $V \ll c$, and show that the time difference between the two round trips of parts **a** and **c** is $\Delta t = LV^2/c^3$. **(e)** If the wavelength of the light is $\lambda$, what is the optical path difference $N$ between the light waves from

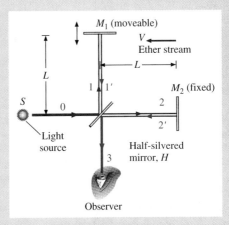

the two arms of the interferometer as they superpose along path 3? Neglect the optical path difference introduced by passage of light through the beam splitter. **(f)** You now rotate the interferometer 90° clockwise, so as to interchange the "upstream-downstream" and "across-the-stream" arms. As you do so, what fringe shift $\Delta N$ do you expect to see in the light of the foregoing analysis? **(g)** There is no way to estimate the speed of the earth with respect to the ether. For the sake of argument, use the speed of the earth relative to the sun, which is about $3.0 \times 10^4$ m/s. In the Michelson-Morley interferometer, the effective length of the arms was 11.0 m. Assuming that $\lambda = 550$ nm, show that the analysis predicts a shift of 0.4 fringe on rotating the apparatus. **(h)** In fact, Michelson and Morley found no fringe shift. They knew that they could detect a fringe shift as small as 0.1 fringe. But suppose they were very unlucky and happened to make their measurement just at an instant when the earth was at rest with respect to the ether. If they repeated the experiment six months later, what fringe shift does the classical analysis predict? **(i)** What bearing does the negative conclusion of the Michelson-Morley experiment have on the second principle of relativity? on the first principle of relativity?

**38.33** *(G) Fitzgerald to the rescue!* The negative result of the Michelson-Morley experiment (see Problem 38.32) required explanation. Not surprisingly, the first attempts at explanation were carried out in a classical framework. Fitzgerald made the following bold suggestion: The interferometer arm facing into the ether stream is shortened, compared with the length it would have in the absence of an ether stream. When the interferometer is rotated through 90°, that arm lengthens as it is turned across the stream while the arm originally perpendicular to the ether stream shortens as it is turned into the stream. Suppose that the arms are adjusted to have equal lengths $L$ before rotation. **(a)** What is the proportional length change $\Delta L/L$ of either arm needed to account for the zero fringe shift when the interferometer is rotated? Does your result look familiar? **(b)** Fitzgerald's suggestion gives a satisfactory *mathematical* explanation for the negative result of the Michelson-Morley experiment. Why is it not a satisfactory *physical* explanation?

**38.34** *(G) Sagnac interferometer.* The device shown (p. 1066) is often used as part of a gyroscope or an angular accelerometer, especially in inertial guidance systems. Light travels

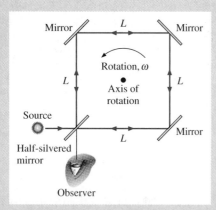

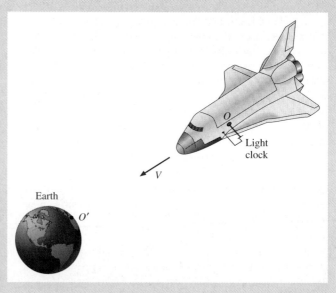

both ways around the square pathway of sides $L$. When the entire system is set into rotation about an axis at its center, as shown, the light path once around is greater in the direction of rotation than in the opposite direction. (You can think of the setup as a rotational version of the boxcar experiment.) As a result, the observer sees a fringe shift whose value depends on the angular speed $\omega$. The instantaneous speed of a point on the pathway depends on its position but can be approximated by a mean speed $\langle v \rangle$. **(a)** Show that the fringe shift is given by

$$\Delta N = \frac{4L\langle v \rangle}{c\lambda} \frac{1}{1 - \langle v \rangle^2/c^2} \simeq \frac{2\omega A}{c\lambda}$$

for small angular speeds $\omega$, where the term $A$ is the area enclosed by the pathway. **(b)** If $L = 1$ m and visible light is used, find the approximate value of $\omega$ that will result in 1 fringe shift; $\Delta N = 1$.

This device was invented by the French physicist G. M. M. Sagnac (1869–1928) in 1913. In modern versions, the light source is often a laser incorporated directly into one or all of the arms of the closed pathway. In order to increase the sensitivity to rotation, the effective enclosed area is sometimes increased by letting the pathway be an optical fiber coiled into a many-turn coil.

**38.35** *(G) Relativistic Doppler shift, I.* As you learned in Section 22.8, the wavelength of sound (or any wave traveling through a medium) is shifted if the source, the observer, or both are moving with respect to the medium. This problem concerns the relativistic Doppler shift, which affects light as it travels through empty space.

In the figure shown here, a light clock is mounted on the side of a spaceship so that its light path is perpendicular to the direction of the ship's motion. (This method of mounting makes it unnecessary to consider any Lorentz contraction of the clock itself.) The light clock "ticks" with frequency $\nu = 1/\Delta t$, as seen by an observer on the spaceship. The spaceship is moving toward an observer on earth at speed $V$. **(a)** Taking into account only the time dilation, determine what frequency would be observed on earth. **(b)** There is another effect that influences the earth observer's measurement of frequency. Because the spaceship is traveling toward him, each flash has less far to go than the preceding one. According to the earth observer, what distance does the spaceship cover between flashes? **(c)** Show that the earth observer sees light flashes

arriving with frequency

$$\nu' = \nu \sqrt{\frac{1 + \dfrac{V}{c}}{1 - \dfrac{V}{c}}}.$$

**(d)** Compare this with the classical result given for a moving source, a stationary medium, and a stationary observer by Equation 22.34, which here assumes the form $\nu' = \nu c/(c - V)$. Show that the two results agree if $V \ll c$. (Note: The relativistic case is simpler than the classical case, in that the only relevant velocity is that of the observer relative to the source. There is no way to make a distinction between a moving source and a moving observer, and there is no medium, moving or otherwise.) **(e)** What is the value of $\nu'$ if the spaceship is moving away from the earth at speed $V$? **(f)** The relativistic Doppler shift is defined as $\Delta \nu \equiv (\nu' - \nu)$. If the spaceship moves away from the earth at speed $V = 0.75c$, find the ratio $\Delta \nu/\nu$. This *red shift* is the basis for determining the rate of expansion of the universe. The most distant objects yet observed appear to be receding from us at speeds in excess of $0.9c$. Interestingly, Doppler himself suggested in 1842 that red and blue shifts might account for the observed differences among the spectra of different stars. But Doppler's suggestion soon proved far too simplistic to account for the wide variations in stellar spectra. In any case, the Doppler shifts for stars within our own galaxy are quite small.

**38.36** *(G) Relativistic Doppler shift, II.* The physicist R. W. Wood (1868–1955) is said to have beaten a traffic ticket he got for passing a red light. He told the judge that as he approached the light his speed relative to the light resulted in a Doppler shift that made the red light appear green. Seeing a green light, he did not stop. The judge, knowing that Wood was a famous physicist and unwilling to expose his own total ignorance of physics, acquitted Wood. Assuming that great physicists never lie, how fast was Wood going? Take "red" to imply $\lambda \simeq 670$ nm and "green" to imply $\lambda' \simeq 525$ nm.

**38.37** *(G) Dop-cop.* Police radars transmit microwave radiation at a fixed frequency. Some of the radiation is reflected off the surface of a moving car, and some of the reflected radiation is detected by a sensitive detector. The speed of the car can be deduced from the Doppler shift ratio $\Delta\nu/\nu$ of the reflected radiation (see part **f** of Problem 38.35). **(a)** What is the value of $\Delta\nu/\nu$ if an approaching car is traveling at 120 km/h? (Hint: There are *two* Doppler shifts to consider. Alternatively, imagine that the car is a mirror, and the radar receiver "sees" a mirror image of the radar transmitter approaching it. How fast is the image moving?) **(b)** The Doppler shift is measured in the detector by superposing the amplified return signal on the transmitted signal and measuring the beat frequency (see Section 22.6). Police radars are expected to be sensitive enough to record speeds within $\pm 1$ km/h. If the transmitted signal frequency is 10 GHz, what is the corresponding change in beat frequency that must be detected?

## GROUP C

**38.38** *(4) The twin paradox, III.* In Problem 38.8, you found that Oliver had aged much more than Olivia. The following argument has been made to show that the two cannot have different ages: From Oliver's point of view, Olivia's time is dilated. Thus, if she returns home after 14 years of Oliver's time have passed, she will have aged more than he, which contradicts the calculation of Problem 38.8. What is wrong with this argument? (Hint: To show that the argument fails, you must find an asymmetry between the two points of view.)

**38.39** *(6) Two Lorentzes in tandem.* Observer $O_0$ measures the speed of observer $O_1$ and finds it to be $V_{10}$. Meanwhile, observer $O_1$ measures the speed of observer $O_2$ and finds it to be $V_{21}$. Show that, when $O_0$ measures the velocity of $O_2$, she will find it to be

$$V_{20} = \frac{V_{21} + V_{10}}{1 + \dfrac{V_{21}V_{10}}{c^2}}.$$

**38.40** *(6) In Lorentz's footsteps, III.* Derive Equation 38.15b or, what amounts to the same thing, Equation 38.15c.

**38.41** *(G) Invariance of the spacetime interval.* Observer $O$ watches a flying saucer as it passes. At time $t_1$, a light on the saucer flashes and she observes its position to be $(x_1, y_1, z_1)$. At time $t_2$, there is another flash, and $O$ locates the saucer at $(x_2, y_2, z_2)$. She thus finds that, in the time interval $\Delta t \equiv t_2 - t_1$, the saucer has moved through the displacement $(\Delta x, \Delta y, \Delta z)$. Observer $O'$ sees the same events at times $t_1'$ and $t_2'$. According to his measurements, the saucer has moved through the displacement $(\Delta x', \Delta y', \Delta z')$ in the time interval $\Delta t'$. Show that

$$c^2(\Delta t)^2 - (\Delta x)^2 - (\Delta y)^2 - (\Delta z)^2 =$$
$$c^2(\Delta t')^2 - (\Delta x')^2 - (\Delta y')^2 - (\Delta z')^2.$$

The quantity on either side of this equation is called the *spacetime interval*. It is called an *invariant* because its value is the same for all observers. What would be the corresponding invariant(s) in Newtonian kinematics?

**38.42** *(G) Proper-time interval.* Observer $O$ sees event 1 occur at $(x_1, t_1)$ and event 2 at $(x_2, t_2)$. Observer $O'$ moves past $O$ at just such a speed $V$ that he sees both events occur at the same place; $x_1' = x_2'$. **(a)** Show that the time interval $\Delta t'$ between the events for $O'$ must be

$$\Delta t' = \sqrt{(\Delta t)^2 - \frac{(\Delta x)^2}{c^2}},$$

where $\Delta t = t_2 - t_1$ and $\Delta x = x_2 - x_1$. The quantity $\Delta t'$ is called the *proper-time interval*. Show that $O'$ can make the observation just described *only* if $\Delta x < c\,\Delta t$. What is the physical meaning of this condition? **(b)** Show that, if $\Delta x > c\,\Delta t$, the time order of events depends on the frame of reference chosen. That is, event 1 occurs before, simultaneously with, or after event 2, depending on the reference frame chosen. What is the physical meaning of the condition $\Delta x > c\,\Delta t$?

**38.43** *(G) Unambiguity of causal sequences.* You learned in Section 38.1 that it is impossible in general to order a sequence of events in an unambiguous way. But consider a pair of *causally related* events. Suppose, for example, that observer $O$ sees light flash 1 and that flash 1 triggers a device that produces flash 2. The positions and relative speeds of the two flash lamps are arbitrary (but $V \leq c$). Show that there is no reference frame $O'$ in which flash 2 *precedes* flash 1. (Hint: See Problem 38.42.)

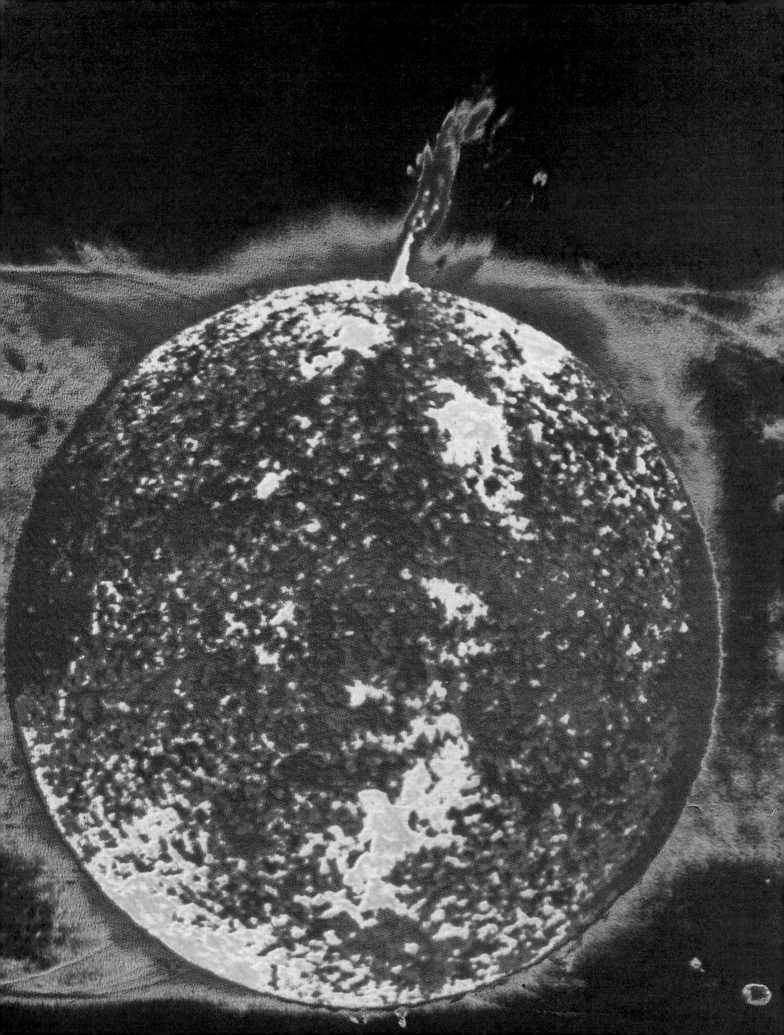

# Relativistic Dynamics

———— The observed mass of a body depends on its speed with respect to the observer.

———— The rest energy, $m_0c^2$, may be regarded as a form of potential energy; the total energy of the body is $mc^2$.

———— Mass and energy are not distinct in this view but constitute the quantity mass-energy. The principle of mass-energy conservation is broader than, and supersedes, the separate principles of mass conservation and energy conservation.

———— In chemical reactions and everyday processes, the transformations of energy from one form into another do not significantly change the mass of the system. In nuclear reactions, however, the mass changes, and the transformations of mass into other forms of energy are considerable.

*Left:* The sun has emitted radiant energy copiously for billions of years and will continue to do so for billions more, because it is fueled by nuclear fusion processes. A little mass is equivalent to a lot of energy because $c^2$ is a big number!

*The very fact that the totality of our sense experiences is such that by means of [scientific] thinking it can be put in order—this fact is one which leaves us in awe, but which we shall never understand. One may say "the eternal mystery of the world is its comprehensibility."*

—ALBERT EINSTEIN

## SECTION 39.1    Introduction

We now develop the theory of **relativistic mechanics**, building on the kinematics developed in Chapter 38. This is parallel to the procedure we adopted in developing the theory of classical mechanics at the beginning of the book. Not surprisingly, there will be significant differences in detail between the classical and relativistic developments. Nevertheless, we will be interested in the same dynamical quantities—notably mass, force, momentum, and energy.

Today there exists a large and varied body of experimental evidence to support the theory of relativistic mechanics. When Einstein began his work there was little or no evidence—certainly none that was clear. As in the kinematic development, he relied on thought experiments. We will follow Einstein's development fairly closely.

Section 39.2 is a study of *mass* and *momentum*. In Section 39.3, we consider *energy* and *force*. In Section 39.4, we develop the renowned concept of *relativistic mass-energy*. Section 39.5 deals with the application of these ideas to nuclear energy.

## SECTION 39.2    Mass and Momentum

When we want to measure the mass of a body under everyday circumstances, we almost always weigh it. That is, we depend on the *gravitational manifestation* of mass—the property that bodies exhibit when they are attracted gravitationally by other bodies (Section 14.6). But it is not possible to put a body on the pan of a scale if the body is moving past the pan at a speed $V$ comparable to the speed of light $c$. Fortunately, there is another way to determine mass, which we discussed in Section 9.6. We can determine the mass $m_2$ of a frictionless glider by making it collide with another glider of known (or standard) mass $m_1$. If we know the initial and final velocities of both gliders, we have all the information needed to determine $m_2$. We use the principle of conservation of linear momentum in the form of Equation 9.34,

$$m_1 \, \Delta v_1 = -m_2 \, \Delta v_2. \tag{39.1}$$

In this equation, $\Delta v_1$ and $\Delta v_2$ are the changes in velocity experienced by bodies 1 and 2. The quantities $m_1$ and $m_2$ represent the *inertial manifestation* of mass.

A particularly simple form of the experiment is sketched in Figure 39.1. We arrange for the collision to be head on and thus one-dimensional. For further simplification, we make the initial velocities equal and opposite; $v_{2i} = -v_{1i}$. Finally, we use two gliders that collide elastically.

Suppose that we find the final velocities to be

$$v_{1f} = -v_{1i} \quad \text{and} \quad v_{2f} = -v_{2i} = v_{1i}.$$

Using Equation 39.1, we write

$$m_1(-2v_{1i}) = -m_2(2v_{1i}).$$

We conclude that the two masses are equal and call their common value $m_0$.

$$m_1 = m_2 \equiv m_0.$$

We now modify the experiment, placing it in a relativistic context as shown in Figure 39.2. Observer $O$ has glider $m_1$, and $O'$ has glider $m_2$. They have performed the experiment just described and have found that both gliders have mass $m_0$. Now $O$ and $O'$ separate, leaving their helper $H$ with a flash lamp. They set up their coordinate systems so that the $x$ and $x'$ axes are collinear. Observer $O$ positions her glider on the $y$ axis at $y = a$. Observer $O'$ positions his glider on the $y'$ axis at $y' = -a$. On a signal from $H$, $O$ and $O'$ approach each other, bringing their gliders with them. Observers $O$ and $O'$ each see the other moving at speed $V$ along the $x$ (or $x'$) axis. From the point of view of helper $H$, $O$ and $O'$ approach at equal speeds and are always equidistant from him. When all is ready, $H$ sets off a flash lamp. Seeing the flash, $O$ propels her glider in the $-y$ direction with an agreed-upon speed $v_\perp$ very much smaller than $V$. Similarly, $O'$ propels his glider in the $y'$ direction with the same speed $v_\perp$.

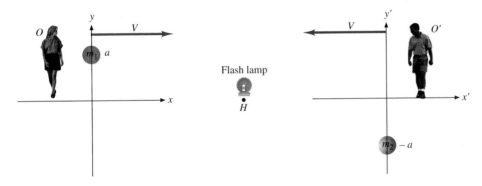

Observers $O$ and $O'$ pass $H$ simultaneously. (Note that "simultaneous" is meaningful here because the positions of all three coincide.) At that instant the two gliders collide, and they subsequently rebound. Figure 39.3 shows what the experiment looks like from the point of view of $H$. Because we have taken care to symmetrize all the experimental conditions from his point of view, the paths of the gliders are symmetrical; $\theta_{1i} = \theta_{2i} = \theta_{1f} = \theta_{2f}$. (These angles are exaggerated for clarity; they are actually very small because $v_\perp \ll V$.) We conclude from the symmetry that the collision does not affect the velocity components of the gliders along the $x$ (or $x'$) axis. The components along the $y$ (or $y'$) axis are reversed.

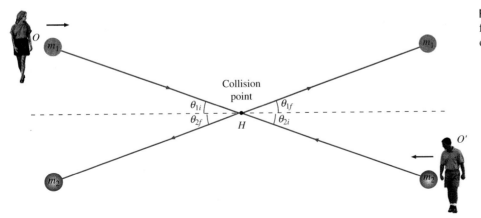

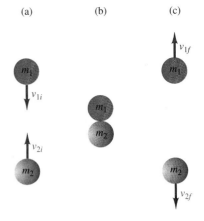

(a)     (b)     (c)

FIGURE 39.1 A particularly simple form of the Mach experiment. (a) Two bodies approach each other. Their velocities are equal and opposite and very much less than $c$. (b) The bodies collide. (c) The bodies move apart. Measurement shows that their directions are reversed and their speeds are unchanged. We conclude that their masses are equal.

FIGURE 39.2 The Mach experiment modified so that it can be performed by observers $O$ and $O'$, who move with relative speed $V$. Observer $O$ carries $m_1$ with her, and $O'$ carries $m_2$ with him. The situation is shown from the point of view of helper $H$, who is always located halfway between $O$ and $O'$. At a signal from $H$, $O$ and $O'$ propel their gliders toward the origins of their own coordinate systems, each with $v_\perp$ parallel to the $y$ and $y'$ axes. The gliders collide just as $O$ and $O'$ pass $H$, and then they rebound.

FIGURE 39.3 The glider paths seen from the symmetrical point of view of $H$.

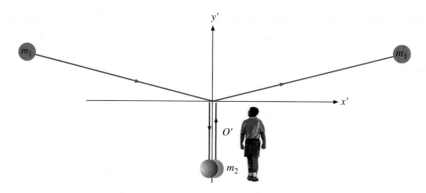

**FIGURE 39.4** The glider paths seen from the asymmetrical point of view of $O'$. Glider 2, which he has propelled, moves up and back along the $y'$ axis. Glider 1, which moves with observer $O$, moves in a shallow, symmetric V.

Figure 39.4 shows what the experiment looks like as $O'$ sees it. From his point of view, the situation is not symmetrical at all. His own glider, $m_2$, travels along the $y'$ axis with initial speed $v_\perp$ and rebounds along the $y'$ axis with the same speed. Its velocity change in the collision is thus

$$\Delta v_{2y'} = -2v_\perp.$$

But $O'$ sees glider $m_1$ traveling along the V-shaped path shown. What is its velocity component $v_{1y'}$ along the $y'$ axis? The value of this component is *not* $v_\perp$, because $O'$ must take into account the Lorentz velocity transformation. He must use Equation 38.15b, the second of the velocity-transformation equations, to find $v_{1y'}$. With minor changes in notation to conform to the present situation, that equation is

$$v_{1y'} = \frac{\sqrt{1 - V^2/c^2}}{1 - Vv_{1x}/c^2} v_{1y}.$$

Now, $O'$ knows that $O$ has propelled $m_2$ with speed $v_\perp$ along the $y$ axis, so he sets $v_{1y} = v_\perp$ and $v_{1x} = 0$. He thus obtains

$$v_{1y'} = \sqrt{1 - V^2/c^2}\, v_\perp.$$

The rebound speed is the same, and thus the velocity change is

$$\Delta v_{1y'} = 2\sqrt{1 - V^2/c^2}\, v_\perp.$$

Now that $O'$ has the $y'$ components of the velocity of both gliders, he can apply the momentum-conservation principle in the form of Equation 39.1. He has $m_1 \Delta v_{1y'} = -m_2 \Delta v_{2y'}$, or

$$2m_1\sqrt{1 - V^2/c^2}\,v_\perp = 2m_2 v_\perp.$$

Canceling the factor $2v_\perp$ common to both sides, $O'$ obtains

$$m_1\sqrt{1 - V^2/c^2} = m_2. \qquad (39.2)$$

This equation cannot be true if $m_2 = m_1 = m_0$. But we know from wide experience with bodies moving at *low* speeds that the mass is independent of the speed of the body with respect to the observer. The speed of $m_2$ with respect to $O'$ is always small. Thus it must be correct to set $m_2 = m_0$, because that is the value $O$ and $O'$ measured when everything was at rest. Equation 39.2 can thus be solved for $m_1$ to obtain

$$m_1 = \frac{m_0}{\sqrt{1 - V^2/c^2}}.$$

That is, $O'$ sees the mass of the glider moving with $O$ to be greater than that of his glider.

There is just one more step to our analysis of the collision experiment. We could repeat the argument from the point of view of $O$, but we need not. The first principle of relativity guarantees that she will see $m_2$, the mass of the glider moving with $O'$, to be greater than that of her glider, whose value is $m_1 = m_0$. Recognizing that her point of view and that of $O'$ must be on equal footing, $O$ will write

$$m_2 = \frac{m_0}{\sqrt{1 - V^2/c^2}}.$$

We thus conclude that *the mass of a body increases with increasing speed relative to the observer*. This generalization allows us to drop the subscripts 1 and 2 and to substitute the speed $v$ of the body with respect to an arbitrary observer for the speed $V$ of either of two observers with respect to one another. We thus write the **relativistic mass equation**:

$$m = \frac{m_0}{\sqrt{1 - v^2/c^2}}.$$ 

(39.3)

In this equation, $m_0$ is called the **rest mass**, and $m$ is called the **relativistic mass**.

EXAMPLE **39.1**

Find the mass of a 1-kg body when it moves with respect to the observer at speed (a) $0.1c$, (b) $0.8c$, and (c) $0.9999c$.

**SOLUTION:**

(a) For $v = 0.1c$ and $m_0 = 1$ kg, Equation 39.3 gives you

$$m = 1 \text{ kg} \times \frac{1}{\sqrt{1 - (0.1)^2}} = 1.005 \text{ kg}.$$

Even though $0.1c$ ($= 3 \times 10^7$ m/s) is a very considerable speed by everyday standards, the relativistic mass differs from the rest mass by only $\frac{1}{2}$%.

(b) For $v = 0.8c$, you calculate

$$m = 1 \text{ kg} \times \frac{1}{\sqrt{1 - (0.8)^2}} = 1.67 \text{ kg}.$$

The relativistic mass is about 70% more than the rest mass.

(c) For $v = 0.9999c$, you have

$$m = 1 \text{ kg} \times \frac{1}{\sqrt{1 - (0.9999)^2}} = 70.7 \text{ kg}.$$

As you have seen before for quantities involving the Lorentz factor $\sqrt{1 - v^2/c^2}$, the difference between the relativistic and classical values is quite small until $v$ becomes an appreciable fraction of the speed of light.

As its speed with respect to the observer approaches the speed of light, the relativistic mass of a body increases without limit. What would happen to such a body if you tried to accelerate it indefinitely by pushing on it with a constant force?

## *Momentum Conservation*

Our derivation of the relativistic mass equation (Equation 39.3) is based on the assumption that momentum is conserved in all circumstances. The price we pay for holding on to this assumption is the abandonment of mass as a quantity independent of motion. That is, we have asserted that *momentum conservation is more fundamental than mass conservation*. Experiment validates this assertion. It therefore makes sense to define the **relativistic momentum p** of a body:

$$\mathbf{p} \equiv m\mathbf{v}.$$ 

(39.4)

In this equation, $m$ is the relativistic mass given by Equation 39.3; that is, $m = m_0/\sqrt{1 - v^2/c^2}$.

## Force and Energy

The definition of the **relativistic force** is the same as that given by Equation 9.30 for the classical force—namely,

$$\mathbf{F} = \frac{d\mathbf{p}}{dt}.$$ 

(39.5)

We have to be careful, however, in expressing relativistic force in terms of acceleration because the mass is not independent of the speed. When we substitute the value $\mathbf{p} = m\mathbf{v}$

given by Equation 39.4 into Equation 39.5 and use the chain rule for differentiation, we obtain

$$\mathbf{F} = \frac{d}{dt}(m\mathbf{v}) = m\frac{d\mathbf{v}}{dt} + \mathbf{v}\frac{dm}{dt}. \qquad (39.6a)$$

By definition, we have $d\mathbf{v}/dt = \mathbf{a}$, the acceleration. So we write

$$\mathbf{F} = m\mathbf{a} + \mathbf{v}\frac{dm}{dt}. \qquad (39.6b)$$

Classically, the mass of a particle does not change as it is acted on by a force; we have $m = m_0 =$ constant. Consequently, we have $dm/dt = 0$, and Equation 39.6b reduces to the familiar classical form $\mathbf{F} = m_0\mathbf{a}$. But, in general, you must take the factor $dm/dt$ into consideration in applying either of Equations 39.6a and 39.6b.

There is one important situation, however, in which $dm/dt = 0$ for relativistic particles. Consider a charged particle moving with instantaneous velocity $\mathbf{v}$ in a uniform magnetic field. The particle is acted on by a force of constant magnitude $F = qv\mathscr{B}$, which is always directed perpendicular to $\mathbf{v}$ (Chapter 28). Thus the particle moves in a circle with constant speed $v$, and the relativistic mass does not change with time. Equation 39.6b then simplifies to

$$\mathbf{F} = m\mathbf{a} \quad \text{for } m = \text{constant}, \qquad (39.7)$$

where $m$ is the relativistic mass given by Equation 39.3.

A charged particle moving at high speed in a magnetic field provides a basis for experimental verification of Equation 39.3. As early as 1901, the German physicist Walter Kaufmann (1871–1947) had repeated J. J. Thomson's $e/m$ experiment, using $\beta$ particles (the energetic electrons expelled from radioactive nuclei) in place of the less energetic electrons produced in earlier cathode-ray tube experiments (Section 28.8). Some of these $\beta$ particles had speeds in excess of $0.8c$. Kaufmann found evidence that the most energetic $\beta$ particles (electrons) exhibited values of $e/m$ as small as one-third the value $e/m_0$ obtained for electrons at lower energies.

Before Einstein's work of 1905, several theories were proposed for interpreting Kaufmann's results, but the results were not precise enough to provide a crucial test among the theories. In 1908, Bucherer performed an experiment whose results were much more precise, at least for electrons having speeds up to about $0.7c$. The electrons were made to pass through a velocity selector (Problem 28.23) and then into a region of uniform magnetic field. The mass was then determined from the curvature of the trajectory. Kaufmann's and Bucherer's results, and those of a third investigation, are shown in Figure 39.5. Later experiments, using electrons having still greater speeds and employing more precise measuring techniques, furnish further confirmation.

**The Commanding Voice of Experiment**

The German physical chemist and physicist Alfred H. Bucherer (1863–1927) was nearly unique among European scientists of his day in that he spent a major part of his graduate-student years in the United States. At Johns Hopkins, he developed a process important for aluminum manufacture; at Cornell, he was among the first to appreciate the usefulness of vector analysis in physics. Bucherer had his own theory to explain Kaufmann's observations and originally designed the improved experiment described in the text to test it. However, his results convinced him of the validity of the relativistic explanation.

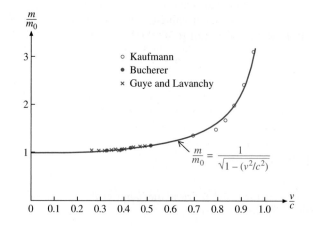

**FIGURE 39.5** Electron mass as a function of speed. Speed is plotted as the ratio $v/c$ and mass as the ratio $m/m_0$. The points represent the results of various experimenters; the solid curve is obtained by using Equation 39.3.

## Relativistic Work

When a force acts on a particle, work is done. This is true relativistically as well as classically, and work is defined identically in both cases. In the simple case where motion is confined to the $x$ direction and the force is parallel to the $x$ direction, the definition has the form

$$W \equiv \int_{x_i}^{x_f} F \, dx. \qquad (39.8)$$

The limits of the integral are the initial position $x_i$ and the final position $x_f$ of the particle.

To evaluate the work integral, we use the value of $F$ given by Equation 39.6a,

$$F = m \frac{dv}{dt} + v \frac{dm}{dt}.$$

But we must express this value as a function of $x$. There is a neat way to do this simultaneously for both terms. Beginning with Equation 39.3, $m = m_0 / \sqrt{1 - v^2/c^2}$, we square both sides and rearrange to obtain

$$m^2 c^2 - m^2 v^2 - m_0^2 c^2 = 0.$$

Next, we take the derivative of both sides of this equation with respect to time:

$$c^2 \frac{d(m^2)}{dt} - m^2 \frac{d(v^2)}{dt} - v^2 \frac{d(m^2)}{dt} = 0.$$

We carry through the differentiation to obtain

$$2mc^2 \frac{dm}{dt} - 2m^2 v \frac{dv}{dt} - 2mv^2 \frac{dm}{dt} = 0.$$

We divide through by $-2mv$ and transpose the first term to the right side of the equation:

$$m \frac{dv}{dt} + v \frac{dm}{dt} = c^2 \frac{1}{v} \frac{dm}{dt}.$$

We are almost there. We substitute into the right side of this equation the definition $v \equiv dx/dt$, and we have the desired result

$$F = m \frac{dv}{dt} + v \frac{dm}{dt} = c^2 \frac{dt}{dx} \frac{dm}{dt}$$

$$= c^2 \frac{dm}{dx}.$$

We can now substitute this value of $F$ into the integrand of Equation 39.8. This gives us

$$W = \int_{x_i}^{x_f} c^2 \frac{dm}{dx} \, dx = c^2 \int_{m_i}^{m_f} dm.$$

The limits of integration $m_i$ and $m_f$ are the initial and final masses of the particle, corresponding to the positions $x_i$ and $x_f$. Evaluating the integral, we obtain

$$W = c^2 (m_f - m_i).$$

Suppose that we begin with the particle at rest at $x_i$. Then its initial mass is its rest mass; $m_i = m_0$. We have placed no restrictions on the final position, and hence the final mass is the relativistic mass; $m_f = m$. The work done on a particle accelerated from rest can thus be expressed in the general form

$$W = (m - m_0) c^2. \qquad (39.9)$$

This equation tells us that the work done on the particle is equal to its change in mass, $m - m_0$, multiplied by $c^2$.

### Relativistic Kinetic Energy

If no other forces act on the particle, the work $W$ serves solely to change the kinetic energy $K$ of the particle. That is, we can use the work-energy theorem in its simplest form, $W = K$ (Section 7.4). Equation 39.9 then gives us the **relativistic kinetic energy**

$$K = mc^2 - m_0c^2. \tag{39.10}$$

We are used to expressing the kinetic energy as a function of $v$, the speed of the particle with respect to the observer. To do this, we use the value of $m$ given by Equation 39.3. Equation 39.10 then becomes

$$K = \frac{m_0c^2}{\sqrt{1 - v^2/c^2}} - m_0c^2, \tag{39.11a}$$

or

$$K = m_0c^2\left(\frac{1}{\sqrt{1 - v^2/c^2}} - 1\right). \tag{39.11b}$$

We must make sure that this relativistic equation reduces to the familiar classical equation $K = \frac{1}{2}mv^2$ when $v \ll c$. In this case, the first term in parentheses can be replaced by the approximate value $(1 + \frac{1}{2}v^2/c^2)$. Equation 39.11b then becomes

$$K = m_0c^2\left(1 + \frac{1}{2}\frac{v^2}{c^2} - 1\right) = \frac{1}{2}m_0v^2. \tag{39.12}$$

This is indeed the classical value for the kinetic energy.

Figure 39.6 is a plot of kinetic energy versus speed, according to Equation 39.11b. As in Figure 39.5, the speed is plotted as the ratio $v/c$. The kinetic energy is plotted as the ratio $K/m_0c^2$. The two crosses represent values obtained in a very direct experiment performed in 1964 by the American physicist William Bertozzi. An accelerator produced a beam of electrons of precisely controlled energy. The speed $v$ of the electrons was measured by timing the passage of the electrons between two detectors. The electrons were then stopped by collision in a metal target. The collision transformed the kinetic energy of the electrons into thermal energy, and the temperature rise of the target was used to measure the total energy lost by the electrons. At the same time, the electric charge flowing from the target was used to measure the number of electrons in the beam. Thus the energy per electron could be calculated.

Equation 39.11b furnishes a dynamical reason why no particle having a nonzero rest mass can be accelerated to a speed $c$, let alone a speed greater than $c$. To accelerate the particle requires work. But as the speed $v$ approaches $c$, the kinetic energy of the particle increases without limit, becoming mathematically infinite at $v = c$. Thus infinite work would be required to accelerate a particle, no matter how small its (nonzero) rest mass, to the speed of light. However, there is no theoretical restriction on accelerating a particle to a speed *approaching* $c$ as closely as desired.

**FIGURE 39.6** Electron energy as a function of speed. Speed is plotted as the ratio $v/c$ and kinetic energy as the ratio $K/m_0c^2$. The red curve is obtained using Equation 39.11b, and the lower curve is the classical prediction. Bertozzi's experimental values are shown as crosses.

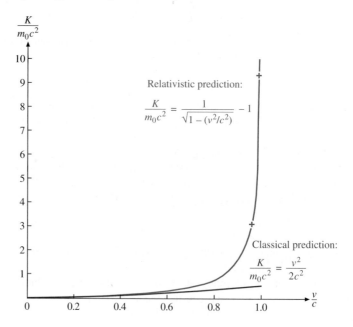

EXAMPLE **39.2**

A cyclotron (Section 28.4) is used to accelerate deuterons of rest mass $m_0 = 3.36 \times 10^{-27}$ kg. Show that the cyclotron will certainly fail to accelerate deuterons to energies in excess of 100 MeV.

**SOLUTION:** The key to the operation of the cyclotron is the *synchrony* of the orbits of the particles being accelerated. Regardless of the energy attained by a particular particle of charge $e$ and mass $m$, its angular frequency of revolution in the magnetic field $\mathscr{B}$ is given by Equation 28.10b, $\omega_c = e\mathscr{B}/m$. On account of this synchrony, all particles arrive at the gap between the dees (Figure 28.11) just in time to acquire additional energy from the alternating electric field that is imposed across the dees.

But the isochrony condition fails if $m$ is not constant. To see what happens to deuterons in the cyclotron at 100 MeV, suppose that a deuteron has attained that energy. Call the angular frequency of this deuteron $\omega_r$. To allow for the possibility that relativistic effects are significant, use the relativistic mass $m$ of the deuteron to write

$$\omega_r = \frac{e\mathscr{B}}{m} = \frac{e\mathscr{B}}{m_0}\sqrt{1 - \frac{v^2}{c^2}}.$$

Now, compare this angular frequency with the classical angular frequency $\omega_c$, which is certainly the angular frequency of deuterons that have just begun the acceleration process. You have the ratio

$$\frac{\omega_r}{\omega_c} = \frac{e\mathscr{B}/m}{e\mathscr{B}/m_0} = \sqrt{1 - \frac{v^2}{c^2}}. \qquad \textbf{(39.13)}$$

You now need to express $\sqrt{1 - v^2/c^2}$ in terms of the kinetic energy $K$ of a 100-MeV deuteron. From Equation 39.11b, you have

$$\sqrt{1 - \frac{v^2}{c^2}} = \frac{1}{\dfrac{K}{m_0 c^2} + 1}.$$

You insert this value into Equation 39.13 to obtain

$$\frac{\omega_r}{\omega_c} = \frac{1}{\dfrac{K}{m_0 c^2} + 1}.$$

Now you can calculate numerical values. Remembering that 1 MeV $= 1.6 \times 10^{-13}$ J, you have

$$\frac{\omega_r}{\omega_c} = \frac{1}{\dfrac{100 \text{ MeV} \times 1.6 \times 10^{-13} \text{ J/MeV}}{3.36 \times 10^{-27} \text{ kg} \times (3 \times 10^8 \text{ m/s})^2} + 1} = 0.95.$$

That is, the 100-MeV deuterons revolve in their magnetic orbits with only 95% of the angular speed of the low-energy deuterons and thus take about 5% longer to make a circuit around the dees. Imagine that the 100-MeV deuterons start around their orbits together with the low-energy deuterons. In five revolutions, they will be out of phase by one-quarter of a cycle. That is, they will arrive at the gap between the dees just when the electric field is zero and will receive no further acceleration. When a deuteron arrives at the gap with the proper phase, each passage across the gap adds a few thousand electron-volts of energy. Thus the cyclotron will certainly not further accelerate deuterons whose energy is 100 MeV and may well fail to accelerate deuterons with considerably smaller energies.

---

# Conservation of Mass-Energy

We now proceed to a famous result: Mass is a form of energy. Let us return to Equation 39.10, $K = mc^2 - m_0 c^2$, and rearrange it in the form

$$mc^2 = K + m_0 c^2. \qquad \textbf{(39.14)}$$

Each of the three terms in this equation has the dimensions of energy. Indeed, the equation looks very much like Equation 8.17, which expresses the total energy $E$ of a system as the sum

$$E = K + U.$$

It looks as though the energy of the system has two parts: a kinetic energy $K$, which it possesses by virtue of its motion, and a **rest energy** $m_0 c^2$, which it possesses when at rest. The sum of the two is called the **relativistic energy** $mc^2$. The relativistic energy is usually represented by the symbol $E$—thus Einstein's famous equation

$$E = mc^2. \qquad \textbf{(39.15)}$$

According to this equation, *the relativistic energy of a particle is the product of its relativistic mass and the square of the speed of light.*

The rest energy is usually represented by the symbol $E_0$; thus we have

$$E_0 = m_0 c^2. \qquad \textbf{(39.16)}$$

This tells us that *mass is a form of energy*. In form, Equation 39.16 is similar to Equation 17.27, $W = JQ$, which expresses the equivalence between the frictional work $W$ done on a body and the heat $Q$ evolved in the process. Recall the evolution of our understanding of Equation 17.27. At first, the proportionality constant $J$ was just an experimentally determined quantity that linked two quite different quantities, mechanical work $W$ (expressed in joules) and heat $Q$ (expressed in calories). But with further study we came to understand, and to become increasingly confident in, the closeness of the connection between heat and energy. We successively took the following points of view:

1. The proportionality constant connecting the heat evolved and the work done is $J$.
2. There is an equivalence between heat and the mechanical energy required to create it.
3. Heat is a form of energy and can be converted into and from other forms under proper conditions.

We now pass (much more quickly) through the same steps with respect to Equation 39.16:

1'. The proportionality constant between mass (expressed in kilograms) and energy (expressed in joules) is $c^2$.
2'. There is an equivalence between mass and energy.
3'. Mass is a form of energy and can be converted into and from other forms under proper conditions.

We have yet to make an argument for statement 3'. We begin to do so in terms of a thought experiment. In Section 39.5, we will turn to very important "real" versions of the thought experiment.

In Figure 39.7a, a short length of string holds two bodies together against the tendency of a compressed spring to force them apart. The rest masses of the two bodies are $m_{01}$ and $m_{02}$. The potential energy of the compressed spring is $U$. In Figure 39.7b, we burn the string with a match, and the spring propels the bodies apart. The spring now has zero potential energy, but the bodies have kinetic energy $K_1$ and $K_2$. According to the principle of conservation of mechanical energy, we have

$$U = K_1 + K_2.$$

If we are to take the idea of rest energy seriously, we must add the total rest energy of the system, $m_{01}c^2 + m_{02}c^2$, to both sides of this equation. (This is certainly permissible mathematically.) Thus modified, the energy-conservation equation is

$$m_{01}c^2 + m_{02}c^2 + U = (K_1 + m_{01}c^2) + (K_2 + m_{02}c^2).$$

According to Equation 39.14, we can write this equation in the form

$$m_{01}c^2 + m_{02}c^2 + U = m_1c^2 + m_2c^2. \tag{39.17}$$

(a)

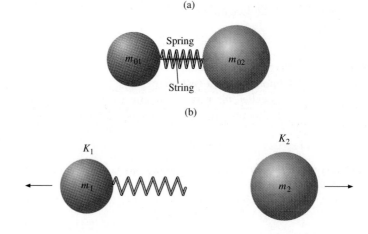

(b)

FIGURE 39.7 (*a*) Two bodies of rest mass $m_{01}$ and $m_{02}$ are held together by a string. Between the bodies is a compressed spring whose potential energy is $U$. (*b*) The string is burned and the spring expands, propelling the two bodies apart with kinetic energies $K_1$ and $K_2$.

As you can see from this equation, the energy originally in the form of potential energy $U$ of the compressed spring has been converted into mass-energy and is now part of the total mass-energy $(m_1 + m_2)c^2$ of the system. Neither the total mechanical energy $U$ of the system nor its total rest mass $m_{01} + m_{02}$ is conserved. *But the mass-energy is conserved.*

If we write Equation 39.17 in the form

$$\left( m_{01} + m_{02} + \frac{U}{c^2} \right) c^2 = (m_1 + m_2)c^2,$$

we have inside the parentheses on the left side a sum of three terms, each of which has the dimensions of mass. We can thus consider the term $U/c^2$ as a part of the rest mass of the system in Figure 39.7a. But $U$ is the energy of the compressed spring; thus it is part of the internal energy of the system. Any change in the internal energy of a system changes its rest mass. In this case, the rest mass of the system with the spring compressed is greater than it would be if the spring were not compressed.

This thought experiment is a highly idealized model of what actually happens when an atomic nucleus breaks up into two parts. (We leave the details to Section 39.5.) But we are usually not in a position to observe the ''compressed spring'' separately, let alone evaluate its energy $U$. Rather, we lump the total initial rest energy of the system, given on the left side of Equation 39.17, into a single quantity $M_0c^2$. We then have

$$M_0c^2 = m_1c^2 + m_2c^2, \qquad \textbf{(39.18a)}$$

or
$$M_0c^2 = m_{01}c^2 + m_{02}c^2 + K_1 + K_2. \qquad \textbf{(39.18b)}$$

The second of these two forms makes evident the fact that the final rest energy of the system is smaller than its initial rest energy. This is because some of the initial rest energy has been converted into kinetic energy.

We can readily subject Equation 39.18b to experimental verification. We measure the rest mass $M_0$ on the left side directly—say, with a laboratory balance. We evaluate the two kinetic energies on the right side by measuring the speeds of the two bodies. Then we bring the two bodies to rest (removing the kinetic energy from the system) and measure their rest masses by the same means used to measure $M_0$. Just such experiments are considered in Section 39.5.

Let us generalize Equations 39.18. A system has an initial mass-energy

$$M_i c^2 = M_{0i}c^2 + K_i$$

and a final mass-energy

$$M_f c^2 = M_{0f}c^2 + K_f.$$

According to the **principle of conservation of mass-energy**, the two are equal:

$$M_i c^2 = M_f c^2 \quad \text{or} \quad M_{0i}c^2 + K_i = M_f c^2 + K_f. \qquad \textbf{(39.19a,b)}$$

This principle supplants the separate classical principles of conservation of mass and conservation of energy.

## Momentum and Energy

We now explore the relations among four important relativistic quantities: total mass-energy $E = mc^2$, rest energy $m_0c^2$, kinetic energy $K$, and momentum $p = mv$. We begin by taking the square of both sides of Equation 39.15, $E = mc^2$. This gives us

$$E^2 = m^2c^4.$$

We express $m$ in terms of $m_0$ by means of Equation 39.3, $m = m_0/\sqrt{1 - v^2/c^2}$:

$$m^2c^4 = \frac{m_0^2 c^4}{1 - v^2/c^2}.$$

We multiply through by the denominator on the right side of this equation:

$$m^2c^4 - m^2v^2c^2 = m_0^2c^4.$$

The first term on the left is $E^2$, and the second term is $-(mv)^2c^2 = -p^2c^2$. So we have

$$E^2 = p^2c^2 + m_0^2c^4.$$

Taking the square root of both sides of the equation gives us the result

$$E = \sqrt{p^2c^2 + m_0^2c^4}. \qquad \textbf{(39.20a)}$$

It is sometimes useful to express this result in a slightly different form. We substitute $mc^2$ for $E$ on the left side and divide both sides through by $c^2$ to obtain

$$m = \sqrt{(p/c)^2 + m_0^2}. \qquad \textbf{(39.20b)}$$

Equations 39.20a and 39.20b express the total relativistic energy $E$ and the relativistic mass $m$ in terms of the momentum $p$ and the rest mass $m_0$.

According to Equation 39.19, the total relativistic energy is the sum of the rest energy and the kinetic energy; $E = m_0c^2 + K$. It follows immediately that $K = E - m_0c^2$. We use the value of $E$ given by Equation 39.20a to obtain the relation

$$K = \sqrt{p^2c^2 + m_0^2c^4} - m_0c^2. \qquad \textbf{(39.21)}$$

Equation 39.20a is a Pythagorean sum; that is, it has the form $h = \sqrt{a^2 + b^2}$. By drawing the corresponding right triangle, as shown in Figure 39.8a, you can readily remember the relation expressed by that equation, as well as the relation expressed by Equation 39.20b. Alternatively, you can divide the lengths of all three sides of this triangle by $c^2$ and obtain the relations in the form given by Equation 39.20b, as shown in Figure 39.8b.

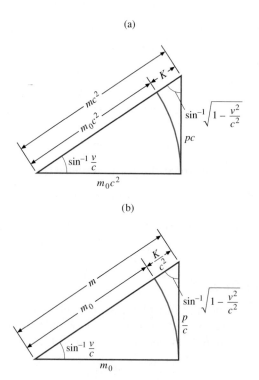

FIGURE 39.8 (a) Right triangle corresponding to the Pythagorean sum expressed in Equation 39.20a. The kinetic energy is given by the difference $K = mc^2 - m_0c^2$. (b) Right triangle corresponding to Equation 39.20b.

EXAMPLE **39.3**

An electron is accelerated through a potential difference of 500 kV. Find **(a)** its relativistic energy, **(b)** its relativistic mass, **(c)** its momentum, and **(d)** its speed.

**SOLUTION:**

**(a)** Find the relativistic energy of an electron whose kinetic energy is 500 keV.

First you must express the kinetic energy in joules:

$$K = 500 \times 10^3 \text{ eV} \times 1.60 \times 10^{-19} \text{ J/eV}$$
$$= 8.00 \times 10^{-14} \text{ J}.$$

Now you are ready to find the total relativistic energy. Using Equation 39.19, you have

$$E = m_0 c^2 + K$$
$$= 9.11 \times 10^{-31} \text{ kg} \times (3.00 \times 10^8 \text{ m/s})^2$$
$$\quad + 8.00 \times 10^{-14} \text{ J}$$
$$= 8.20 \times 10^{-14} \text{ J} + 8.00 \times 10^{-14} \text{ J}$$
$$= 1.62 \times 10^{-13} \text{ J}.$$

The kinetic energy of a 500-keV electron is nearly the same as its rest energy.

**(b)** Find the relativistic mass.

You write Equation 39.3 in the form

$$m = \frac{E}{c^2} = \frac{1.62 \times 10^{-13} \text{ J}}{(3.00 \times 10^8 \text{ m/s})^2} = 1.80 \times 10^{-30} \text{ kg}.$$

This is about twice the rest mass, just as the total energy is about twice the rest energy.

**(c)** Find the momentum.

It is most convenient to solve Equation 39.20b for $p$:

$$p = c\sqrt{m^2 - m_0{}^2}$$
$$= 3.00 \times 10^8 \text{ m/s}$$
$$\quad \times \sqrt{(1.80 \times 10^{-30} \text{ kg})^2 - (9.11 \times 10^{-31} \text{ kg})^2}$$
$$= 4.66 \times 10^{-22} \text{ kg·m/s}.$$

**(d)** Find the speed.

There are several ways to do this, but the easiest is to refer to Figure 39.8b and note that

$$\frac{v}{c} = \sin\frac{p/c}{m} = \sin\frac{p}{mc}.$$

You thus have

$$\frac{v}{c} = \sin\frac{4.66 \times 10^{-22} \text{ kg·m/s}}{1.80 \times 10^{-30} \text{ kg} \times 3.00 \times 10^8 \text{ m/s}} = 0.759.$$

That is, the electron is traveling at about three-quarters the speed of light. You could not have used Newtonian physics to obtain the desired results. You multiply through by $c$ to obtain

$$v = 0.759 \times 3.00 \times 10^8 \text{ m/s} = 2.28 \times 10^8 \text{ m/s}.$$

## The Classical and Extreme Relativistic Limits

The argument leading to Equation 39.12 shows that the relativistic kinetic energy simplifies to the form $K = \frac{1}{2}mv^2$ in the **classical limit** $v \ll c$. The kinetic energy also assumes a simple form in the **extreme relativistic limit** $v \simeq c$.

The kinetic energy is given in general by Equation 39.21, $K = \sqrt{p^2c^2 - m_0{}^2c^4} - m_0c^2$. As $v$ approaches $c$, the relativistic mass $m$ increases without limit. Consequently, the momentum $p = mv$ also increases without limit. The speed-independent term $m_0{}^2c^4$ under the radical becomes negligibly small compared with $p^2c^2$, and the term $m_0c^2$ becomes negligibly small compared with $pc$. We thus have

$$K = pc \quad \text{for } pc \gg m_0c^2 \text{ or } v \simeq c. \tag{39.22}$$

Figure 39.9 shows how $K$ approaches $pc$ as $v/c$ approaches 1.

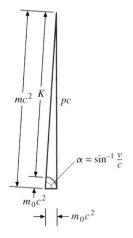

**FIGURE 39.9** The triangle of Figure 39.8a in the case $v \simeq c$. As the angle $\alpha$ approaches $\pi/2$ rad, $K$ approaches $pc$.

We have already noted that, although in principle the speed of a particle of rest mass $m_0$ can approach $c$ as closely as we wish, a particle with even a very small mass cannot achieve the speed of light. This is because *any* nonzero value of $m_0$ will result in a relativistic mass and momentum that increase without limit as $v \to c$. However, there exist particles with *zero rest mass*. The **neutrino** has either zero rest mass or at most a very tiny rest mass.

A particle traveling at the speed of light can carry momentum and energy even if its rest mass is zero. (Indeed, from the relativistic point of view, there is no paradox in a particle's having zero rest mass, precisely because momentum has meaning in the absence of rest mass.) A particle without rest mass must always travel at *exactly* the speed of light, from the point of view of any observer. For any other value of $v$, the relativistic mass equation,

$$E/c^2 = m = m_0/\sqrt{1 - v^2/c^2},$$

yields $E = 0$; this contradicts the assertion that the particle carries energy. For $v = c$, the equation yields the indeterminate value $0/0$. Although this result is not informative, it is at least not contradictory. The indeterminate value of the equality $m = 0/0$ simply means that the relativistic mass $m$ is independent of the rest mass when $m_0 = 0$.

You should not lose sleep over the concept of zero rest mass. A particle that *always* moves with speed $c$ can never be placed on a balance pan and weighed at rest. All we really mean when we say that $m_0 = 0$ is that the relativistic mass and momentum of a neutrino depend entirely on its energy, which is purely kinetic. From Equation 39.20a, we have $p = E/c$, and from Equation 39.20b, we have $m = p/c = E/c^2$.

The significance of the speed of light is thus much broader than the fact that it is the speed at which light travels. It is, rather, the *limiting* speed for any particle of nonzero mass and the *only possible* speed for any particle of zero rest mass. Aside from neutrinos, there are a few other particles whose rest mass is zero. One is the *photon*, a zero-rest-mass particle associated with light, which will be considered in Chapter 40. The *graviton*, whose existence is postulated on the basis of gravitational theory, also has zero rest mass. But gravitons have not yet been detected.

---

# Chemical and Nuclear Reactions

We have already noted that the thought experiment of Figure 39.7, with two particles pushed apart by a spring, has direct counterparts in the "real" world. One such counterpart can be seen in chemical reactions. Let us consider, as a simple example, the chemical reaction in which a zinc atom and an oxygen atom combine to form a zinc oxide molecule:

$$Zn + O \to ZnO + 3.61 \text{ eV}. \tag{39.23}$$

This is a typical *exothermic* reaction—a reaction in which energy is released to the surroundings. (When the reaction takes place in a laboratory vessel, the energy ultimately becomes apparent as heat.) The original potential energy of the "spring" is thus 3.61 eV.* As in all chemical reactions, the energy is electric potential energy, and its conversion into kinetic energy arises from the rearrangement of the electrons near the outer bounds of the two atoms.

A serious question now arises: Nothing is more thoroughly established in chemistry than the conservation of mass in chemical reactions. The mass of the product (here ZnO) is always the same as the mass of the reactants (here Zn and O). How can we reconcile this fact with relativistic principles? The answer lies in the calculation of Example 39.4.

---

*Contrary to the thought experiment of Figure 39.7, where the spring pushes the bodies apart, the spring here pulls the atoms together to form a molecule. But that is immaterial for the present discussion.

EXAMPLE **39.4**

Let the combined initial mass of the reactant Zn and O atoms be $M$. Find the proportional mass change $\Delta M/M$ of the atoms in Reaction 39.23. Assume that the zinc atom is the isotope $^{64}$Zn, whose mass is $m_{Zn} = 63.9291$ u, and the oxygen atom is the isotope $^{16}$O, whose mass is $m_O = 15.99491$ u.

**SOLUTION:** First, you need to express the mass and the energy in SI units. You have for the energy

$$U = 3.61 \text{ eV} \times 1.6 \times 10^{-19} \text{ J/eV} = 5.78 \times 10^{-19} \text{ J}.$$

The corresponding relativistic mass change is a loss, because the energy escapes from the original system. You have

$$\Delta M = \frac{U}{c^2} = \frac{5.78 \times 10^{-19} \text{ J}}{(3 \times 10^8 \text{ m/s})^2} = 6.42 \times 10^{-36} \text{ kg}.$$

For the atomic masses, you use $1 \text{ u} = 1.661 \times 10^{-27}$ kg. You have

$$M = (63.9291 + 15.994\,91) \text{ u} \times 1.661 \times 10^{-27} \text{ kg/u}$$
$$= 1.33 \times 10^{-25} \text{ kg}.$$

The proportional mass loss is

$$\frac{\Delta M}{M} = \frac{6.42 \times 10^{-36} \text{ kg}}{1.33 \times 10^{-25} \text{ kg}} = 4.84 \times 10^{-11}.$$

The ultimate sensitivity of the best modern balances is something like 1 part in $10^8$. You can see why it is so often satisfactory to consider conservation of mass and conservation of energy as separate principles.

## Nuclear Reactions

The situation becomes quite different from that described in Example 39.4 when we consider *nuclear reactions*. There are many types of nuclear reactions, but the most dramatic is **fission**. In this process, the nucleus of an atom—usually one of large atomic mass, comprising many protons and neutrons—splits into two pieces of more or less equal size, often together with a few small fragments. In some cases, the process is *spontaneous*; that is, it occurs without external interference. In other cases, fission is caused by a collision between the nucleus and a neutron.* This latter process, first observed in 1938, is called **neutron-induced fission**.[†] The neutron, which is electrically neutral, does not need to have great initial kinetic energy to approach the positively charged nucleus, which would strongly repel a positively charged particle such as a proton. Thus neutrons collide readily with atomic nuclei.

An atomic nucleus contains only positive charge. In view of the strong electrostatic repulsion among its components—called *nucleons*—what is the ''glue'' that keeps the undisturbed nucleus together? The ''glue'' is the *strong nuclear force*. This force is unfamiliar in everyday experience because, unlike the gravitational and electric forces, its range is very short. Indeed, its strength drops to zero at any distance larger than the radius of a typical atomic nucleus—that is, something less than $10^{-14}$ m. Consequently, we never experience the strong nuclear force on the macroscopic scale. At distances comparable to the nuclear radius, however, the strong nuclear force can act attractively so as to overcome the electrostatic repulsion.

The impact of a neutron disturbs the nucleus. You can picture the disturbance as distortion of an originally spherical liquid drop, as shown in Figure 39.10. If the nucleus is sufficiently distorted, the strong nuclear force can no longer hold it together. The repulsive electric force takes over and pushes the two *fission fragments* apart. (Once they are separated, each of the fission fragments constitutes an independent atomic nucleus, of atomic mass roughly half that of the parent nucleus.)

In its gross operation, the system strongly resembles the idealized system of Figure 39.7. The ''string'' is the strong nuclear force; the ''spring'' is the electric repulsion between the fission fragments; and the two bodies are the fragments themselves. The energy of the ''compressed spring'' is the initial electrostatic potential energy of the two positively charged fragments.

---

*A neutron is a particle having approximately the same mass as a proton but zero electric charge.

[†]The discovery was first made by the German chemist Otto Hahn (1879–1968) and his physicist associate Fritz Strassmann (b. 1902). The work was quickly followed up by many others and led to the development of the first nuclear reactor (1942) and the atomic bomb (1945).

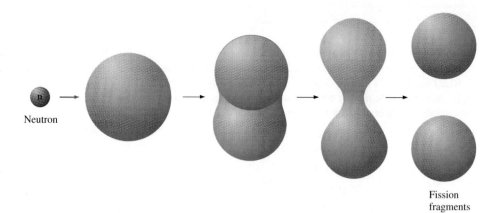

Neutron

Fission
fragments

**FIGURE 39.10** Sequence of drawings of a disturbed nucleus.

## EXAMPLE 39.5

Make a crude calculation of the electrostatic energy released in nuclear fission. Suppose that a uranium nucleus, of charge $+92e$, splits into two fragments each having charge $+46e$. Suppose, moreover, that the electrostatic force begins to dominate when the distance between the centers of the two fragments is $r_0 \simeq 1 \times 10^{-14}$ m. **(a)** How much energy is ultimately released to the surroundings? **(b)** Let the uranium nucleus be the isotope $^{235}$U, of atomic mass 235 u, and suppose that the fission is caused by collision with a neutron of atomic mass 1 u. Find the proportional mass change $\Delta M/M$.

**SOLUTION:**

**(a)** How much energy is ultimately released to the surroundings?

For a crude estimate, it is adequate to treat the fission fragments as point charges. Using Equations 25.14b and 25.5, you have

$$ U = \frac{1}{4\pi\epsilon_0} \frac{q_1 q_2}{r} = \frac{1}{4\pi\epsilon_0} \frac{(46e)^2}{r}. $$

You insert the values of the constants, and the value of $r_0$, to obtain

$$ U = 8.99 \times 10^9 \ \text{N·m}^2/\text{C}^2 \times \frac{(46 \times 1.6 \times 10^{-19} \ \text{C})^2}{1 \times 10^{-14} \ \text{m}} $$

$$ \simeq 5 \times 10^{-11} \ \text{J}. $$

This value corresponds to about 300 MeV—that is, a factor of about $8 \times 10^7$ greater than the typical chemical energy given in Example 39.4, about 4 eV.

**(b)** Find $\Delta M/M$.

The relativistic mass of change is

$$ \Delta M = \frac{5 \times 10^{-11} \ \text{J}}{(3 \times 10^8 \ \text{m/s})^2} \simeq 6 \times 10^{-28} \ \text{kg}. $$

The original mass of the system is that of the uranium nucleus and the neutron:

$$ M = (235 + 1) \ \text{u} \times 1.661 \times 10^{-27} \ \text{kg} \simeq 4 \times 10^{-25} \ \text{kg}. $$

The ratio is of order of magnitude

$$ \frac{\Delta M}{M} \simeq 1 \times 10^{-3}, $$

or about one part per thousand. This is a small but far from trivial change.

As a general rule, it is not possible to refine the calculation of Example 39.5 into an accurate calculation because it is not possible—or at best very difficult—to replace the crude assumptions with accurate descriptions. But we often find it unnecessary to go into such detail. Rather, we make careful measurements of the initial and final rest masses of the system. We then use the difference in these masses to calculate the energy released.

EXAMPLE 39.6

One of the first fission reactions studied was the process

$$n + {}^{238}U \rightarrow {}^{143}Ba + {}^{93}Kr + 3\,n + E,$$

which led Hahn and Strassmann to discover neutron-induced fission. Find the energy released per fission. The atomic masses are as follows:

n: $m_n = 1.008\,665\,u$
${}^{143}Ba$: $m_{Ba} = 142.9205\,u$
${}^{238}U$: $m_U = 238.0508\,u$
${}^{93}Kr$: $m_{Kr} = 92.931\,13\,u$

Find the total energy $E$ released in the process.

**SOLUTION:** The easiest way to proceed is to find the net mass change:

$$\Delta M = m_U - 2m_n - m_{Ba} - m_{Kr}$$
$$= 238.0508\,u - 2 \times 1.008\,665\,u$$
$$- 142.9205\,u - 92.931\,13\,u$$
$$= 0.1818\,u.$$

Note that this change represents a loss of mass. Now, reexpress this result in kilograms:

$$\Delta M = 0.1818\,u \times 1.661 \times 10^{-27}\,kg/u$$
$$= 3.020 \times 10^{-28}\,kg.$$

In the last step, you find the energy to which this mass loss is equivalent. You have

$$E = \Delta M\,c^2 = 2.718 \times 10^{-11}\,J,$$

or $\quad E = 169.9\,MeV.$

## THINKING LIKE A PHYSICIST

If you look back on the calculations of Examples 39.5 and 39.6, you will see that we have obtained important information concerning two complicated systems about whose details we have said very little, by substituting for the actual systems a simple model consisting of nothing more than two masses and a spring, tied together with a string. This is the model first introduced in Figure 39.7 and used as the basis for developing the principle of conservation of mass-energy. Isn't it remarkable that so simple a model can be used both to develop a profound physical principle and to interpret the experimental evidence that verifies the principle? For initial inquiry into an unfamiliar world, the simplest possible model is often the best.

Needless to say, a lot of important information about chemical and nuclear reactions depends on the structural details of molecules and nuclei and will not emerge from a spring-mass model. Nevertheless, the mileage to be gained from simple, apparently artificial models is a never-ending source of wonder to even the most experienced physicist. What other models that you have studied in this book have surprised you with their power?

The neutron-induced reaction of Example 39.6 is of the not-uncommon type in which two or more neutrons are among the fission fragments. This opens the possibility of a **chain reaction**, the process that underlies the operation of **fission reactors**. A stray neutron induces the first fission; if not too many of the product neutrons are lost through absorption or other processes, more than one neutron is available for further fission processes. The average number of neutrons produced by each reaction and surviving to induce a new reaction is called the **multiplication ratio** $\mu$ of the process. If $\mu < 1$, the process will not sustain itself, but will quickly die down. This condition is called **subcritical**. If $\mu = 1$, the fission process continues at a constant rate, and the power output is constant. This condition, called **critical**, is the condition for which nuclear reactors are adjusted. If $\mu > 1$, the rate of the process increases exponentially. This condition is called **supercritical**. The result is a runaway chain reaction, and huge amounts of energy can be released very quickly. This is what happens in "atomic bombs." The three possible processes—subcritical, critical, and supercritical—are shown schematically in Figure 39.11.

Evidently, it is of the utmost importance to control the value of $\mu$ in nuclear reactors. This is often done by using rods of materials such as cadmium, which are effective neutron absorbers. By inserting and withdrawing the rods, the reaction rate can be controlled with great precision.

**Obscure Heroes**

When the first nuclear reactor was allowed to reach criticality on the night of 2 December 1942, precautions had to be taken againt unforeseen events. A "suicide squad" of young (and brave) physicists and chemists was stationed on top of the reactor. Each had a glass carboy, filled with a saturated solution of a cadmium salt, and a sledgehammer. If all else failed, they were to smash the bottles so that the solution would flow into the reactor and stop the reaction. One of them, the chemist James B. Parsons, recalled much later that—at least in retrospect—the worst part of the job was the preparation. Cadmium salts are hard to dissolve, and he spent long hours rolling the heavy jugs up and down the hall, trying to get the stuff into solution.

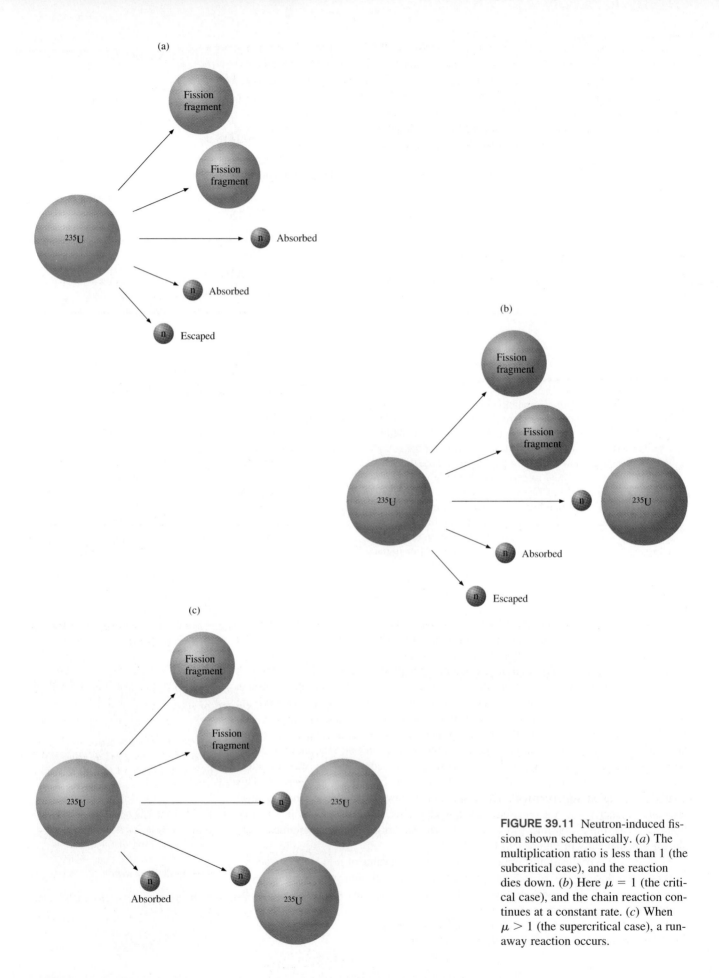

(a)

Fission
fragment

Fission
fragment

$^{235}$U

n   Absorbed

n   Absorbed

n   Escaped

(b)

Fission
fragment

Fission
fragment

$^{235}$U

n   $^{235}$U

n   Absorbed

n   Escaped

(c)

Fission
fragment

Fission
fragment

$^{235}$U

n   $^{235}$U

n

Absorbed   n   $^{235}$U

**FIGURE 39.11** Neutron-induced fission shown schematically. (*a*) The multiplication ratio is less than 1 (the subcritical case), and the reaction dies down. (*b*) Here $\mu = 1$ (the critical case), and the chain reaction continues at a constant rate. (*c*) When $\mu > 1$ (the supercritical case), a runaway reaction occurs.

In nuclear reactors, most of the energy is released in the form of the kinetic energy of the fission fragments. These fragments make repeated collisions with other atoms in the reactor. The kinetic energy is quickly distributed randomly among many atoms and is thus transformed into heat energy. In commercial power reactors, the heat is used to produce steam, which drives a turbine generator, very much as in a fossil-fuel reactor. The dramatic difference between nuclear and fossil-fuel reactors lies in the factor of $8 \times 10^7$ or so that distinguishes the energy of nuclear reactions from that of chemical reactions. One kilogram of nuclear fuel yields about as much energy as 800,000 tonnes of coal and reduces the carbon dioxide burden of the atmosphere by a corresponding amount. However, nuclear fuels have their own problems, mostly having to do with the radioactivity of the waste. Power production is a colossal systems-engineering problem!

## Symbols Used in Chapter 39

| | | | |
|---|---|---|---|
| **a** | acceleration | **p**, $p$ | momentum |
| $\mathscr{B}$ | magnetic field | $U$ | potential energy |
| $c$ | speed of light | **V**, $v$ | velocity, speed of a body |
| $E$ | total relativistic energy | $V$ | speed of a reference frame with respect to an observer |
| **F** | force | | |
| $K$ | kinetic energy | $W$ | work |
| $m$, $m_0$ | relativistic mass, rest mass | $\mu$ | multiplication ratio |
| $M$ | total relativistic mass of a system | $\omega$ | angular speed |

## Summing Up

We repeat the Mach experiment in an adaptation designed for use by two observers moving at speed $V$ with respect to one another. Taking the principle of conservation of momentum as a starting point and using the rules of relativistic kinematics developed in Chapter 38, we find that the mass $m$ of a body of rest mass $m_0$ depends on its speed $v$ with respect to the observer. The dependence is given by Equation 39.3,

$$m = \frac{m_0}{\sqrt{1 - v^2/c^2}}.$$

When a particle of rest mass $m_0$ is accelerated from rest, its kinetic energy can be expressed as a function of the relativistic mass change according to Equation 39.10,

$$K = mc^2 - m_0 c^2.$$

Because $m$ is itself a function of speed $v$, the kinetic energy can also be expressed in terms of $v$ according to Equation 39.11b,

$$K = m_0 c^2 \left( \frac{1}{\sqrt{1 - v^2/c^2}} - 1 \right).$$

The total relativistic energy $E$ of a particle can be expressed as the sum of its kinetic energy $K$ and its rest energy $m_0 c^2$. From this follows Equation 39.15, which expresses $E$

in terms of the relativistic mass according to the relation

$$E = mc^2.$$

The **principle of conservation of mass-energy** is expressed by either of Equations 39.19a and 39.19b:

$$M_i c^2 = M_f c^2 \quad \text{or} \quad M_{0i} c^2 + K_i = M_{0f} c^2 + K_f.$$

According to this principle, the removal of energy from a system must be accompanied by a change in its rest mass. For chemical reactions this change is negligibly small, but for nuclear reactions the change is significant.

Total relativistic energy $E$, momentum $p$, relativistic mass $m$, and rest mass $m_0$ are related by either of Equations 39.20a and 39.20b:

$$E = \sqrt{p^2 c^2 + m_0^2 c^4} \quad \text{or} \quad m = \sqrt{(p/c)^2 + m_0^2}.$$

The kinetic energy $K$ is related to $p$, $m$, and $m_0$ according to Equation 39.21,

$$K = \sqrt{p^2 c^2 + m_0^2 c^4} - m_0 c^2.$$

In the **extreme relativistic limit**, this relation simplifies to the form of Equation 39.22, $K = pc$.

## KEY TERMS

### Section 39.2  Mass and Momentum
rest mass, relativistic mass ▪ relativistic momentum

### Section 39.3  Force and Energy
relativistic force ▪ relativistic kinetic energy

### Section 39.4  Conservation of Mass-Energy
rest energy, relativistic energy ▪ classical limit, extreme relativistic limit ▪ neutrino

### Section 39.5  Chemical and Nuclear Reactions
fission, neutron-induced fission ▪ chain reaction, fission reactor

## Queries and Problems for Chapter 39

## QUERIES

**39.1** *(3) Shoving them along.* A machine that increases the energy of elementary particles (such as protons or electrons) to large values is conventionally called a particle *accelerator*. Why is this name not really descriptive? (Perhaps the machine should be called a *ponderator*!)

**39.2** *(4) Letting go.* The spring in the system of Figure 39.7 is allowed to expand, and the two masses ultimately come to rest owing to frictional interactions with other objects. Is the total rest mass of the system increased or decreased over the entire process?

**39.3** *(4) Rest-mass dependence.* Sketch a graph of relativistic kinetic energy $K$ versus $pc$ for **(a)** a particle with zero rest mass, **(b)** a particle with small rest mass, and **(c)** a particle with larger rest mass. (Hint: What form must curves **b** and **c** have for $v \ll c$? for $v \simeq c$?)

## PROBLEMS

### GROUP A

**39.1** *(2) Tepid electron.* At what speed is the mass of an electron increased by 1% over its rest mass?

**39.2** *(2) Warmish proton.* At what speed is the mass of a proton increased by 0.5% over its rest mass?

**39.3** *(3) Hot electron, I.* What is the value of the ratio $v/c$ for an electron having kinetic energy 100 keV?

**39.4** *(3) Hot electron, II.* What is the value of the ratio $v/c$ for an electron having kinetic energy 1 MeV?

**39.5** *(3) Really hot electron.* An electron has total relativistic energy 1 J. Find **(a)** its speed, **(b)** its relativistic mass, and **(c)** its momentum. (Hint: You will need to use the approximation suggested in Problem 38.9.)

**39.6** *(3) Double negative . . .* An electron has relativistic mass twice its rest mass. **(a)** What is its speed? Express your result as the ratio $v/c$. **(b)** What is its kinetic energy? Express your result in J and in eV.

**39.7** *(3) . . . makes a positive?* A proton has relativistic mass twice its rest mass. What is its kinetic energy? Express your result in J and in eV.

**39.8** *(3) Threshold.* At what speed does the relativistic kinetic energy of a body differ from its classical kinetic energy by 1%?

**39.9** *(3) Dust shield.* A spaceship, traveling at high speed, strikes a cosmic dust particle of rest mass 1 μg. The collision speed is $0.9c$, and the collision is totally inelastic. How much energy is released? Compare with the explosive energy of TNT, which releases $4.2 \times 10^6$ J/kg. You will see why de-

signing a shield for such a spaceship is more difficult than designing a cowcatcher for a locomotive!

**39.10** *(4) Hidden energy.* Find the rest energy of a proton. Express your result in J and in MeV.

**39.11** *(4) Positively hot.* **(a)** What is the total relativistic energy $E$ of a proton whose speed is $0.8c$? **(b)** What is its kinetic energy $K$? Express your results in terms of $m_0$ and also in eV.

**39.12** *(4) Cosmic consciousness.* The most energetic protons found in cosmic rays have energies of order $10^{20}$ eV. **(a)** What is the speed of such a proton? Express your result as the difference $c - v$. **(b)** What is the relativistic mass of the proton?

**39.13** *(4) Gone but not forgotten.* To every elementary particle (such as an electron, a proton, or a neutron) there corresponds an *antiparticle*. When a particle and its antiparticle collide, they annihilate each other totally, producing electromagnetic radiation in the process. Suppose that a particle-antiparticle pair collide at speed $v \ll c$. What is the energy of the radiation if the particles are **(a)** an electron and its antiparticle, a positron? **(b)** a proton and an antiproton? **(c)** Why does the question make no sense for a neutrino-antineutrino pair?

**39.14** *(4) Energy from the sun, I.* The sun radiates energy at a rate of about $10^{26}$ W. **(a)** At what rate, in kg/s, does it lose mass as a result? **(b)** The mass of the sun is $1.99 \times 10^{30}$ kg. What is the fractional mass loss per year due to radiation? **(c)** If the sun lost mass in no other way, how long would it last if it continued to radiate energy at the present rate? (Note:

This must be at best an upper limit, because not all of the sun's mass can be converted into energy; see Problem 39.30.)

**39.15** *(4) Energy from the sun, II.* Sunlight falls on each square meter of the earth's surface normal to the sun–earth direction at a rate of 1.4 kW. (This is called the *solar constant.*) If the earth absorbed all this radiation and did not reemit it (*not* the case), what would be the resulting annual rate of increase of the earth's mass?

**39.16** *(5) Decay.* A free neutron is not stable but decays into a proton (p), an electron (e), and an antineutrino ($\bar{\nu}$):

$$n = p + e + \bar{\nu} + 7.8024 \times 10^5 \text{ eV.}$$

The antineutrino has zero rest mass. The rest mass of the neutron is given in Example 39.6, and the rest mass of the electron is $5.4854 \times 10^{-4}$ u. Find the rest mass of the proton.

**38.17** *(5) Relativistic chemistry, I.* Of all chemical reactions, the one that releases the greatest amount of energy per unit mass of reactant is the reaction

$$H + F \rightarrow HF + 2.79 \text{ eV,}$$

in which hydrogen and fluorine combine to form hydrogen fluoride. The atomic mass of the hydrogen isotope $^1H$ is 1.007 825 u; that of the fluorine isotope $^{19}F$ is 18.998 40 u. Find the proportional mass change $\Delta M/M$ that occurs in the reaction. (The actual overall reaction begins and ends with molecular forms: $H_2 + F_2 \rightarrow 2$ HF. But neglect the comparatively small energy difference between monatomic H and F and their molecular forms, $H_2$ and $F_2$.)

**39.18** *(5) Relativistic chemistry, II.* Hydrogen and oxygen gas are combined to form 1 kmol of water. The reaction is

$$1 \text{ kmol } H_2 + \tfrac{1}{2} \text{ kmol } O_2 \rightarrow 1 \text{ kmol } H_2O + 5.75 \times 10^8 \text{ J.}$$

What is the difference in mass between the original reactants and the final product? Is the mass change a gain or a loss?

**39.19** *(5) Uranium fission.* When $^{235}U$ undergoes fission in a nuclear reactor, the energy released depends on exactly how the nucleus splits. But on the average, the fission process yields about 200 MeV per nucleus. **(a)** What is the mass change when 1 kmol of $^{235}U$ undergoes fission? **(b)** The mass of 1 kmol of $^{235}U$ is close to 235 kg. What fraction of the initial mass is converted into other forms of energy?

## GROUP B

**39.20** *(2) Single variable, I.* Equation 39.4, $\mathbf{p} = m\mathbf{v}$, expresses the relativistic momentum in terms of two variables. Reexpress the magnitude $p$ of the momentum of a body of rest mass $m_0$ in terms of the variable $v$ only, together with any necessary constants.

**39.21** *(2) Single variable, II.* Reexpress the magnitude $p$ of the relativistic momentum of a body of rest mass $m_0$ in terms of the variable $m$ only, together with any necessary constants.

**39.22** *(3) Helping it along, I.* Through what potential difference must an electron pass if it is to acquire a speed $v = 0.95c$?

**39.23** *(3) Helping it along, II.* Through what potential difference must a proton pass if it is to acquire a speed $v = 0.90c$?

**39.24** *(3) Grand tour, I.* In the Fermilab accelerator, protons are subjected to a magnetic field that makes them follow an essentially circular orbit of radius 1 km. They reach a maximum energy of about 1 TeV ($10^{12}$ eV). **(a)** What is the centripetal acceleration of such a proton? **(b)** What is its mass? **(c)** What force must the magnetic field exert on the protons? **(d)** What is the magnitude of the magnetic field?

**39.25** *(3) A short life but a merry-go-round one.* The 450-MeV synchrocyclotron at the University of Chicago was the largest ever built. It remained in service for about a decade, beginning in 1953. In order to circumvent the difficulty described in Example 39.2, the deuterons to be accelerated were injected in bursts. The angular frequency of the electric potential applied to the dee system was varied as a burst of deuterons accelerated, so as to keep the accelerating potential in phase with the relativistic angular speed $\omega_r$, as $\omega_r$ varied with increasing energy. Each cycle of modulation produced a burst of 450-MeV deuterons, and the completion of a cycle required a little more than 1 s. **(a)** Calculate the ratio $\omega_r/\omega_c$ of the final to the initial angular speed of the deuterons. **(b)** Calculate $\omega_r$ and $\omega_c$ on the assumption that the magnetic field had the uniform value 1.6 T.

**39.26** *(3) Constant force.* A particle of rest mass $m_0$ is originally located at $x = 0$, at rest with respect to observer $O$. Using appropriate means (for example, a uniform electric field), she exerts a constant force $\mathbf{F}$ on it, parallel to the $x$ axis. She measures the speed $v$ and position $x$ of the particle at various times $t$. **(a)** Show that the speed of the particle depends on $t$ according to the relation

$$v = \frac{cFt}{\sqrt{m_0^2c^2 + F^2t^2}}.$$

(Hint: Remember that, for a body initially at rest, impulse = momentum.) **(b)** Show that the position of the particle depends on $t$ according to the relation

$$x = \sqrt{\left(\frac{m_0c^2}{F}\right)^2 + (ct)^2} - \frac{m_0c^2}{F}.$$

**39.27** *(4) Constant in a changing world.* Observers $O$ and $O'$ make observations on the same particle. Show that, regardless of their speed $V$ with respect to one another, they measure the *same* value for the quantity $E^2 - p^2c^2$. Such a quantity is said to be *Lorentz invariant*. What is the value of the quantity?

**39.28** *(4) Classical simplicity.* Show that the total relativistic energy $E$ given by Equation 39.20a simplifies in the case $v \ll c$ to the form $E = m_0c^2 + \tfrac{1}{2}mv^2$.

**39.29** *(4) "Classical" self-energy of the electron.* Imagine that you could construct an electron by bringing little bits of charge together from infinity to form a sphere of charge $-e$ and radius $R$. **(a)** Show that the resulting electrostatic potential energy, called the *classical self-energy*, is

$$U = \frac{3}{5}\frac{e^2}{4\pi\epsilon_0 R}.$$

**(b)** Using the known mass of the electron, $9.1 \times 10^{-31}$ kg, estimate the radius of the electron.

**39.30** (5) *Energy from the sun, III.* The process by which the sun produces energy was first suggested in 1939 by the German-American physicist Hans A. Bethe (b. 1906). The process involves a sequence of *nuclear fusion* reactions, in which hydrogen nuclei combine to produce helium nuclei, with the release of energy. The overall process is

$$4 \, {}^1\text{H} \rightarrow {}^4\text{He} + 2 \, \text{e} + 2 \, \nu.$$

In this reaction, e represents an electron and $\nu$ a neutrino. The rest mass of a hydrogen (${}^1$H) atom is 1.007 825 u; that of a helium (${}^4$He) atom is 4.002 60 u, that of an electron is $5.4854 \times 10^{-4}$ u, and that of a neutrino is zero. Because the colliding nuclei must overcome their mutual Coulomb repulsion to combine, a high temperature—about $2 \times 10^7$ K—is required to give them adequate thermal kinetic energy. **(a)** Find the energy released in the reaction. Express your result in MeV. **(b)** The sun radiates energy at a rate of about $10^{26}$ W. Its mass is $1.99 \times 10^{30}$ kg. If the sun were originally made of pure hydrogen and if the reaction could go on until all the hydrogen was fused into helium, how long would the sun "burn" at its current rate? For comparison, the age of the sun is about $5 \times 10^9$ years.

**39.31** (5) *Senile stars and nucleosynthesis.* When all the hydrogen in a star has been converted into helium, as in Problem 39.30, the star can still use other processes to release energy, provided it is sufficiently large that the temperatures and pressures in its core are high enough. The main processes are, successively,

$$3 \, {}^4\text{He} \rightarrow {}^{12}\text{C},$$

$$ {}^{12}\text{C} + {}^4\text{He} \rightarrow {}^{16}\text{O},$$

$$2 \, {}^{12}\text{C} \rightarrow {}^4\text{He} + {}^{20}\text{Ne} \quad \text{and} \quad 2 \, {}^{12}\text{C} \rightarrow {}^{24}\text{Mg},$$

$$2 \, {}^{16}\text{O} \rightarrow {}^4\text{He} + {}^{28}\text{Si} \quad \text{and} \quad 2 \, {}^{16}\text{O} \rightarrow {}^{32}\text{S},$$

and

$$2 \, {}^{28}\text{Si} + 2 \, \text{e} \rightarrow {}^{56}\text{Fe}.$$

Find the energy released in each of these reactions. The atomic masses are as follows: ${}^4$He: 4.002 50 u; ${}^{12}$C: 12.000 00 u; ${}^{16}$O: 15.994 91 u; ${}^{20}$Ne: 19.992 44 u; ${}^{24}$Mg: 23.985 04 u;

${}^{28}$Si: 27.976 93 u; ${}^{32}$S: 31.972 07 u; ${}^{56}$Fe: 55.9349 u.

Each reaction begins as the "fuel" for the preceding one becomes exhausted. Exhaustion spreads from the center of the star outward, and all of the reactions can take place simultaneously in "shells" at different distances from the center. The temperatures required to sustain these reactions are successively higher, as the number of protons in the nuclei increases and the Coulomb repulsion becomes stronger. For the last reaction, a temperature of about $4 \times 10^9$ K is required. Note that the processes produce less energy per unit mass as a star goes up the ladder. Building still larger nuclei is an exothermic process. If the star is big enough, it becomes unstable when all the lighter nuclei are exhausted. The core of the star begins to collapse, the temperature rising so high that some of the fusion products dissociate back to helium—an exothermic process that removes heat from the center of the star. The star collapses violently, blowing off its outer part in a *supernova event*. The blown-off gases are later incorporated into newly forming stars. This is the source of the heavier elements in our solar system. More detailed arguments show that the sun must be at least a third-generation star, some of its matter having passed through two or more supernovae. For an excellent short account of the supernova process, see S. E. Woolsey and M. M. Phillips, *Science* **240**, 750 (6 May 1988).

**39.32** (G) *Merry-go-round.* Suppose that a hydrogen atom consists of a ball-like electron in a circular orbit about a much more massive ball-like proton. When a proton captures an electron, the newly formed atom radiates 13.6 eV of electromagnetic energy. **(a)** What is the classical kinetic energy of the electron as it circles the proton? (Hint: See Section 14.5 and especially Equations 14.20 and 14.21.) **(b)** What is the total relativistic energy of the electron? **(c)** What is its relativistic kinetic energy? **(d)** By what percentage does the relativistic kinetic energy differ from the classical kinetic energy? [Note: The hydrogen atom can be treated nonrelativistically with fairly good accuracy. But the inner electrons in atoms of large atomic number (and nuclear charge) must be treated relativistically.]

## GROUP C

**39.33** (2) *Inelastic collision, I.* A body of rest mass $m_0$ is at rest with respect to an observer. Another body of equal mass, having initial speed $v$, strikes the first body. The collision is totally inelastic, and the two bodies stick together as they move away from the observer with speed $v'$. **(a)** Show that $v'$ and $v$ bear the relation

$$v' = \frac{\gamma v}{\gamma + 1}, \quad \text{where } \gamma \equiv \frac{1}{\sqrt{1 - v^2/c^2}}.$$

**(b)** Show that the final rest mass $M_0'$ of the system is

$$M_0' = m_0\sqrt{2(\gamma + 1)}.$$

**(c)** Express the result of part b in terms of the initial kinetic energy $K_0$ of the system. **(d)** From the result of part b, you can see that $M_0' \geq 2m_0$. What is the physical reason for this?

**39.34** (2) *Inelastic collision, II.* From your point of view, the relativistic mass of a particle is three times its rest mass; $m = $

$3m_0$. It makes a totally inelastic collision with another particle of equal rest mass. **(a)** What is the final rest mass $M_0'$ of the system? **(b)** Just after the collision, you measure the speed $v'$ of the system. What ratio $v/c$ do you expect? (Hint: Use the results of Problem 39.33.)

**39.35** (3) *Grand tour, II.* The now abandoned SSC (superconducting supercollider) accelerator was designed to accelerate protons to energies of about 20 TeV—about ten times the energy of the Fermilab accelerator. The radius of the ring was to be 13.5 km. What is the required value of the magnetic field?

**39.36** (3) *Not always parallel.* **(a)** Using the Lorentz transformation equations (Section 38.6), show that, when a force **F** acts in an arbitrary direction on a particle moving in an arbitrary direction with velocity **v**, the acceleration **a** of the particle is not in general parallel to the force. **(b)** Show that **a** *is* parallel to **F** for two special cases: **F** ∥ **v** and **F** ⊥ **v**.

**39.37** *(4) Photon rocket.* A galactic explorer ship makes use of a rocket engine that expels a stream of photons from the tail of the ship. **(a)** During a certain time interval, the engine expels photons having total energy $E$ and total momentum $p$. Express $E$ in terms of the ship mass $m_i$ at the beginning of the interval and the mass $m_f$ at the end of the interval. **(b)** Express $p$ in terms of $m_i$ $m_f$, and the speed change $dv$ of the ship. **(c)** An observer watches the ship start from rest and achieve speed $v$. Using the results of parts **a** and **b**, show that, from her point of view, $v$ bears the following relation to the initial and final rest masses:

$$v = c \, \frac{m_{0i}^2 - m_{0f}^2}{m_{0i}^2 + m_{0f}^2}.$$

**(d)** If the rest mass of the ship is reduced to $\frac{1}{5}$ its initial value, find $v/c$. **(e)** Compare the result of part **c** with the Tsiolkovskii formula for chemical rockets (Equation 9.14). Explain the much greater efficacy of photons as a propellant compared with chemical fuels.

**39.38** *(G) Conservative systems.* Observer $O$ measures the mass $m_i$ and the velocity $v_i$ of each of $N$ particles moving along the $x$ axis. The particles make many collisions with one another. Nevertheless, $O$ finds that

$$\sum_{i=1}^{N} m_i = M_c \quad \text{and} \quad \sum_{i=1}^{N} m_i v_i = P_c,$$

where $M_c$ and $P_c$ are constants. (The subscript "c" stands for "classical.") The constancy of $M_c$ expresses the classical conservation of mass. The constancy of $P_c$ expresses the conservation of momentum. **(a)** Another observer $O'$ moves with velocity $V$ with respect to $O$. Show that, in the classical domain, where the Galilean transformation holds, $O'$ also observes that momentum is conserved but finds a different value for $P_c$. **(b)** Show that, if $O'$ uses the Lorentz velocity transformation, he does *not* obtain a constant value for $P_c$. **(c)** Now consider the following quantities:

$$\sum_{i=1}^{N} \frac{m_i}{\sqrt{1 - v^2/c^2}} \equiv M_r \quad \text{and} \quad \sum_{i=1}^{N} \frac{m_i v_i}{\sqrt{1 - v^2/c^2}} \equiv P_r.$$

(The subscript "r" stands for "relativistic.") Show that both $O$ and $O'$ will find that $M_r$ and $P_r$ are conserved, though the two observers will find different values for these quantities.

**39.39** *(G) Lorentz transformation for momentum and energy.* Observer $O$ finds that a particle has momentum $\mathbf{p} = (p_x, p_y, p_z)$ and total relativistic energy $E$. Observer $O'$ is moving with speed $V$ in the $x$ direction with respect to $O$. Show that for $O'$ the corresponding quantities $\mathbf{p}' = (p_x', p_y', p_z')$ and $E'$ for the particle are

$$p_x' = \frac{p_x - VE/c^2}{\sqrt{1 - V^2/c^2}}, \tag{1}$$

$$p_y' = p_y, \quad p_z' = p_z, \tag{2,3}$$

and

$$E' = \frac{E - Vp_x}{\sqrt{1 - V^2/c^2}}. \tag{4}$$

**39.40** *(G) Classical check.* Show that Equations 1 and 4 in Problem 39.39 simplify to familiar classical transformation equations in the case $V \ll c$.

**39.41** *(G) Colliding beams.* An *antiproton* is a particle having the same rest mass $m_0$ as a proton but opposite charge. In the original 1955 Bevatron experiment used to produce antiprotons at the University of California, Berkeley, a beam of fast-moving protons ($p^+$) was made to collide with a metal target containing many protons essentially at rest. If the incident protons are energetic enough, new particles are created and a shower of protons and antiprotons ($p^-$) emerges. Because electric charge must be conserved, new matter is created in the form of proton-antiproton pairs, and the overall reaction is of the form

$$p^+ + p^+ \rightarrow 3\,p^+ + p^-.$$

**(a)** What is the minimum initial kinetic energy $K_0$ of the protons in the incident beam needed to produce the reaction? Express your result in terms of $m_0 c^2$ and also in eV. **(b)** What is the efficiency of the process? That is, what fraction of $K_0$ goes into the creation of new matter? What fraction of $K_0$ appears as kinetic energy of the four product particles? **(c)** Whenever possible, this method is supplanted in modern practice by *colliding-beam* experiments, in which the collision takes place between two beams of particles moving at equal speed in opposite directions. What is the efficiency of this process? (Hint: See Problem 39.40.)

# Part *XII*

# The Quantum World

Without doubt, the advent of relativity theory in the first two decades of the twentieth century would have resulted, by itself, in the most profound change of perspective on the physical world since Newton. But the *relativistic revolution* was not the only revolution that took place in these years. More or less simultaneously, it became clear that the world is not ultimately "smooth," but "grainy," and that this graininess permeates every aspect of nature. This fundamental change of viewpoint, called the *quantum revolution*, is the main subject of the last five chapters of this book. You may not be surprised to learn that Einstein made outstanding contributions to the quantum revolution as well as to the relativistic revolution. However, the quantum revolution was much less a one-man show than the relativistic revolution was.

Chapter 40 concerns the birth of the quantum revolution, which began with the discovery by Planck, Einstein, and others that light is *quantized*—grainy. That is to say, light has essential properties that can be understood only if we think of it as exhibiting properties of a stream of particles *at the same time* that it exhibits continuous, wavelike properties.

In Chapter 41, we study how quantized light interacts

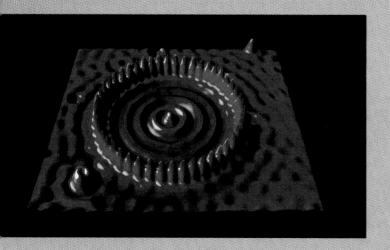

with matter. Out of this study comes our first clear picture of the structure and behavior of atoms—in particular, of hydrogen atoms.

The view of nature that arises out of Chapters 40 and 41, called the *old quantum mechanics*, is fine as far as it goes. But it has two serious shortcomings. First, attempts to expand on its initial spectacular successes soon turn out to have definite limits. Second, those limits are clearly related to the presence in the theory of *ad hoc*, unjustified assumptions. The solution to these difficulties, called the *new quantum mechanics* or *wave mechanics*, is the subject of Chapter 42.

In Chapters 43 and 44, we consider a few of the most important applications of quantum mechanics: the structure of the atomic nucleus, the electronic properties of crystalline solids, and the spectacular phenomena called superfluidity and superconductivity. All of these have important engineering applications, which we address briefly.

Chapter 45 deals with the structure of matter on the finest and most fundamental level we can study. Here we face the same questions that have motivated physicists since classical Greek times: Can we explain the entire universe on the basis of a small number of fundamental particles that interact by means of a small number of fundamental forces? What do we mean by "fundamental particle" and "fundamental force"? Unlike the plot of a detective novel, the "plot" of a physics book cannot be sandwiched between a pair of covers. The questions we have raised are by no means solved, even in principle. We close this book with a survey of the present state of affairs.

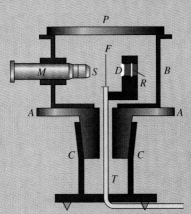

# 40

# The Quantum of Light

——— The electromagnetic spectrum emitted by a blackbody can be accounted for only if one assumes that energy is emitted into the radiation field and absorbed by the blackbody in discrete amounts $nh\nu$, called quanta.

——— The photoelectric effect, in which light of sufficiently short wavelength produces emission of electrons from metal surfaces, can be explained only if one assumes that the quanta, once emitted into the radiation field, continue their existence as quanta called photons.

——— The Compton effect, which involves the use of X-ray photons much more energetic than those used in the photoelectric effect, demonstrates that photons carry momentum and transfer it to other particles in collisions, just as familiar particles do.

——— Is light a stream of particles or a wave?

*Left:* It's hard to make out detail in the firebox of a furnace because it approximates a blackbody in which the radiation flux is uniform.

*What I did can be described as simply an act of desperation. . . . A theoretical interpretation* had *to be found at any cost, no matter how high. It was clear to me that classical physics could offer no solution. . . . The two laws [of thermodynamics], it seems to me, must be upheld under all circumstances. For the rest, I was ready to sacrifice every one of my previous convictions about physical laws.*

—MAX PLANCK (1931)

# Introduction

Is nature ultimately smooth or grainy? This question, which has intrigued thoughtful persons since classical Greek times, is easier to ask than to answer. In this chapter, we will think mostly about light and other forms of electromagnetic radiation.* In Chapters 34 through 37, we dealt with the *propagation* of light. We did so almost entirely in terms of a "smooth" theory—a theory of wave trains. We noted in passing that physicists (including Newton) have at various times thought of light as a stream of particles—that is, in terms of a "grainy" theory. That theory did not provide a satisfactory explanation of the properties of light discussed in Chapters 34 through 37. Nevertheless, a particulate theory of light does have an important appeal: If light were grainy, we could unify the treatment of particle mechanics with the treatment of electromagnetism and thus explain the entire universe in terms of the mechanics we developed in the first twenty-two chapters of this book.

In Chapters 35 and 36, we discarded the particulate view of light because it could not account for numerous phenomena that are satisfactorily explained by the wave picture. But it sometimes happens in physics that an apparently dead-end theme is revived in a more sophisticated form. In this chapter, a particulate view of light emerges from a study of the way in which matter emits and absorbs light. In Section 40.2, we consider this emission and absorption in a very general way. We conclude that the energy exchange must take place in little "packets," each one called a quantum (plural: quanta). In Section 40.3, we study the *photoelectric effect* and conclude that quanta are not merely the medium of exchange between matter and the surrounding *radiation field* but exist in the radiation field as discrete "particles" called *photons*. In Section 40.4, we study the *Compton effect*, in which photons clearly demonstrate their behavior as relativistic particles having energy and momentum.

# Blackbody Radiation

Macroscopic bodies continually emit radiation to their surroundings and absorb radiation from their surroundings. When a body is warmer than its surroundings, it radiates more energy than it absorbs. You have probably noted that a heated body—a toaster

---

*For convenience, we will use the terms "light" and "radiation" interchangeably when there is no possibility for confusion.

coil, for example—begins to glow as its temperature is increased. At first, the glow is not visible, but infrared radiation can be felt as heat. With increasing temperature, the glow becomes barely visible; it is dim and deep red. With further temperature increase, the glow becomes much brighter and whiter. For example, the tiny filament of an electric light bulb produces much more and much whiter light than a far bigger toaster element.

We focus our attention on radiation and absorption by a special kind of body called a *blackbody*. To define precisely what is meant by "blackbody," we begin with the everyday meaning of *black*. In common experience, a black body absorbs most of the visible radiation incident on it. Let us idealize and consider a body so black that it absorbs *all* of the radiation incident on it—visible, infrared, ultraviolet, or otherwise.

In particular, consider such a body situated in an enclosure in which the temperature is uniform and constant. If the temperature of the body is to remain constant, the body must emit energy at exactly the same rate at which it absorbs energy. We say that such a body is in *thermal equilibrium* with its surroundings.

Now, suppose that we place in the same enclosure a second body that is not black in the everyday sense—a shiny block of metal, say. In order to remain in thermal equilibrium with its surroundings, this second body must behave exactly as the first body does—it must emit energy at exactly the same rate at which it absorbs energy. In the enclosure, therefore, *all bodies are ideally black*. If the inside walls of the enclosure are thermally insulated from the outside world, the same argument applies to them as well. Thus the entire enclosure is ideally black, and this property does not depend on the material of which the enclosure (or its contents, if any) is made. With this in mind, we call the entire enclosure—the walls and the cavity they enclose—a **blackbody** (Figure 40.1). The absorption and emission of energy by a blackbody has a universal character, and this universality underlies the importance of studying blackbody radiation.

Suppose that the temperature of one part of the wall increases, disturbing the equilibrium. That rate at which that part of the wall emits energy will then increase and thus exceed the rate at which it absorbs energy. As a result, any temperature variation will soon smooth out. Thus the equilibrium temperature $T$ is a property of the blackbody as a whole.

Not only must every part of the wall emit and absorb total energy at the same rate per unit area as every other part, but it must do so for every range of wavelength. That is, for the range between any wavelength $\lambda$ and a neighboring wavelength $\lambda + \Delta\lambda$, the energy emitted and absorbed per unit (time · area) is the same for all parts of the blackbody wall. This situation is called *detailed equilibrium*. Why must the equilibrium be detailed? There are many ways to answer this question, but the one that will prove most fruitful for us involves the thermodynamic *principle of equipartition of energy* (Section 18.3).

Imagine the cavity to be filled with standing waves. It is difficult to visualize standing waves in three dimensions, but nothing is lost in principle by considering instead the one-dimensional analogue shown in Figure 40.2. As you learned in Section 21.7, the only wavelengths $\lambda_n$ are those that satisfy the *standing-wave condition*

$$\frac{n\lambda_n}{2} = L, \quad n = 1, 2, 3, \ldots; \qquad \textbf{(40.1)}$$

compare with Equation 21.56a. Each of these possibilities is called a **standing-wave mode**. According to the equipartition theorem, the energy of the system must be distributed equally (on the average) among all the standing-wave modes.

Suppose that some area element $dA$ in the wall absorbs more energy than it emits in one wavelength range $\lambda_1 \leftrightarrow \lambda_1 + \Delta\lambda$ and nevertheless keeps its total energy constant by emitting more energy than it absorbs in some other wavelength range $\lambda_2 \leftrightarrow \lambda_2 + \Delta\lambda$. This has the effect of "pumping" energy from one mode to another in the radiation field and thus violates the equipartition principle. *The system must therefore be not merely in overall equilibrium but in detailed equilibrium.*

It is not possible to realize an ideal blackbody. But it is possible to build a well-insulated oven that approximates the ideal blackbody quite closely. An electric heater

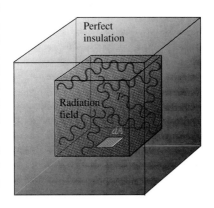

**FIGURE 40.1** An ideal blackbody. The evacuated box is surrounded by perfect thermal insulation. The inner walls are at temperature $T$. The cavity is filled with electromagnetic radiation of various wavelengths. In thermal equilibrium, any radiant energy incident on the area element $dA$ must be balanced by emission of an equal amount of energy. Thus the entire blackbody—radiation field and cavity walls—is in equilibrium.

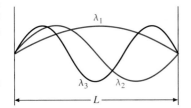

**FIGURE 40.2** A one-dimensional "cavity" of length $L$ filled with standing waves of many wavelengths. Only those wavelengths that satisfy the condition $n\lambda/2 = L$ are allowed. The first three *modes* are shown.

in the walls of the oven supplies energy to replace the energy lost to the surroundings because the device is not perfectly isolated. A precise thermostat keeps the temperature very close to any desired value. A small hole in the oven wall allows us to sample the radiation with a spectrometer, without allowing enough radiation to escape from the cavity to disturb the detailed equilibrium significantly.

Blackbodies are often operated at red heat. Because of the uniformity of the radiation inside the blackbody, it is impossible to distinguish any features inside; all you can see is an overall red glow. If you have ever looked into the firebox of an operating furnace, you have probably observed this. Even in a pile of hot coals in a fireplace (which is far from equilibrium and not a very good blackbody), the detailed form of the individual coals often disappears into a shimmering red glow.

### The Blackbody Spectrum

Starting about 1880, physicists began measuring the **blackbody spectrum**. Like all spectra, the blackbody spectrum is simply a plot of radiation intensity as a function of wavelength. By 1899, blackbody spectra like those shown in Figure 40.3 had been measured with good accuracy, and still better measurements followed. In Figure 40.3, the *spectral intensity* $I_\lambda$ is plotted as a function of wavelength $\lambda$. To understand the meaning of "spectral intensity," consider a surface illuminated by light falling normally on it with total intensity $I$ (Equation 34.38b) expressed in the SI unit $W/m^2$. We are now interested in the way in which the total intensity $I$ is distributed over the spectrum.

**FIGURE 40.3** Blackbody spectra, showing spectral intensity $I_\lambda$ as a function of wavelength $\lambda$ for temperatures $T = 4000$ K, 5000 K, and 5600 K. (The last temperature is the effective temperature of the sun, as seen from the surface of the earth.)

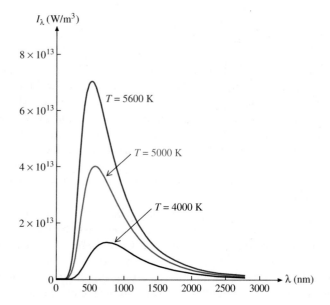

The **spectral intensity** $I_\lambda$ is defined as the intensity per unit wavelength range; that is,

$$I_\lambda \equiv \frac{\Delta I}{\Delta \lambda}, \qquad (40.2)$$

whose value depends on the particular value of $\lambda$. The SI unit of $I_\lambda$ is $(W/m^2)/m$, or $W/m^3$.

As you can see from any one of the curves of Figure 40.3, the value of $I_\lambda$ is zero for $\lambda = 0$, rises to a peak value $I_{\lambda\,max}$ at some wavelength $\lambda_{max}$, and then tails off asymptotically to zero at $\lambda = \infty$.

### The Stefan-Boltzmann Law and Wien's Displacement Law

Figure 40.3 bears out the qualitative impressions you have gathered from everyday observations of radiation from heating coils. As the temperature increases, the total intensity $I$ (the area under the curve) increases rapidly. Also, the maximum of the curve,

$\lambda_{max}$, shifts to shorter wavelengths—''toward the blue''—with increasing temperature. These observations were made quantitative in the late nineteenth century. As early as 1879, the Austrian physicist Josef Stefan (1835–1893) suggested, on the basis of the experimental evidence then available, that the total intensity of a blackbody emitter was proportional to the fourth power of the absolute temperature:

$$I = \sigma T^4, \qquad (40.3a)$$

where the proportionality constant $\sigma$ is called the **Stefan-Boltzmann constant** (not to be confused with Boltzmann's constant, $k$). The value of $\sigma$ is now known to be

$$\sigma = 5.670\ 51 \times 10^{-8}\ \mathrm{W/(m^2 \cdot K^4)}. \qquad (40.3b)$$

In 1884, Boltzmann derived Equation 40.3a by means of a beautiful thermodynamic argument in which he carried a blackbody through an infinitesimal Carnot cycle; see Problem 40.62. The relation $I \propto T^4$ is therefore called the **Stefan-Boltzmann law**. The fourth-power relation of the law certainly bears out the impression that the energy radiated by a hot surface increases rapidly with increasing temperature.

The shift of $\lambda_{max}$ to smaller values as temperature increases is expressed quantitatively by the simple relation

$$\lambda_{max} T = \epsilon, \qquad (40.4a)$$

where the constant $\epsilon$ is now known to have the value

$$\epsilon = 2.897\ 756 \times 10^{-3}\ \mathrm{m \cdot K}. \qquad (40.4b)$$

Equation 40.4a is known as **Wien's displacement law**.

Although the Stefan-Boltzmann law and Wien's displacement law described important features of the blackbody spectrum, the task remained of understanding why the spectrum has the shape it has. Various efforts were made in this direction with limited success. Steadily improving experimental results imposed increasingly stringent tests on proposed theoretical forms of the function $I_\lambda = f(\lambda, T)$ displayed in Figure 40.3. In particular, a function derived in 1896 by Wien fit the experimental results not badly. Unfortunately, Wien's derivation rested on an implausible and rather vague assumption, which we need not consider here. Much worse, it was clear by 1900 that Wien's ''law'' did not agree with the improved experimental results, especially at the long-wavelength end of the spectrum.

In 1900, Lord Rayleigh tried to explain the blackbody spectrum on the basis of an argument involving no *ad hoc* assumptions at all. He began by applying the equipartition theorem to the wave modes in the cavity, as we did in the discussion after Equation 40.1. Assigning to each mode an energy $kT$ and carrying out the argument in detail, he succeeded in deriving a distribution ''law.'' This law, as modified slightly by James Jeans, has the form

$$I_\lambda = \frac{2\pi c k T}{\lambda^4}, \qquad (40.5)$$

where $c$ is the speed of light and $k$ is Boltzmann's constant. Equation 40.5 is called the **Rayleigh-Jeans law**. Because the argument on which it is founded involves no *ad hoc* assumptions, the relation contains no adjustable constants. As you can see in Figure 40.4, the fit of the Rayleigh-Jeans law to the experimental curve is good at long wavelengths. But Equation 40.5 is completely wrong at short wavelengths, where it blows up—that is, the Rayleigh-Jeans law predicts that the spectral intensity of a blackbody increases without limit at short wavelengths. If that were so, any blackbody, no matter how small or how cold, would emit infinite power! This total breakdown of the ''law'' is called the **ultraviolet catastrophe**.

The reason for the ultraviolet catastrophe lies in the standing-wave condition of Equation 40.1, $n\lambda_n/2 = L$. We can use this equation to write the wavelength of the $n$th mode as

$$\lambda_n = \frac{2L}{n}.$$

**Wien's Displacement Law**

Equation 40.4a was first proposed by the German physicist Wilhelm Carl Werner Otto Fritz Franz Wien (1864–1928), whose name was mercifully shortened to Willy Wien by almost all of his contemporaries. Wien was awarded the Nobel Prize in 1911 for his work in blackbody radiation; he also did important work in electromagnetism and pioneered in ac measurement technique.

**James Jeans**

James (later Sir James) H. Jeans (1877–1946), British physicist, astronomer, mathematician, and science writer, was noted for the extraordinary clarity of his writing, and his textbooks were used for decades.

**FIGURE 40.4** Comparison of a blackbody spectrum with the Rayleigh-Jeans "law" (Equation 40.5). Part *b* extends the comparison to a range of wavelengths much longer than that of part *a*.

(a)

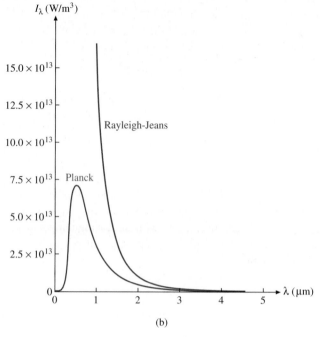

(b)

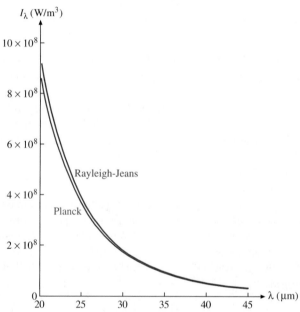

The number of modes is infinite because $n$ can have any integral value. But as $n$ becomes large and $\lambda_n$ thus becomes small, the modes are crowded closer and closer together. This becomes clear if we differentiate the equation for $\lambda_n$, obtaining

$$\frac{d\lambda_n}{dn} = -\frac{2L}{n^2};$$

the spacing between modes goes as $1/n^2$. As $n$ approaches infinity, the crowding increases without limit. For an arbitrarily short wavelength $\lambda$, the wavelength range between $\lambda$ and $\lambda + \Delta\lambda$ contains an indefinitely large number of modes—all of which have equal energy $kT$. Thus $I_\lambda$ increases without limit. A detailed analysis leads to Equation 40.5, plotted in Figure 40.4.

## Planck's Radiation Law

By 1900, Max Planck had applied himself single-mindedly to the blackbody spectrum problem for six years. Early in 1900, he found by trial and error that an excellent fit to experiment could be made by means of a slight modification of the incorrect formula Wien had proposed in 1896. With this modification, Planck proposed the formula

$$I_\lambda = \frac{b}{\lambda^5 (e^{a/\lambda T} - 1)},$$ (40.6)

where $b$ and $a$ are constants whose values are adjusted to produce the best fit to experiment. The largely trial-and-error basis of Equation 40.6 is made clear by the modest title of the paper in which Planck reported it: "On an Improvement on Wien's Spectral Equation." Superficially, Planck's modification was small; the only difference between Equation 40.6 and Wien's formula (stated explicitly in Query 40.3) is the $-1$ in Planck's denominator. However, with further intensive work along the lines of the statistical arguments made in Section 20.7, Planck found that he could derive Equation 40.6 from elementary principles if he made an *ad hoc* assumption. He was able to replace the constants $b$ and $a$ with fundamental constants, together with a single adjustable constant $h$, and write the blackbody radiation law in the form

$$I_\lambda = \frac{2\pi c^2 h}{\lambda^5 (e^{ch/\lambda kT} - 1)}.$$ (40.7)

Equation 40.7 is called **Planck's radiation law**.

As you will see in Section 40.3, the "adjustable" constant $h$ is in fact a universal constant now called **Planck's constant**:

$$h = (6.626\ 075\ 5 \pm 0.000\ 000\ 60) \times 10^{-34}\ \text{J·s}.$$ (40.8a)

We will usually use the value

$$h = 6.63 \times 10^{-34}\ \text{J·s}.$$ (40.8b)

**A New Century and a New World**

Planck announced the radiation law at a meeting of the Prussian Academy of Sciences on 14 December 1900, a date often regarded as the birthday of modern physics. Subsequent refinement of blackbody spectrum measurements has not revealed any discrepancy between experiment and the prediction of Equation 40.7. As you will see, Planck's radiation law has become so deeply embedded in the fabric of modern physics that no such discrepancy is expected.

---

**Max Planck: Conservative Revolutionary**

The German physicist Max Planck (1858–1947) personifies the conservatism of science; old, well-established theories are not abandoned for new ones except in the face of dire necessity. It took some time before the importance of Planck's work was fully appreciated; he did not receive the Nobel Prize until 1918. But Planck came to be acknowledged as the leader of the German physics community, an eminence he held until his death. The semipublic chain of national research institutes originally called the Kaiser Wilhelm Institute was renamed the Max Planck Institute after the Second World War. Planck's moral rectitude led him to take bold political positions on matters affecting the autonomy of the scientific community; at various times, he criticized the Minister of Education for holding up the appointment of a Jew to a professorship, protected political dissenters in the scientific community, and allowed women to attend his lectures. Planck struggled—eventually unsuccessfully—to minimize the influence of Nazis in the scientific community. Unfortunately, he consistently underestimated the malevolence of the Nazis; ironically, his son was executed for participating in the attempted assassination plot against Hitler in 1943. Distraught and disillusioned by the catastrophes of his old age, he spent his last years as an itinerant preacher. See J. L. Heilbron, *The Dilemma of an Upright Man* (Berkeley: University of California Press, 1986).

---

## Planck's Quantum Hypothesis

Let us now look closely at the assumption Planck made in deriving Equation 40.7. Like Rayleigh and others before him, Planck assumed that the electrically charged particles (electrons) in the walls of a blackbody cavity absorb energy as they are driven into oscillatory motion by the incident electromagnetic radiation. For simplicity, he assumed that the charges act like one-dimensional oscillators, each oscillating at some natural frequency $\nu$. (He could make this simplification because the details of the absorption-radiation process cannot be significant. After all, the radiation of a blackbody is independent of the material of which the walls are made.) An oscillator having any particular frequency $\nu$ is excited by light waves of frequency $\nu$ and thus of wavelength $\lambda = c/\nu$. The process works both ways, and so the oscillator emits radiation of the same frequency $\nu$.

But Planck had to make an *ad hoc* assumption he could not justify. He had to assume that an oscillator may emit energy in the form of electromagnetic radiation *only* when its total energy has one of the exact values

$$E_n = nh\nu, \quad n = 1, 2, 3, \ldots \tag{40.9}$$

In this equation, $\nu$ is the oscillation frequency of the particular oscillator and $h$ is the constant appearing in Equation 40.7. When the oscillator does emit energy, this equation says, it must emit *all* of its energy at once. The "packet" of emitted energy is called a **quantum**, a Latin word meaning "this much."

This assumption is called **Planck's quantum hypothesis**. Beginning with this hypothesis, Planck used a statistical thermodynamic argument (which we will not consider) to deduce Equation 40.7.

Planck originally intended his quantum hypothesis to be a temporary, purely calculational crutch. It led to the right answer—an equation that accurately described the blackbody spectrum. It also had the virtue that the apparently unjustifiable quantum hypothesis fit in a clearly defined way into an otherwise very solid classical theory. Planck fully expected that with further work he would be able to circumvent the quantum hypothesis and derive Equation 40.7 on the basis of some other, more justifiable condition. But, although he devoted great effort to this project over a period of years, he was never able to do so. (In hindsight, we cannot be surprised!)

Why does the quantum hypothesis avoid the ultraviolet catastrophe? The density of modes still increases without limit at short wavelengths. But now consider the *smallest* amount of energy that can be emitted by an oscillator into a mode of wavelength $\lambda$. We have for this energy $E_1 = h\nu = hc/\lambda$. For small values of $\lambda$, the minimum energy $E_1$ is large. Consider the extreme case of a mode whose wavelength is so short that the value of $E_1$ exceeds the total energy of all the oscillators in the blackbody. This mode must remain unexcited; there is not enough energy in the whole system to provide energy $hc/\lambda$ to it. As a less extreme case, consider a mode for which $E_1 = hc/\lambda \gg kT$. The minimum amount of energy that can be emitted into this mode by the walls of the blackbody is much greater than $kT$, the average energy per mode dictated by the equipartition theorem. That is, $kT$ is much closer to zero than it is to $E_1$. If the energy of the mode, averaged over time, is to be $kT$, the mode must almost always be unexcited.

We can now understand the general shape of the blackbody spectrum. At the long-wavelength end of the blackbody spectrum, where $E_1 \ll kT$, the considerations of the preceding paragraph do not enter. Consequently, the curve rises rapidly with decreasing $\lambda$, in accordance with the Rayleigh-Jeans law (Figure 40.4). But, when the value of $\lambda$ becomes small enough that $E_1 \simeq kT$, the probability of full excitement of the modes begins to drop off; the curve flattens out and then declines. At still smaller $\lambda$, the modes are empty nearly all the time, and the blackbody spectrum falls to zero (Figures 40.3 and 40.4a).

If the temperature of the blackbody is increased, the total available energy also increases, and it is possible to populate more of the short-wavelength modes with the necessary energetic quanta. As a result, the peak in the spectrum moves toward shorter wavelengths in accordance with Wien's displacement law. Also, more quanta of every energy are available, and the intensity increases for every wavelength, as shown in Figure 40.3.

Planck's intensive quest for an understanding of the black-body radiation spectrum followed a pattern that you have seen before in your study of physics. Planck's first step was a trial-and-error one, and his initial goal was limited to finding a mathematical function that would fit the experimental results. His search was motivated in large part by the steady improvement in the precision of the available measurements—a precision that gave good hope that only one plausible function would fit the data well.

Planck did not embark on even this relatively modest effort without background provided by others. He knew that the desired function had to be consistent with Wien's formula at short wavelengths and with the Rayleigh-Jeans function (Equation 40.5) at long wavelengths (because those functions *did* fit the experimental results in those ranges).

Planck could not be content merely with having found the function given by Equation 40.7. He had to achieve an understanding of *why* this particular function fit the data. In searching for this understanding, he relied on his profound knowledge of statistical thermodynamics. As the famous chemist and bacteriologist Louis Pasteur put it, "In the fields of observation, chance favors only the prepared minds."

Planck's search was successful—to a point. As we have seen, he was able to show that the function was not merely a mathematical description of the experimental data but could be derived from fundamental principles of statistical thermodynamics—except that the derivation required a single assumption that he could not justify. The search for justification is the subject of much of the rest of this chapter. But, though Planck did not himself succeed in finding the justification, he made the crucial contribution of *isolating* what he could not understand on the basis of accepted physical principles, thus making it possible for others to focus their efforts.

# The Photoelectric Effect

Planck's quantum hypothesis requires only that the exchange of energy between the oscillators in the walls of the blackbody and the radiation field within the cavity be quantized. It does not require that the energy, once injected into the radiation field, retain a "grainy" form. Figure 40.5a shows a dripping-water analogy. In 1905, however, Einstein showed that it was necessary to generalize the quantum hypothesis and

FIGURE 40.5 (*a*) Analogy to Planck's quantum hypothesis. Water passes from the funnel to the basin in the form of discrete drops (particles). In both the funnel and the basin, however, the water is in the form of a continuous fluid. (*b*) Analogy to Einstein's extension of the quantum hypothesis. Sand passes from the funnel to the basin in the form of discrete grains. But it retains its grainy character both in the funnel and in the basin.

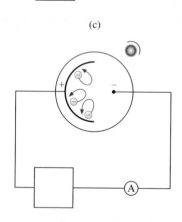

(a)

Light source of
variable wavelength
and intensity

Photoelectrons

Anode

Photocathode

Variable
voltage
source

(A)

(b)

(A)

(c)

(A)

**FIGURE 40.6** (*a*) Apparatus for experiments on the photoelectric effect. Photoelectrons are emitted by the photocathode when light (usually ultraviolet) shines on it. When some or all of the photoelectrons reach the anode, a photocurrent flows through the circuit. The flow of photocurrent can be aided or opposed by the imposition of a potential difference in the external circuit. (*b*) Arrangement for measuring the total photocurrent as a function of light intensity. (*c*) Arrangement for measuring the maximum kinetic energy of the photoelectrons.

argue that energy retains its quantized character in the radiation field. Figure 40.5b shows a flowing-sand analogy.

Einstein based his argument on the *photoelectric effect*. We now consider the major features of this effect, originally discovered accidentally in 1887 by Heinrich Hertz as he worked to improve the spark detectors he used in his work on electromagnetic radiation (Section 34.5). Many other physicists were attracted to the study of this new effect. By 1902, the following salient facts had been established.

**1.** The photoelectric effect is the light-induced emission—**photoemission**—of electrons from a metal surface, called the **photocathode**. The emitted electrons are called **photoelectrons**.

If the photocathode is placed together with an anode in an evacuated glass container, as shown in Figure 40.6a, the photoelectrons can be collected at the anode. The external wiring completes a circuit of which the vacuum tube is part. The current through the circuit is called a **photocurrent**. All the photoelectrons will contribute to the photocurrent if the external voltage source is used to hold the anode at a positive potential with respect to the cathode, as shown in Figure 40.6b. Under these circumstances, any electron emitted by the cathode will be accelerated by the electric field between the cathode and the anode and will be collected at the anode.

**2.** The photocurrent is proportional to the intensity of the light. There is no threshold intensity; reduction of the light intensity results in reduction, but never cutoff, of the photocurrent. (In practice, the proportionality has been confirmed experimentally over more than a millionfold range of light intensities.)

**3.** Photoemission is *prompt*; there is no significant time delay between the instant the light is turned on and the beginning of photoemission.

The energy of the photoelectrons can be measured by using the external voltage source to hold the anode at a negative potential $V$ with respect to the cathode—that is, by imposing a *retarding potential*, as shown in Figure 40.6c. The electrons must then move "uphill" toward the anode. The only electrons able to reach the anode are those whose kinetic energy, just after emission from the cathode, exceeds the potential energy $eV$ they acquire in moving from the cathode to the anode. As the magnitude of $V$ increases, fewer and fewer electrons reach the anode, while more and more fall back to the cathode and do not contribute to the photocurrent. At some potential $V_0$, called the **stopping potential**, the photocurrent just vanishes. The potential energy $eV_0$ is equal to the initial kinetic energy of the most energetic photoelectrons.

**4.** The greater the frequency of the light, the greater is the maximum energy of the photoelectrons. There is a threshold frequency $\nu_0$ for photoemission. Light of frequency less than $\nu_0$ does not produce photoemission, no matter how intense the light. The value of $\nu_0$ depends on the metal of which the photocathode is made. Figure 40.7 shows schematically the results of an experiment in which the stopping potential is measured as a function of the frequency of the incident light.

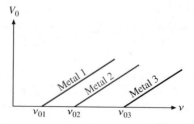

$V_0$

Metal 1

Metal 2

Metal 3

$\nu_{01}$ $\nu_{02}$ $\nu_{03}$ $\nu$

**FIGURE 40.7** Schematic plot of the stopping potential $V_0$ as a function of the frequency $\nu$ of the incident light for photocathodes made of three different metals. For each metal, the threshold frequency $\nu_0$ is shown.

## Theory of the Photoelectric Effect

Electrons are bound to the metals of which they are part. (If they were not, they would spill out spontaneously!) In nearly all metals, the binding energy for the electrons that carry electric current lies between 1 eV and 5 eV. If an electron is to leave the metal, it must acquire at least this much energy. It is not surprising that light incident on the metal surface is a possible source of this energy; from the classical point of view, the electric field vector of a light wave will exert a force on an electron. But any classical explanation of the photoelectric effect faces a key difficulty connected with items 2 and 3 in our list of facts: Photoemission is prompt and can be produced by extremely faint illumination (provided the frequency is great enough). How, then, can an electron acquire sufficient energy in a sufficiently short time? Example 40.1 illustrates the difficulty.

### EXAMPLE 40.1

Light of intensity $I = 10^{-13}$ W/m$^2$ is barely bright enough to be seen, but it can produce a clearly detectable photocurrent in a well-designed apparatus. As you saw in Example 27.4, electrons in metals make collisions roughly every $10^{-14}$ s. Because the collisions randomize the motion of the electrons, an electron must gain the energy necessary to escape the metal within a time interval comparable to this collision time interval $\tau$. Imagine that a single electron somehow accumulates all the energy falling on an area $A$ of metal surface during the interval $\tau$. Estimate the necessary value of $A$.

**SOLUTION:** As noted in the preceding text, the binding energy $\phi$ of a typical electron is of order 1 eV, or $10^{-19}$ J. This energy must be gained from the light whose intensity—energy per unit (area · time)—is $I$. So you have

$$\phi = I\tau A,$$

or $\quad A = \dfrac{\phi}{I\tau} \simeq \dfrac{10^{-19}\ \text{J}}{10^{-13}\ \text{W/m}^2 \times 10^{-14}\ \text{s}} \simeq 10^8\ \text{m}^2.$

It is inconceivable that a single electron could accumulate all the energy falling on a square 10 km on a side, even assuming that the photocathode were that big!

Einstein's solution to the difficulty illustrated by Example 40.1 was a bold extension of Planck's quantum hypothesis. Einstein asserted that *the quanta, in which energy is injected by matter into the electromagnetic radiation field, persist as quanta in the field and can leave the field by interacting individually with particles of matter.* In modern terminology, we say that *the radiation field is quantized.* Even though the radiation is very faint, these quanta, called **photons** (from the Greek word meaning ''light''), must strike the metal surface somewhere. Wherever a photon strikes, there are plenty of electrons with which to interact.

*The photon is created when it is emitted into the radiation field by matter*—say, matter in the walls of a blackbody. *The photon is annihilated in the inverse process, when it is absorbed by matter,* as in the photoelectric effect. Thus all the energy of the photon,

$$E = h\nu = \frac{hc}{\lambda}, \tag{40.10}$$

is transferred to the electron. If the light is faint, the density of photons is small. But *the energy of each photon depends only on its frequency and is independent of the light intensity.* Even a single photon can produce a photoelectron. This deduction from Einstein's photon hypothesis is consistent with the observed absence of an intensity threshold.

When is the photon energy adequate to produce photoelectrons? The minimum energy required to remove an electron from a metal is equal to the binding energy of the electron and is called the **work function** $\phi$ of that metal. We thus have a minimum requirement on the photon energy:

$$h\nu_0 = \phi. \tag{40.11a}$$

What if the energy of the photon exceeds the required minimum—that is, if $\nu > \nu_0$?

*The operation of this photographer's light meter depends on the excitation of electrons across an energy gap by the photons incident on the white hemisphere.*

The excess energy appears as the kinetic energy $K$ of the ejected photoelectron:

$$h\nu = \phi + K. \tag{40.11b}$$

Subtracting Equation 40.11a from Equation 40.11b, we obtain

$$h(\nu - \nu_0) = K. \tag{40.12}$$

Now, we have already noted that the maximum kinetic energy of electrons as they emerge from the photocathode is equal to $eV_0$, where $V_0$ is the stopping potential. So we have $eV_0 = K = h(\nu - \nu_0)$. Substituting $h\nu_0 = \phi$ into this equation for $eV_0$, we have **Einstein's photoelectric equation**,

$$eV_0 = h\nu - \phi. \tag{40.13}$$

This equation is indeed the equation of a straight line that intercepts the abscissa at $\nu_0 = \phi/h$; it is the equation of each of the curves shown in Figure 40.7. This in itself is a strong argument for the validity of Einstein's argument. But we can make a stronger statement yet. The curves of Figure 40.7 are mutually parallel; all have the same slope $h/e$. Because $e$ is known, the photoelectric effect provides a means for measuring Planck's constant $h$. We noted in Section 40.2 that Planck's constant can also be determined by adjusting its value to make the best fit between Planck's radiation law (Equation 40.7) and the experimentally determined blackbody spectrum. The two methods of evaluating $h$ are completely independent; yet, within experimental error, they yield the same result. On this basis, we can assert with confidence that Planck's constant is indeed a universal constant and not merely a number used to "fix up" Planck's law.

## EXAMPLE 40.2

You perform a photoelectric experiment, using a calcium photocathode. With an ultraviolet spectrometer, you illuminate the photocathode successively with light of three wavelengths, and you obtain the following data:

|         | Wavelength (nm) | Stopping Potential ($V_0$) |
|---------|-----------------|----------------------------|
| $\lambda_1$: | 400        | 0.40                       |
| $\lambda_2$: | 350        | 0.84                       |
| $\lambda_3$: | 300        | 1.43                       |

**(a)** Evaluate the threshold frequency $\nu_0$ and threshold wavelength $\lambda_0$ for calcium. **(b)** Use your data to obtain a value for Planck's constant. **(c)** Evaluate the work function $\phi$ for calcium.

**SOLUTION:**

**(a)** Evaluate $\nu_0$ and $\lambda_0$ for calcium.

When you plot a curve like those of Figure 40.7, it is most convenient to plot $V_0$ versus $\nu$ rather than versus $\lambda$, because Equation 40.13 is linear in $\nu$ but the corresponding equation expressed in terms of wavelength, $eV_0 = hc/\lambda - \phi$, is not linear in $\lambda$. Using the relation $\nu = c/\lambda$, you find frequencies for the three wavelengths:

$$\nu_1 = \frac{c}{\lambda_1} = \frac{3.00 \times 10^8 \text{ m/s}}{400 \times 10^{-9} \text{ m}} = 7.50 \times 10^{14} \text{ Hz};$$

similarly,

$$\nu_2 = 8.57 \times 10^{14} \text{ Hz} \quad \text{and} \quad \nu_3 = 10.0 \times 10^{14} \text{ Hz}.$$

You plot the corresponding stopping potentials versus these frequencies, as in Figure 40.8. The points do indeed lie on a straight line, and the intercept of the line with the $\nu$ axis gives you the value

$$\nu_0 = 6.54 \times 10^{14} \text{ Hz}.$$

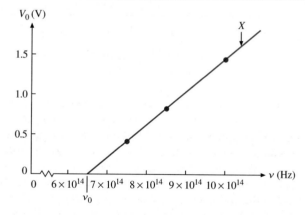

**FIGURE 40.8**

The corresponding wavelength is

$$\lambda_0 = \frac{3.00 \times 10^8 \text{ m/s}}{6.54 \times 10^{14} \text{ Hz}} = 4.59 \times 10^{-7} \text{ m} = 459 \text{ nm}.$$

**(b)** Evaluate Planck's constant.

The slope of the curve is $h/e$. If in Figure 40.8 you pick the point $X$ and the $\nu$-axis intercept in order to evaluate the slope, you obtain

$$\frac{h}{e} = \frac{(1.60 - 0) \text{ V}}{(10.42 - 6.54) \times 10^{14} \text{ Hz}} = 4.12 \times 10^{-15} \text{ V/Hz}.$$

So you have

$$h = 4.12 \times 10^{-15} \text{ V/Hz} \times 1.60 \times 10^{-19} \text{ C}$$

$$= 6.60 \times 10^{-34} \text{ J·s}.$$

This numerical value agrees within 0.5% with the accepted value.

How do you get from the unit C·V/Hz in the preceding calculation to the standard unit for $h$, J·s? Note that 1 C·V = 1 J; remember the relation $W = qV$ (Section 25.2). It requires 1 J of work to transport a 1-C charge across a 1-V potential difference. In addition, you have 1 Hz $\equiv$ 1 s$^{-1}$, and therefore 1 C·V/Hz = 1 J·s.

(c) Evaluate the work function for calcium.

You insert the numerical results of parts **a** and **b** into Equation 40.11a, $\phi = h\nu_0$, and obtain

$$\phi = 6.60 \times 10^{-34} \text{ J·s} \times 6.54 \times 10^{14} \text{ Hz}$$
$$= 4.32 \times 10^{-19} \text{ J},$$

or $\quad \phi = \dfrac{4.32 \times 10^{-14} \text{ J}}{1.60 \times 10^{-19} \text{ J/eV}} = 2.70 \text{ eV}.$

# The Compton Effect

If we grant that the electromagnetic radiation field is grainy, in the sense that it comprises individual photons, we are implying that photons have a distinctly particulate character. As with a mass of sand, we can ignore the graininess as long as we deal with large quantities. But on the microscopic level (for example, in the case of the photoelectric effect), we must deal with light photon by photon. In the sand analogy, when we deal with a pinch of sand, we cannot ignore the fact that sand consists of small grains.

Photons—quanta of light—travel at exactly the speed of light. As you learned in Section 39.4, a particle that travels at the speed of light must behave in extreme relativistic fashion. In particular, it must have zero rest mass. Its energy must be purely kinetic, $E = K$, and its energy and momentum must be related by Equation 39.22, which we write as

$$E = pc. \tag{40.14}$$

When we combine this relation with Equation 40.10, $E = h\nu$, we have

$$p = \frac{h\nu}{c} \quad \text{or} \quad p = \frac{h}{\lambda}. \tag{40.15a,b}$$

Momentum is certainly among the most basic properties of a particle and, if we wish to assert that a photon is a particle in any meaningful sense, we ought to be able to demonstrate experimentally the transfer of momentum from a photon to another particle—preferably a more familiar, nonrelativistic one. If we succeed in this demonstration, we will have considerable confidence in the particulate nature of photons.

Transfer of momentum from photons to electrons was first observed in 1922. The effect is now called the **Compton effect**.

As we will soon see, momentum is always exchanged in collisions between photons and microscopic particles. However, the effect is much easier to see with X-ray photons than with visible-light photons because of the much greater momentum of X-ray photons. (Note that the wavelength of X rays is of order 0.1 nm. This is something like $\frac{1}{5000}$ the wavelength typical of visible light. According to Equation 40.15b, the momentum of the X-ray photon is thus some 5000 times as great.)

The Compton experiment is shown in Figure 40.9. A narrow beam of monochromatic X rays (X rays having a very narrow range of wavelengths) is directed at a graphite target. The detector is an *X-ray spectrometer*, a device that can measure the wavelength of the X rays incident on it. The detector can be rotated around the target so as to detect the X rays scattered from the target at any angle $\theta$.

Most of the incident X-ray beam passes straight through the target, but some of the radiation is scattered. Compton found that the wavelength $\lambda'$ of the scattered X rays is slightly greater than the wavelength $\lambda$ of the incident X rays and that the change in wavelength, $\Delta\lambda \equiv \lambda' - \lambda$, depends on the angle $\theta$. Compton suggested that the effect was due to transfer of momentum from the photons incident on the target to the electrons with which they collided in the target. According to Equation 40.15b, $\lambda = h/p$, a

**The Compton Effect**
The Compton effect is named after its discoverer, the American physicist Arthur Holly Compton (1892–1962). A distinguished authority on X rays, Compton devoted most of his academic career at the University of Chicago to their study and won the Nobel Prize in 1927. With his student Samuel K. Allison, he wrote what remained for decades the standard work on X rays. Compton was the son of the president of the College of Wooster (Ohio), and he and his brothers all had distinguished academic careers. The Compton family furnishes a remarkable example of an active, intellectually oriented life-style in small-town turn-of-the-century America; see James R. Blackwood, *The House on College Avenue: The Comptons at Wooster, 1891–1913* (Cambridge, Mass.: MIT Press, 1968).

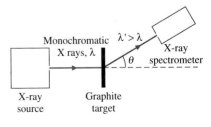

**FIGURE 40.9** The Compton effect.

decrease in photon momentum requires an increase in wavelength, and the scattered photon has a longer wavelength than the incident photon.

The analysis of the collision is much simplified if we assume that (1) the target electron is free and (2) the collision is elastic. The free-electron approximation is a good one *if* the impulse is imparted to the electron in the form of a large force of short duration. If this force is large compared with the force that binds the electron to its atom, we can safely neglect the latter force.* We also assume (3) that the electron is initially at rest. (We can determine whether the assumptions are correct or not when we have finished our analysis, by comparing the theoretical prediction with experimental results.)

Figure 40.10 shows the collision in the form of a momentum vector diagram. The incident X-ray photon has known wavelength $\lambda$, and thus its frequency is $\nu = c/\lambda$. According to Equation 40.15a, the incident photon momentum is $h\nu/c$. In the collision, the electron, whose rest mass is $m_0$, recoils with momentum $\mathbf{p} = m\mathbf{v}$. The recoil direction is denoted by the angle $\psi$. The emerging photon strikes the detector. Its frequency is $\nu' = c/\lambda'$, and its direction is expressed by the angle $\theta$.

**FIGURE 40.10** (a) Momentum vector diagram of a Compton collision between an X-ray photon of momentum $h\nu/c$ and an electron of mass $m$, initially at rest. (b) Because momentum is conserved in the collision, the sum of the final momentum vectors must be equal to the initial momentum vector.

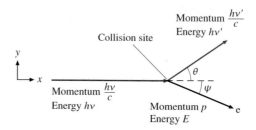

Because the collision is elastic, we can employ the principles of mass-energy conservation and momentum conservation. (We must consider the collision relativistically, conserving total mass-energy rather than kinetic energy alone, because photons are extreme relativistic particles whose speed is always $c$. We have already expressed the momentum of the photon in relativistic terms in Equation 40.15a.)

Equating the $x$ components of the initial and final total momenta, we have

$$\frac{h\nu}{c} + 0 = \frac{h\nu'}{c} \cos \theta + p \cos \psi. \tag{40.16a}$$

Equating the $y$ components of the initial and final total momenta, we have

$$0 + 0 = \frac{h\nu}{c} \sin \theta - p \sin \psi. \tag{40.16b}$$

Mass-energy conservation gives us

$$h\nu + m_0 c^2 = h\nu' + E, \tag{40.17}$$

where $E = mc^2$ is the total relativistic energy of the electron as it recoils.

Experiment gives us the values of $\nu$, $\nu'$, and $\theta$. But it is harder to observe the recoiling electron and thus determine $\psi$. In the simplest form of the Compton scattering experiment, the electron is not observed. Instead, we eliminate $\psi$ from the pair of simultaneous equations (Equations 40.16a and 40.16b). We solve Equation 40.16a for $p \cos \psi$ and Equation 40.16b for $p \sin \psi$ and then square each equation:

$$p^2 \cos^2 \psi = \frac{h^2 \nu^2}{c^2} + \frac{h^2 \nu'^2}{c^2} \cos^2 \theta - 2 \frac{h\nu}{c} \frac{h\nu'}{c} \cos \theta$$

and

$$p^2 \sin^2 \psi = \frac{h^2 \nu'^2}{c^2} \sin^2 \theta.$$

---

*In like manner, we can neglect the gravitational force exerted on a baseball during the brief interval when it is in contact with the bat; see Section 9.2.

We add these two equations to obtain

$$p^2 = \frac{h^2 v^2}{c^2} + \frac{h^2 v'^2}{c^2} - 2 \frac{h^2 v v'}{c^2} \cos \theta. \qquad \textbf{(40.18)}$$

Next, we solve Equation 40.17 for $E$ and obtain $E = h(v - v') + m_0 c^2$. Squaring both sides of this equation, we get

$$E^2 = h^2(v - v')^2 + m_0^2 c^4 + 2h m_0 c^2(v - v').$$

Expanding the factor $(v - v')^2$ and dividing both sides by $c^2$ yields

$$\frac{E^2}{c^2} = \frac{h^2 v^2}{c^2} + \frac{h^2 v'^2}{c^2} - 2 \frac{h^2 v v'}{c^2} + m_0^2 c^2 + 2h m_0(v - v'). \qquad \textbf{(40.19)}$$

We now have expressions for both $E^2/c^2$ and $p^2$. We take advantage of Equation 39.20a, $E = \sqrt{p^2 c^2 + m_0^2 c^4}$, which we cast in the form

$$\frac{E^2}{c^2} - p^2 = m_0^2 c^2.$$

We subtract Equation 40.18 from Equation 40.19 and obtain

$$m_0^2 c^2 = -2 \frac{h^2 v v'}{c^2} + m_0^2 c^2 + 2h m_0(v - v') + 2 \frac{h^2 v v'}{c^2} \cos \theta,$$

which simplifies to

$$m_0(v - v') = \frac{h}{c^2} v v'(1 - \cos \theta).$$

This equation looks something like what we have been after; it is a relation between the frequency $v$ of the incident photon, the frequency $v'$ of the scattered photon, and the scattering angle $\theta$. To make the relation more transparent, we multiply through by $c/m_0 v v'$:

$$c \frac{v - v'}{v v'} = \frac{h}{m_0 c}(1 - \cos \theta). \qquad \textbf{(40.20)}$$

But the quantity on the left side of this equation is just

$$\frac{c}{v'} - \frac{c}{v} = \lambda' - \lambda.$$

Making this substitution, we have the final result

$$\Delta \lambda \equiv \lambda' - \lambda = \frac{h}{m_0 c}(1 - \cos \theta). \qquad \textbf{(40.21)}$$

The change in wavelength, called the **Compton shift**, depends only on the scattering angle $\theta$ and not on the wavelength (or frequency) of the incident photon or on the material of which the target is made. The Compton shift has its maximum value for *backscattered* photons—that is, photons scattered through the angle $\theta = 180°$. For this angle, we have

$$\Delta \lambda_{max} = \frac{2h}{m_0 c}. \qquad \textbf{(40.22)}$$

The factor $h/m_0 c$ in Equations 40.21 and 40.22 is a constant. You can see from Equation 40.22 that it has the dimensions of length. It is called the **Compton wavelength of the electron**, $\lambda_c$, and its value is

$$\lambda_c \equiv \frac{h}{m_0 c} = 2.426\,310\,58 \times 10^{-12} \text{ m} = 2.426\,310\,58 \text{ pm}. \qquad \textbf{(40.23)}$$

In terms of the Compton wavelength, Equation 40.21 can be written in the compact form

$$\Delta \lambda = \lambda_c(1 - \cos \theta). \qquad \textbf{(40.24)}$$

# EXAMPLE 40.3

Directing 2.086-nm X rays at a target, you observe photons that are Compton scattered from electrons at an angle $\theta = 52°$. Find (a) the Compton wavelength shift $\Delta\lambda$ to three significant figures, (b) the fractional difference $\Delta\lambda/\lambda$ between the wavelengths of the incident and scattered photons, (c) the frequency shift $\Delta\nu$, and (d) the kinetic energy $K$ of the recoil electrons.

**SOLUTION:**

**(a)** Find $\Delta\lambda$.

Using Equation 40.24, you have

$$\Delta\lambda = 2.42 \text{ pm} \times (1 - \cos 52°) = 0.933 \text{ pm}.$$

**(b)** Find $\Delta\lambda/\lambda$.

You write

$$\frac{\Delta\lambda}{\lambda} = \frac{0.933 \times 10^{-12} \text{ m}}{2.086 \times 10^{-9} \text{ m}} = 4.47 \times 10^{-4}.$$

This small value is typical for Compton shifts at X-ray wavelengths.

**(c)** Find $\Delta\nu$.

The basic relation between wavelength and frequency is $\nu = c/\lambda$. To express the frequency change $\Delta\nu$ corresponding to a small wavelength change $\Delta\lambda$, you differentiate both sides of the basic relation, obtaining

$$d\nu = -\frac{1}{\lambda^2} d\lambda \qquad \textbf{(40.25a)}$$

or approximately

$$\Delta\nu = -\frac{1}{\lambda^2} \Delta\lambda. \qquad \textbf{(40.25b)}$$

For a small shift $\Delta\lambda$, it does not matter whether you use $\lambda$ or $\lambda'$ in Equation 40.25b. Using $\lambda$, you have

$$\Delta\nu = -\frac{1}{(2.086 \times 10^{-9} \text{ m})^2} \times 0.933 \times 10^{-12} \text{ m}$$

$$= -2.14 \times 10^5 \text{ Hz}.$$

The minus sign denotes a frequency decrease.

You can see that $\Delta\nu$ is small compared with the frequency of the photons:

$$\nu = \frac{c}{\lambda} \simeq \frac{3.0 \times 10^8 \text{ m/s}}{2.1 \times 10^{-9} \text{ m}} = 1.4 \times 10^{17} \text{ Hz}.$$

Although the wavelength shift $\Delta\lambda$ is independent of the wavelength $\lambda$ of the incident photon, the frequency shift $\Delta\nu$ is *not* independent of the frequency $\nu$ of the incident photon. (This point is made clear in Equation 40.20.)

**(d)** Find the kinetic energy $K$ of the recoil electrons.

You begin with Equation 40.10, $E = h\nu$. To find the mass-energy gained by the electron, you need only find the energy difference between the incident and scattered photons. So you differentiate Equation 40.10 to obtain the approximate relation

$$\Delta E = h \, \Delta\nu. \qquad \textbf{(40.26)}$$

You could determine the kinetic energy $K$ of a recoil electron by a relativistic calculation using Equation 39.21, $K = \sqrt{p^2c^2 + m_0^2c^4} - m_0^2c^2$. But the calculation is easier if you assume that the electron does not recoil at relativistic speed. The answer will tell you if the assumption is correct.

From the classical point of view, the recoil energy of the electron is purely kinetic, and you write

$$K = -\Delta E = -h \, \Delta\nu$$

$$= -6.63 \times 10^{-34} \text{ J·s} \times (-2.14 \times 10^5 \text{ Hz})$$

$$= 1.42 \times 10^{-28} \text{ J}.$$

For the check, you rewrite the classical kinetic energy $K = \frac{1}{2}m_0v^2$ in the form

$$v = \sqrt{\frac{2K}{m_0}} = \sqrt{\frac{2 \times 1.42 \times 10^{-28} \text{ J}}{9.1 \times 10^{-31} \text{ kg}}} = 17.7 \text{ m/s}.$$

Indeed you have $v \ll c$, and the classical approximation is well justified.

---

It remains to note that the predictions of the Compton-scattering equation (Equation 40.21 or 40.24) agree very well with experimental results. The photon really does act like a particle!

## Is Light a Stream of Particles or a Wave?

If we do not think carefully, we find ourselves in a logical dilemma. We have developed strong evidence in this chapter of the particle-like behavior of light. At the same time, we cannot ignore the strong evidence, presented in Chapters 34 through 37, that light exhibits wavelike behavior.

But a thing cannot be a stream of particles and a wave at the same time, because particles and waves have conflicting properties. For example, no two particles can occupy the same space at the same time; that is a fundamental property of particles. But waves obey the superposition principle, and any number of waves can occupy the same space at the same time. Thus, how can light be both a stream of particles and a wave?

As often happens, the dilemma is self-imposed. The terms *particle* and *wave* represent idealizations we have contrived. Light is the basic physical reality, not particles or waves, and so it is simply not correct to state such an identity as "Light *is* a stream of particles" or "Light *is* a wave." We must make more limited statements: "Light behaves in some ways *like* a stream of ideal particles," and "Light behaves in some ways *like* an ideal wave." Or, for short, we can say, "Light is both particle-*like* and wave*like*." Such a statement will not get us into trouble. We need only describe the particle-like and wavelike *aspects* of light.

---

| | | | |
|---|---|---|---|
| $c$ | speed of light | $T$ | absolute temperature |
| $e$ | quantum of charge | $V, V_0$ | potential of the anode with respect to the photocathode, stopping potential |
| $E$ | total relativistic energy | | |
| $h$ | Planck's constant | $\epsilon$ | Wien's displacement constant |
| $I$ | total intensity | $\theta$ | Compton scattering angle |
| $I_\lambda$ | spectral intensity | $\lambda$ | wavelength |
| $k$ | Boltzmann's constant | $\lambda_c$ | Compton wavelength |
| $K$ | kinetic energy | $\nu$ | frequency |
| $m_0$ | rest mass of the electron | $\sigma$ | Stefan-Boltzmann constant |
| $p$ | momentum | $\phi$ | work function |
| $P$ | power | $\psi$ | Compton electron recoil angle |

---

## Summing Up

The total intensity of the electromagnetic radiation emitted by a blackbody depends only on its temperature. This dependence is described by Equation 40.3a, the **Stefan-Boltzmann law**,

$$I = \sigma T^4.$$

The blackbody spectrum may be represented as a plot of the **spectral intensity** $I_\lambda$ versus wavelength $\lambda$. For any temperature $T$, the wavelength $\lambda_{max}$ at which $I_\lambda$ has its maximum value is given by Equation 40.4a, the **Wien displacement law**:

$$\lambda_{max}T = \epsilon,$$

where $\epsilon$ is a constant whose value is given by Equation 40.4b.

Blackbody spectra, such as those shown in Figure 40.3, are described excellently by **Planck's radiation law** (Equation 40.7),

$$I_\lambda = \frac{2\pi c^2 h}{\lambda^5 (e^{ch/\lambda kT} - 1)},$$

in which $h$ is a universal constant called **Planck's constant**. Derivation of Planck's radiation law from first principles requires the assumption that light is absorbed and emitted by the blackbody walls in **quanta** whose energy is restricted to the values given by Equation 40.9,

$$E_n = nh\nu, \quad n = 1, 2, 3, \ldots.$$

This assumption is called **Planck's quantum hypothesis**.

In order to explain the **photoelectric effect**, we must expand Planck's quantum hypothesis and argue that the quanta persist in the radiation field until they are annihilated when absorbed by matter. That is, *the radiation field is quantized*. The quanta in the radiation field are called **photons**. It follows that a photon of frequency $\nu$ and wavelength $\lambda$ must have energy given by Equation 40.10,

$$E = h\nu = \frac{hc}{\lambda}.$$

On the basis of this argument, we can derive a relation between the frequency $\nu$ of photons incident on a photocathode and the **stopping potential** $V_0$ required to reduce the photocurrent to zero. The relation, given by Equation 40.13, is called **Einstein's photoelectric equation**:

$$eV_0 = h\nu - \phi,$$

where $\phi$ is a property of the photocathode surface called its **work function**. The value of $h$ derived from photoelectric experiments is in accord with the value derived from blackbody experiments, thus strengthening the assertion that $h$ is a universal constant.

Special relativity predicts that a photon must have momentum given by Equations 40.15a and 40.15b:

$$p = \frac{h\nu}{c} \quad \text{or} \quad p = \frac{h}{\lambda}.$$

The particle-like properties of photons are made evident in the **Compton effect**, in which X-ray photons collide with electrons and transfer momentum to them. In the collision, the wavelength $\lambda'$ of the scattered photon is greater than the wavelength $\lambda$ of the incident photon. The **Compton shift** is given by Equation 40.21,

$$\Delta\lambda \equiv \lambda' - \lambda = \frac{h}{m_0 c}(1 - \cos\theta).$$

There is no conflict between the particle-like and wave-like properties of electromagnetic radiation. We must bear in mind that the suffix "-like" does not imply identity.

## KEY TERMS

### Section 40.2 Blackbody Radiation

blackbody ▪ standing-wave mode ▪ blackbody spectrum, spectral intensity ▪ Stefan-Boltzmann law, Stefan-Boltzmann constant ▪ Wien's displacement law ▪ Rayleigh-Jeans law ▪ ultraviolet catastrophe ▪ Planck's radiation law, Planck's constant ▪ quantum, Planck's quantum hypothesis

### Section 40.3 The Photoelectric Effect

photoemission ▪ photocathode ▪ photoelectron ▪ photocurrent ▪ stopping potential ▪ photon ▪ work function ▪ Einstein's photoelectric equation

### Section 40.4 The Compton Effect

Compton shift ▪ Compton wavelength of the electron

---

## Queries and Problems for Chapter 40

## QUERIES

**40.1** *(2) Planck thermometer.* The temperature of a glowing hot body can be measured with an *optical pyrometer*. In this instrument, lenses and mirrors are used to superimpose the image of the filament of an electric light bulb on the image of the body whose temperature is to be measured. The observer uses a variable resistor to adjust the current through the filament until the filament cannot be seen against the background. Explain how the instrument works. What are the likely sources of error in such an instrument? Suppose there is no adjustment for which the filament disappears; what can you then infer about the hot object?

**40.2** *(2) Your light or mine?* There are two ways to use infrared radiation to "see" in the dark. In the active method, you illuminate the scene with an infrared lamp equipped with a filter that blocks visible light, and you observe the light reflected from the scene with an infrared-sensitive viewer. In the passive method, you use a viewer to observe the infrared radiation emitted by the objects of interest. Discuss the advantages and disadvantages of each method.

**40.3** *(2) Not so good in the long run.* In 1896, Wien proposed to describe the blackbody spectrum with the expression $I_\lambda = b/\lambda^5 e^{a/\lambda T}$, which describes the short-wavelength part of the spectrum quite well but deviates from experimentally measured values in the middle-wavelength region and fails completely in the long-wavelength region. Explain.

**40.4** *(3) Brownian motion?* The energy of a photon can be written $E = h\nu$. In view of the fact that a photon has particle-like properties, what is the meaning of the frequency $\nu$?

**40.5** *(3) For peaches-and-cream complexions.* In photographic film, light initiates a chemical reaction in which, later during development, a silver salt (such as AgBr) is reduced to form elemental silver. The silver aggregates into grains, clumps of which appear black to the eye. The reduction reaction is endothermic and cannot occur without the energy provided by the light. Explain in a general way why on orthochromatic film (the old-fashioned black-and-white kind) pink cheeks and red lips show up as dark and why the film can be developed in red light.

**40.6** *(3) Too much reflection?* Why could you not use a pure metal surface as the sensitive element in an infrared detector?

**40.7** *(4) Fair bounce?* In principle, the Compton effect is not restricted to X rays but occurs with all electromagnetic radiation. Why do we not see the Compton effect in our everyday observations of objects around us?

**40.8** *(4) Detective work.* In the Compton effect, a photon of wavelength $\lambda$ is incident on an electron, and a photon scattered through an angle $\theta$ emerges from the target with a different wavelength $\lambda'$. Is the scattered photon the same photon with its wavelength shifted or is it a photon different from the incident photon? If the latter is the case, what happens to the initial photon? In pondering this question, ask yourself how you would go about identifying a particular photon.

**40.9** *G) Well, myopic anyway!* In a famous poem, the Scottish poet Robert Burns says, "My love's like a red, red rose." Discuss the distinction between this statement and the statement "My love's a red, red rose." Consider specific cases, not necessarily having to do with physics, in which analogy is useful and cases in which analogy pushed too far leads to difficulties.

# PROBLEMS

## GROUP A

**40.1** *(2) Dishing it out, I.* The temperature of a blackbody is 3200 K. What is the intensity of the radiation it emits?

**40.2** *(2) Dishing it out, II.* The temperature of a blackbody is 6000 K. What is the intensity of the radiation it emits?

**40.3** *(2) Thank you, Spencer Tracy!* The glowing tungsten filament of a 200-W incandescent lamp approximates a blackbody. In normal operation, the filament temperature is 3200 K. Estimate the surface area of the filament.

**40.4** *(2) Color temperature, I.* What is the value of $\lambda_{max}$ for a 3200-K blackbody? What color does the blackbody appear to be?

**40.5** *(2) Color temperature, II.* What is the value of $\lambda_{max}$ for a 6000-K blackbody? What color does the blackbody appear to be?

**40.6** *(2) Color temperature, III.* If you stand unclothed in a room that is neither chilly nor warm, your skin temperature is something like 32°C. What is the approximate wavelength of the most intense radiation you emit? Where in the electromagnetic spectrum does this radiation lie? This is the radiation on which some night-vision detectors (including the ''pits'' on the heads of rattlesnakes and other pit vipers) operate.

**40.7** *(2) Hotter than hell.* The fireball of a detonating hydrogen bomb briefly achieves temperatures of order of magnitude $10^8$ K. **(a)** What is the value of $\lambda_{max}$? Consult Figure 34.12 to find out what the predominant type of electromagnetic radiation is. **(b)** What is the energy of the most numerous photons emitted by the fireball?

**40.8** *(2) Relic and type of our ancestors' worth.* At a certain time in the early history of the universe, as it expanded during the big bang, the mean free path of photons changed quickly from a quite small value to a very large one. That is, the universe made a rapid transition from essentially ''opaque'' to essentially ''transparent.'' The universe today is filled with photons emitted during that brief transition period. Because these photons have not interacted with matter since, they display the blackbody spectrum characteristic of the universe at the time. However, the tremendous expansion of the universe since then has resulted in a very large Doppler shift of the blackbody radiation, so its temperature is now a very cold 3.2 K. At what wavelength is the blackbody radiation most intense? Consult Figure 34.12 to see in what region of the electromagnetic spectrum the predominant radiation lies. What sort of telescope do you need to observe this radiation?

**40.9** *(3) Cesium photoelectrode.* Cesium has the smallest work function for any pure metal: $\phi = 1.9$ eV. **(a)** Find the threshold frequency $\nu_0$ and the threshold wavelength $\lambda_0$. In what region of the spectrum do these values lie? **(b)** If a cesium photoelectrode is illuminated with violet light of frequency $7.5 \times 10^{14}$ Hz, what is the stopping potential for the photoelectrons?

**40.10** *(3) High-sodium diet.* A sodium photoelectrode is illuminated with ultraviolet light of wavelength $\lambda = 375$ nm. The stopping potential is $V_0 = 1.26$ V. Find the work function of sodium.

**40.11** *(3) Checking Planck.* You use a photoelectric experiment to evaluate Planck's constant. You find that, when the photocathode is illuminated with light of frequency $2.2 \times 10^{15}$ Hz, the stopping potential is 6.6 V. With light of frequency $4.6 \times 10^{15}$ Hz, the stopping potential is 16.5 V. **(a)** Find $h$. **(b)** What is the work function of the photocathode?

**40.12** *(3) Roses are red . . .* Find the frequency and the energy of a photon of red light for which $\lambda = 675$ nm. Express the energy both in J and in eV.

**40.13** *(3) . . . and violets are violet.* Find the frequency and the energy of a photon of violet light for which $\lambda = 420$ nm. Express the energy both in J and in eV.

**40.14** *(3) A handy number worth remembering.* **(a)** Show that, when the wavelength of a photon is expressed in nanometers, the photon energy in electron-volts is given by the relation

$$E \text{ (in eV)} = \frac{1240.7}{\lambda \text{ (in nm)}}.$$

**(b)** What are the dimensions of the constant in this equation?

**40.15** *(3) Going to extremes, I.* What is the energy of the photons emitted by an extremely low frequency (ELF) transmitter of the kind used to communicate with submerged submarines? The frequency is typically of order 1 Hz.

**40.16** *(3) Going to extremes, II.* A radioactive source emits $\gamma$ rays of energy 5.5 MeV. What are the frequency and wavelength of the $\gamma$-ray photons?

**40.17** *(3) Getting some rays.* The work function of magnesium is 3.68 eV. A magnesium surface is illuminated with ultraviolet light of wavelength 250 nm. **(a)** What is the maximum kinetic energy of the photoelectrons? **(b)** What is their maximum speed as they emerge from the surface?

**40.18** *(3) Commercial break.* A manufacturer of underarm deodorant buys 1 minute of time on a 50-kW radio station that broadcasts at a frequency of 1030 kHz. How many photons are used to sell the product?

**40.19** *(3) Copious, I.* The solar constant, 1353 W/m², is the intensity of the sunlight reaching the top of the earth's atmosphere. Make the approximation that the light is monochromatic, with wavelength $\lambda = 520$ nm, equal to $\lambda_{max}$ for actual sunlight. How many photons per second fall on each square meter of the upper atmosphere?

**40.20** *(3) Thresholds, I.* The work function of tungsten is 4.5 eV. Find **(a)** the threshold frequency and **(b)** the threshold wavelength.

**40.21** *(3) Thresholds, II.* The work functions of the alkali metals are as follows: Li, 2.3 eV; Na, 2.3 eV; K, 2.2 eV; Rb, 2.1 eV; Cs, 1.9 eV. For each metal, find **(a)** the threshold frequency and **(b)** the threshold wavelength.

**40.22** *(3) Stop everything!* A potassium surface is illuminated with light of wavelength 330 nm. What is the stopping potential for the photoelectrons? Let $\phi = 2.2$ eV.

**40.23** (3) *Lighting the way, I.* You illuminate a metal surface with light of wavelength 450 nm and find that the stopping potential for photoelectrons is 0.50 V. **(a)** What is the maximum kinetic energy of the photoelectrons? Express your result in eV and in J. **(b)** What is the work function of the surface? **(c)** What are the threshold frequency and threshold wavelength?

**40.24** (3) *Lighting the way, II.* When a metal surface is illuminated with light of wavelength $\lambda$, the stopping potential for photoelectrons is 3.0 V. The threshold wavelength for this surface is 500 nm. What is the work function?

**40.25** (3) *No shorter.* In an X-ray tube, the X-ray photons are produced by bombarding a metal target with electrons that have been accelerated through a large potential difference $V$. (Don't worry about the details; that's a subject for the next chapter.) **(a)** Show that no photon can be emitted with a wavelength less than

$$\lambda_{min} = \frac{hc}{eV}.$$

This statement is called the *Duane-Hunt law*. **(b)** Find $\lambda_{min}$ for $V = 90$ kV.

**40.26** (4) *Clout, I.* A visible-light photon has wavelength 550 nm. Find **(a)** its energy and **(b)** its momentum.

**40.27** (4) *Clout, II.* An X-ray photon has wavelength 0.12 nm. Find **(a)** its energy and **(b)** its momentum.

**40.28** (4) *Deck the halls with Arthur Holly.* X rays of wavelength 70.8 pm strike a graphite target. What is the wavelength of the X rays scattered at **(a)** 45°? **(b)** 125°?

**40.29** (4) *Shifty and devious.* X rays are incident on a graphite target. The Compton shift is 2.0 pm. What is the scattering angle?

**40.30** (4) *Deduction from the increase.* Compton-scattered X rays are observed at $\theta = 90°$. The wavelength is 236 pm. What is the wavelength of the incident X rays?

**40.31** (4) *Energy drain.* X rays of 0.72 nm are Compton scattered. What is the maximum fraction of the initial energy that can be lost?

**40.32** (4) *Pick-me-up.* An electron initially at rest recoils from a collision with an incident photon of energy 6.0 keV. What is the maximum energy the electron can acquire? What are the directions of motion of the electron and of the scattered photon?

**40.33** (4) *Special photon.* Consider a photon whose wavelength is equal to the Compton wavelength $\lambda_c$ of the electron. Show that the energy of such a photon is equal to the rest energy of an electron.

**40.34** (4) *Compton scattering by protons.* Compton scattering is not limited to photons incident on electrons, even though the Compton effect was first discovered with electrons. **(a)** What is the Compton wavelength of a proton? **(b)** What is the Compton shift $\Delta\lambda$ when a photon is scattered from a proton through an angle of 90°? **(c)** If the incident photon is a $\gamma$-ray photon of energy $7.5 \times 10^8$ eV, what is the energy of the Compton-scattered photon?

## GROUP B

**40.35** (2) *Intensity distribution, I.* The effective temperature of the sun is 5600 K, and the corresponding value of $\lambda_{max}$ is 520 nm, which is in the green region of the spectrum. What is the ratio of the spectral intensity $I_{675}$ at $\lambda = 675$ nm, in the red region of the spectrum, to the intensity $I_{520}$ at $\lambda_{max}$?

**40.36** (2) *Intensity distribution, II.* Using the data given in Problem 40.35, find the solar blue-to-green spectral intensity ratio $I_{425}/I_{520}$.

**40.37** (2) *Intensity distribution, III.* Using the effective solar temperature given in Problem 40.35, find the solar spectral intensity ratios **(a)** $I_{1000}/I_{520}$ for a wavelength in the infrared; **(b)** $I_{200}/I_{520}$ for a wavelength in the ultraviolet; **(c)** for a wavelength $\lambda = 1.2$ cm in the microwave region. (Hint: For long wavelengths, the Rayleigh-Jeans "law" works just fine.)

**40.38** (2) *Solar constant, I.* The energy flux that reaches the top of the earth's atmosphere from the sun is 1353 W/m². (This quantity is called the *solar constant.*) **(a)** What is the total power radiated by the sun? Take the earth–sun distance to be $1.50 \times 10^{11}$ m. **(b)** Find the temperature of the surface of the sun. Take the radius of the sun to be $6.96 \times 10^8$ m. **(c)** Show that, neglecting the (important) effects of atmospheric heat retention, the absolute temperature of rotating planets is inversely proportional to the square root of their distance from the sun.

**40.39** (2) *Solar constant, II.* If the earth did not reradiate all the energy that falls on it from the sun, it would soon become unbearably hot. (Indeed, that is what happens locally, for a few hours, on a hot summer day.) Assume (1) that the earth radiates like a blackbody; (2) that, although the earth receives solar energy only on one-half of its surface at any time, it radiates from its entire surface; and (3) that, to a crude approximation, the entire surface of the earth is at the same temperature. **(a)** What must the mean surface temperature of the earth be? Does this value sound reasonable? **(b)** What is the wavelength $\lambda_{max}$ for the reradiated energy? In what region of the spectrum does $\lambda_{max}$ lie?

**40.40** (2) *Vacation on the moon?* The moon has no atmosphere, and it rotates on its axis only once a month. The complicating effects of atmosphere and rotation can thus safely be neglected. **(a)** What is the value of the solar constant at the moon? **(b)** Assuming that the moon acts as a blackbody, find the temperature at a location on the moon where the sun appears overhead.

**40.41** (2) *Connections.* A blackbody radiates with total intensity $I$. **(a)** Derive an expression for $\lambda_{max}$. **(b)** If $I = 19.0 \times 10^4$ W/m², find $\lambda_{max}$. **(c)** What is the temperature of the blackbody?

**40.42** (2) *Something extra needed.* A resting human being generates metabolic heat at a rate of about 100 W. Suppose that, as in Problem 40.6, you are standing unclothed in a room, and your skin temperature is 32°C. Suppose also that the room is dimly lighted and draft free and that the temperature is a uniform 20°C. If your body surface area is 1.6 m² and radiates approximately like a blackbody, **(a)** how much power do you radiate to your surroundings? **(b)** how much power do you absorb from your surroundings? **(c)** You have mechanisms other than radiation for shedding excess heat, principally conduction to the air, convection, exhalation of water-vapor–saturated air at 37°C, and evaporation of sweat. Under what circumstances are these mechanisms likely to come into play?

**40.43** *(2) Little stars, big stars.* The Stefan-Boltzmann law can be used to estimate the radius of stars. The relatively nearby, bright star Capella is known to lie at a distance 45 light-years ($4.2 \times 10^{17}$ m) from the earth. The intensity of the light reaching us from Capella is 12 nW/m². Like most stars, Capella radiates in fairly good approximation to a blackbody. **(a)** What is the total power radiated by Capella? **(b)** The wavelength at which $I_\lambda$ has its maximum value is $\lambda_{max} = 560$ nm. What is the temperature of the surface of Capella? **(c)** What is the surface area of the star? **(d)** What is the radius of Capella? **(e)** Compare Capella's radius with that of the sun. You can see why stars like Capella are called *red giants*.

**40.44** *(2) Classical simplicity.* Show that Planck's radiation law (Equation 40.7) reduces to the Rayleigh-Jeans "law" (Equation 40.5) in the long-wavelength limit $hc/\lambda \ll kT$.

**40.45** *(2) Wien's radiation law.* As noted in the text, Planck arrived at the empirical form of his radiation law (Equation 40.6) by trial-and-error adjustment of Wien's "law." Wien's law fits the experimental results quite well at short wavelengths; its form is given in Query 40.3. **(a)** Show that, for small $\lambda$, Planck's law simplifies to the form shown in Query 40.3. **(b)** Using Equation 40.7, evaluate the constants $a$ and $b$ in the equation of Query 40.3. (Note that Wien's law fails just where the Rayleigh-Jeans law works best and vice versa.)

**40.46** *(2) And Planck begat Stefan and Boltzmann.* Show that the Stefan-Boltzmann law, given by Equation 40.3, is a consequence of Planck's radiation law. (Hint: Express the total intensity $I$ in terms of an integral of $I_\lambda$ over all possible values of $\lambda$. Then insert Equation 40.7 into this integral, and carry out the integration by making the substitution of variable $x \equiv hc/\lambda kT$.)

**40.47** *(2) Peak value, I.* **(a)** Beginning with Planck's radiation law (Equation 40.7), show that, at a given temperature $T$, $\lambda_{max}$ is the particular value of $\lambda$ that yields a solution to the transcendental equation

$$e^{-a/\lambda} + \frac{a}{5\lambda} - 1 = 0,$$

where $a \equiv ch/kT$. **(b)** Solve this equation by trial and error for $T = 5600$ K, and show that your result for $\lambda_{max}$ agrees with the corresponding blackbody spectrum of Figure 40.3.

**40.48** *(2) Wien's displacement law.* Planck's radiation law can be cast in the form

$$I_\lambda = G \, \frac{x^5}{e^x - 1},$$

where $G$ is a constant and $x = ch/\lambda kT$. Find the value $x = x_{max}$ for which $I_\lambda$ has its maximum value, and show that this leads to Wien's displacement law (Equation 40.4).

**40.49** *(3) Putting a stop to it.* A metal sphere of radius $r$ and wave function $\phi$ is isolated in vacuum. Monochromatic light of frequency $\nu$ shining on the sphere produces photoelectrons. How many photoelectrons can be liberated from the sphere before the process comes to a stop?

**40.50** *(3) Electron and photon.* A photon has wavelength 520 nm. **(a)** What is the speed of an electron whose kinetic energy is equal to that of the photon? **(b)** What is the speed of an electron whose momentum is equal to that of the photon?

**40.51** *(3) Working backward.* When a metal surface is illuminated with light of wavelength $\lambda_1$, the stopping potential for the emitted photoelectrons is 3.5 V. When the same surface is illuminated with light of wavelength $\lambda_2 = 2\lambda_1$, the stopping potential is 0.8 V. What is the work function of the surface?

**40.52** *(3) Photoelectric twins.* The figure below shows a vacuum tube containing electrodes made of different metals, 1 and 2, whose work functions are $\phi_1$ and $\phi_2$. The electrodes are illuminated simultaneously. Show that the photoelectric equation takes the form $eV_0 = h\nu - (\phi_2 - \phi_1)$.

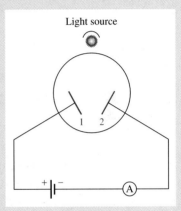

**40.53** *(3) Copious, II.* What is the number of photons contained in a 1-m³ volume at the top of the earth's atmosphere? (Hint: Solve Problem 40.19, and remember that photons move with speed $c$.)

**40.54** *(4) Lebedev's experiment, I.* We derived Equation 40.14, $E = pc$, relativistically. However, Maxwell obtained the same result in 1871, based on a classical analysis. The equation predicts that light falling on a surface should impart an impulse to the surface. The first experimental confirmation of this prediction was provided in 1898 by the Russian physicist Peter N. Lebedev (1866–1912), using the sensitive torsion balance shown below. The entire apparatus is suspended

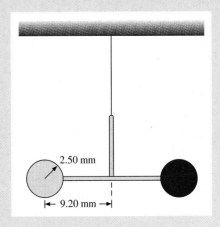

in a vacuum chamber. The glass crossarm carries two platinum disks, each of radius $r = 2.50$ mm. The distance from the center of either disk to the vertical axis of the balance is 9.20 mm. One disk is polished, and the other is blackened. (a) Express the force $F_b$ exerted on the black disk in terms of $r$ and the incident light intensity $I$. (b) Express the force $F_p$ on the polished disk in terms of $F_b$. (c) The torsion constant of Lebedev's torsion fiber was $\kappa = 2.20 \times 10^{-11}$ m·N/rad. When he illuminated the apparatus with light of intensity 775 W/m$^2$, what was the deflection angle of the torsion balance?

**40.55** (4) *Lebedev's experiment, II.* In one of Lebedev's experiments (see Problem 40.54), he used light of wavelength 560 nm and intensity 425 W/m$^2$. (a) How many photons per second fell on each disk? (b) What was the net torque exerted on the balance? (c) What was the deflection angle of the balance?

**40.56** (4) *Space sail.* It has been proposed that spacecraft could be propelled through the solar system—at least the part not too far from the sun—by means of huge sails made of very thin material. Suppose you have such a sail in the form of a square 10 km on a side. (a) If the sail is located in space near the orbit of the earth, what is the total power of the sunlight incident on it? Take the solar constant to be 1353 W/m$^2$. (b) What is the force exerted by the sunlight on the sail? Assume that the sail is perfectly reflective.

**40.57** (4) *Comet tails.* Most comets spend most of their time far from the sun and are therefore very cold. As it approaches the sun, a comet warms enough to cause the ejection of surface matter, both as vapor and as dust particles. This ejected matter forms the tail, whose shape is determined by the force exerted on the matter by the sunlight. (a) If a black, essentially spherical dust particle has density $1 \times 10^3$ kg/m$^3$, what is its radius if the force exerted on it by the sunlight is equal to and opposite the gravitational force exerted on it by the sun? Assume that the sun–comet distance is equal to the sun–earth distance, $1.50 \times 10^{11}$ m, and take the solar constant to be 1353 W/m$^2$. (b) Show that the particle size obtained in part a is independent of the distance of the comet from the sun.

**40.58** (4) *An energetic view of Compton scattering.* Show that the Compton-scattering equation (Equation 40.21) can be written in the alternative form

$$\frac{1}{\epsilon'} - \frac{1}{\epsilon} = \frac{1}{E_0} (1 - \cos \theta),$$

where $\epsilon$ is the energy of the incident photon, $\epsilon'$ is the energy of the scattered photon, and $E_0 = m_0 c^2$ is the rest energy of the electron.

**40.59** (4) *Big loss.* In a Compton backscattering collision, the scattered photon has one-half the energy of the incident photon. What is the energy of the incident photon? (Hint: See Problem 40.58.)

**40.60** (4) *Double Compton scattering.* An X-ray photon is scattered successively from two electrons before it is detected. The scattering angle is 38° in the first collision, and the scattering angle is 67° in the second collision. (a) What is the Compton shift $\Delta\lambda_2$ for the double-scattering process? (b) What is the Compton shift $\Delta\lambda_1$ for a photon detected at the same total scattering angle, 105°, after a single-scattering event? (c) How can you tell whether the observed photon has suffered a single or a double scattering?

---

## GROUP C

**40.61** (2) *Radiative cooling.* You have a metal sphere of mass $M$, density $\rho$, and specific heat capacity $\chi$. (We use $\chi$ here instead of the standard $c$ in order to prevent confusion with the speed of light.) At time $t = 0$, the temperature of the sphere is $T_i$. The sphere is mounted in vacuum in a chamber whose walls are cooled to a temperature very much less than $T_i$. As the surface of the sphere radiates energy, its temperature falls. However, the thermal conductivity of the metal is large enough that the temperature remains nearly uniform throughout the sphere. Assume that the sphere radiates like a blackbody. (a) Find a general expression for the temperature $T$ as a function of the initial temperature $T_i$ and the time $t$. (b) Find the time $t$ required for the temperature of the sphere to fall to one-half its initial value. (c) Suppose the sphere is made of copper ($\chi = 3.85 \times 10^3$ J/kg·K and $\rho = 8.92 \times 10^3$ kg/m$^3$) and has mass 50 kg. The container is cooled by liquid helium to $T = 4.2$ K. If the initial temperature of the sphere is 300 K, how long will it take to cool to 150 K?

**40.62** (2) *Boltzmann's derivation of the Stefan-Boltzmann law.* You learned in Section 18.2 that the pressure $p$ of an ideal gas is related to its (purely kinetic) energy density $u$ by Equation 18.10b, which can be written $p = \frac{2}{3}u$. A similar relation can be deduced from Maxwell's equations for electromagnetic radiation. Only the constant is different; the result is $p = \frac{1}{3}u$. (The proof of this relation is the subject of Problem 40.65.) Imagine a Carnot engine in the form of an ideal blackbody with a moveable wall, as shown in figure $a$. The Carnot cycle is shown in figure $b$. Initially, the pressure is $p$, the

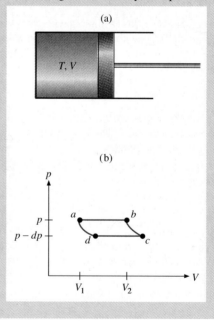

---

volume is $V_1$, and the temperature is $T$. The engine is allowed to expand isothermally to a larger volume $V_2$ in the process $a \rightarrow b$. This is followed by the infinitesimal adiabatic expansion $b \rightarrow c$, during which the temperature drops to $T - dT$ and the pressure drops to $p - dp$. The process $c \rightarrow d$ is an isothermal compression, and the system is restored to its original state by the infinitesimal adiabatic compression $d \rightarrow a$.

(a) Show that the work done by the engine during the process $a \rightarrow b$ is $-W = \frac{1}{3}u(V_2 - V_1)$. [Hints: (1) The energy density in a blackbody depends on the temperature only and therefore remains constant in an isothermal process. (2) See Problem 20.42.]

(b) Show that the change in internal energy of the engine during the process $a \rightarrow b$ is $E_2 - E_1 = u(V_2 - V_1)$. (c) Use the first law of thermodynamics to find the total heat $Q_{in}$ that flows into the engine during the entire cycle. (d) Show that the net work $-dW$ done by the engine during the cycle is $-dW = dp(V_2 - V_1)$. (e) Using the definition of thermodynamic efficiency given by Equation 20.2, $\eta = -W_{cyc}/Q_{in}$, together with the fact that the engine is a Carnot engine, show that

$$\frac{du}{u} = 4\,\frac{dT}{T}.$$

(f) Integrate this equation, and show that $u = \text{constant} \times T^4$.

**40.63** *(2) Planck's radiation law in terms of frequency.* When you use a spectrometer to study the spectrum of a blackbody, the intensity of the light you detect at wavelength $\lambda$ depends on the wavelength range, or "window," $\Delta\lambda$ you choose to resolve. That is, the intensity registered by your recording device is $I_\lambda\,\Delta\lambda$. (This is the area under the blackbody spectrum curve between the limits $\lambda$ and $\lambda + \Delta\lambda$.) Alternatively, you could scan the spectrum in terms of frequency $\nu = c/\lambda$ and determine the blackbody spectrum as a plot of intensity $I_\nu$ versus $\nu$. In this case, you would scan the spectrum with a window of width $\Delta\nu$. The intensity registered by your recording device at frequency $\nu$ would be $I_\nu\,\Delta\nu$. For every possible observation $I_\lambda\,d\lambda$, there exists a corresponding identical observation $I_\nu\,d\nu$. Assuming arbitrarily small windows ($\Delta\lambda \rightarrow d\lambda$ and $\Delta\nu \rightarrow d\nu$), show that the Planck law can be written

$$I_\nu = \frac{2\pi h\nu^3}{c^2(e^{h\nu/kT} - 1)}.$$

**40.64** *(2) Peak value, II.* (a) Beginning with the form of Planck's radiation law given in Problem 40.63, derive a transcendental equation, corresponding to the one of Problem 40.47, whose solution gives the frequency $\nu_{max}$ at which $I_\nu$ has its maximum value for any given temperature $T$. (b) Solve this equation by trial and error, and find $\nu_{max}$ for $T = 5600$ K. Compare your answer with that for Problem 40.47, and note that $\nu_{max} \neq c/\lambda_{max}$. This inequality is due to the fact that $\Delta\nu = -1/\lambda^2\,\Delta\lambda$. Thus, if you equate $I_\lambda\,\Delta\lambda = I_\nu\,\Delta\nu$, you will *not* have $I_\lambda = I_\nu$. The two functions $I_\nu$ and $I_\lambda$ are related but not the same.

**40.65** *(4) Light pressure in terms of energy density.* Suppose

a cubic box contains $n$ photons per unit volume. Let the mean energy of the photons be $\langle\epsilon\rangle$. (a) What is the total momentum per unit time transferred to each wall of the box if the photons are reflected (or absorbed and immediately reemitted) from the walls? (Hint: Assume, as for ideal-gas molecules in Section 18.2, that the photons move with equal likelihood in all directions.) (b) What is the pressure $\Pi$ exerted on each wall? (Use $\Pi$ for pressure so as to prevent confusion with momentum $p$.) (c) Express the energy per unit volume contained in the box—the energy density $u$—in terms of $\langle\epsilon\rangle$. (d) Show that $\Pi = u/3$. Compare with Equation 18.11, the analogous result for ideal gases. $\Pi V = 2N\langle\epsilon\rangle/3$.

**40.66** *(4) Stellar interiors.* A container of ideal gas in thermal equilibrium must also be a blackbody containing photons whose distribution in energy is described by Planck's law. Consider an ideal gas at very high temperature. If the temperature is high enough, the pressure exerted by photons may exceed the ordinary kinetic-theory gas pressure that arises from molecular collisions. To see this, consider a hydrogen *plasma*. This plasma consists of hydrogen completely dissociated into free protons and electrons, and at high enough temperatures it behaves like an ideal gas. (a) Show that, within a small constant factor, the temperature at which the radiation pressure is equal to the kinetic gas pressure is

$$T \simeq \left(\frac{kc}{\sigma}\,\frac{N}{V}\right)^{1/3},$$

where $N/V$ is the number of plasma particles per unit volume. (b) Suppose the mass density of the plasma is $\rho = 100\ \text{kg/m}^3$, or about 2200 times the density of atomic hydrogen at STP. (This is not an unusual density to find in the interior of a star.) Estimate the temperature at which the two contributions to the total pressure are equal. (c) The pressure at the center of the sun is about $10^9$ atm, and the temperature is about $10^7$ K. Show that, even under these conditions, kinetic gas pressure, not radiation pressure, is dominant. (In very massive stars, however, radiation pressure bears the main burden of keeping the star from collapsing under its huge gravitational self-attraction.)

**40.67** *(4) Impossible.* An electron initially at rest is struck by a photon of energy $h\nu$. (a) Write the relativistic energy-conservation equation for the collision. (b) Write the momentum-conservation equation for the collision. (c) Show that the collision cannot result in complete absorption of the photon by the electron.

**40.68** *(4) Compton recoil electrons.* In a Compton scattering event, the scattered photon emerges at an angle $\theta$ with respect to the direction of the incident photon of frequency $\nu$. Show that the electron recoils with energy

$$E = h\nu\,\frac{\chi}{1 + \chi}, \quad \text{where } \chi \equiv \frac{h\nu(1 - \cos\theta)}{m_0c}.$$

Compton recoil electrons were first detected in 1923 by the British physicist C. T. R. Wilson (1869–1959) and the German physicist Walther Bothe (1891–1957).

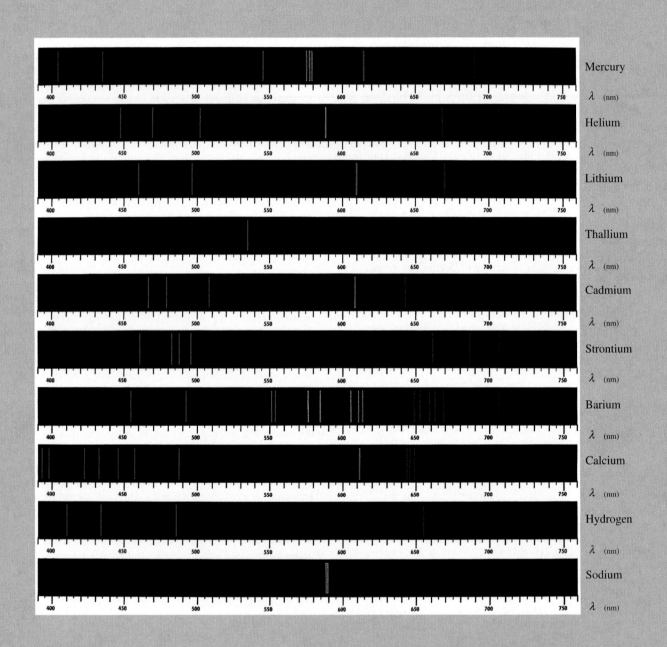

Mercury

$\lambda$ (nm)

Helium

$\lambda$ (nm)

Lithium

$\lambda$ (nm)

Thallium

$\lambda$ (nm)

Cadmium

$\lambda$ (nm)

Strontium

$\lambda$ (nm)

Barium

$\lambda$ (nm)

Calcium

$\lambda$ (nm)

Hydrogen

$\lambda$ (nm)

Sodium

$\lambda$ (nm)

# The Quantized Atom

———— The spectrum of light emitted by individual atoms consists of an array of discrete lines, each representing a specific wavelength.

———— The line spectrum of hydrogen is much simpler than those of most chemical elements. The wavelengths of all the lines can be accounted for by means of the empirical Rydberg formula.

———— The scattering of $\alpha$ particles by atoms in the Rutherford-Geiger-Marsden experiment can be accounted for only if nearly all of the mass of an atom is concentrated in a very small positively charged region, called the nucleus.

———— The planetary model of the atom is inconsistent with the prediction of Maxwell's equations—that an accelerated charge must radiate electromagnetic energy.

———— The Bohr model is based on the assumption that electrons in specific orbits (stationary states) do not radiate energy. Energy in the form of photons is absorbed or emitted by atoms only when an electron makes a transition from one stationary state to another.

———— A crude extension of the Bohr theory to atoms more complicated than hydrogen agrees with the results of Moseley's X-ray experiments and establishes atomic number as the underlying principle of the periodic law of the elements.

---

*Left:* Atoms radiate energy only at certain wavelengths, as these line spectra show. This is strong evidence that the atoms can exist only in specific quantized energy states and can make transitions from one to another only by emitting radiant energy in quantized amounts.

*[Although] the efforts of thinkers have always been
bent upon the "reduction of all physical processes to
the motions of atoms," . . . this is a chimerical ideal.
This ideal has played an effective part in popular
lectures, but in the workshop of the serious inquirer it
has discharged scarcely the least function.*

—ERNST MACH (1894)

*One of the things which distinguishes ours from all
earlier generations is this, that we have seen our
atoms.*

—K. K. DARROW (1936)

---

SECTION 41.1

# Introduction

This chapter concerns two closely linked subjects: the structure of atoms and the light
they emit and absorb. As matters developed historically, there was a great deal of
empirical evidence regarding *atomic spectra* long before there was any model of the
atom that could make possible a satisfactory explanation of the spectra. Atomic spectra
are the subject of Section 41.2. In Section 41.3, we describe the experimental basis for
the discovery of the atomic nucleus, and Section 41.4 concerns the *Bohr atom*. The
Bohr atom was the first model to yield reasonably good quantitative explanations for
at least some observed spectra, and we still have occasion to use this model today. In
Section 41.5, we extend the Bohr model in a semiquantitative fashion and use the
extended model to understand the emission and absorption of X rays. At the same time,
we begin to understand the regularities of the periodic table of the elements. Having
developed a useful if incomplete picture of the process by which atoms absorb and emit
light, we apply this picture in Section 41.6 to a description of the *laser*, an important
optical tool.

---

SECTION 41.2

# Atomic Spectra

In studying *spectra*—the mixture of wavelengths emitted by luminous bodies—we have
concentrated so far on *continuous spectra*. The blackbody spectrum is one kind of
continuous spectrum, but there are others as well. In general, continuous spectra are
characteristic of light emission by dense, macroscopic bodies. Planck made this point
explicit in imagining the walls of a blackbody to be a collection of many harmonic
oscillators.

But individual atoms, as well as such small collections of atoms as molecules, emit
a quite different kind of spectrum. We do not usually observe the light emitted by a
single atom. But we can observe the emission of very many identical atoms, each acting
essentially independently of all the others, by observing the light emitted by a luminous

gas. As you learned in Chapter 18, the atoms or molecules of a gas at ordinary temperatures and pressures are separated from one another by something like ten times their own diameters on the average. The separation is even greater at lower pressures. In view of this separation, it is not surprising that each atom or molecule of a gas interacts with light independently of all the others.

There are many ways to make a gas emit light. A simple and convenient device for doing so is the **gaseous discharge tube**, shown in Figure 41.1. The tube contains a gas, usually pure, at low pressure (typically about $10^2$ Pa, or $10^{-3}$ atm). A high voltage applied across the electrodes results in ionization of the gas, and some of the electrical energy supplied to the tube is emitted as radiation.

Figure 41.2 shows several atomic spectra produced by gaseous discharge tubes. There is a dim continuous background, but most of the light appears as a **discrete**, or **line**, **spectrum**. Each line corresponds to a quite narrow range of wavelengths, and it is often satisfactory to characterize each line as a single wavelength.

Each chemical element emits a unique spectrum. The brightness of the various lines in a spectrum varies with the conditions under which the substance is excited; some lines even appear or disappear as the conditions are varied. But the wavelength of any particular line does not change. The wavelengths of many lines have been painstakingly tabulated. By determining the wavelengths of a number of lines, a spectroscopist can make a reliable determination of the chemical composition of a mixture and can even specify from the relative brightness of the lines the rough proportion of each component in the mixture—a process called *spectrographic analysis*.

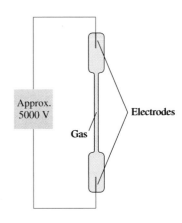

**FIGURE 41.1** A gaseous discharge tube.

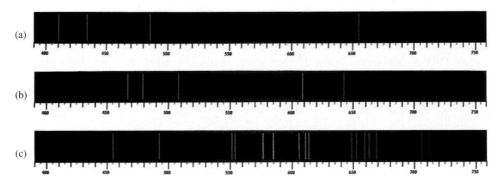

**FIGURE 41.2** Atomic line spectra of (*a*) hydrogen, (*b*) cadmium, and (*c*) barium. Note the difference in complexity among the spectra.

The systematic study of atomic spectra dates from 1859, when Bunsen and Kirchhoff* first developed spectrometers and techniques suitable for the task. An early and spectacular application was to the analysis of the chemical composition of stars. To this day, most of what we know about the stars is based on observations of their spectra.

Although applications of atomic spectra were quick in coming, understanding of the spectra turned out to be quite another matter. The spectra of most elements have many lines, and there is no obvious order in the wavelengths of the lines. Nevertheless, it is possible to set forth some rules of thumb:

1. Hydrogen, the lightest (and presumably the simplest) element, has the simplest spectrum (Figure 41.2a).

2. Elements that have similar chemical properties have, among the lines in their spectra, families of lines that appear related, though they are shifted in wavelength from one element in the group to another. Such related families of lines are particularly evident in the alkali metals—lithium, sodium, potassium, rubidium, cesium, and francium.

3. Elements having several chemical valences (which implies complex chemical properties) often have especially complicated spectra. A good example is barium (Figure 41.2c).

4. As the temperature of the substance whose line spectrum is being observed is increased, some lines appear and others fade out. A family of lines often appears and fades out together.

---

*Robert W. E. Bunsen (1811–1899), German chemist, and Gustav R. Kirchhoff, German physicist. See the margin note in Section 27.4 for a discussion of Kirchhoff's role in the development of Ohm's law.

## Fraunhofer Lines and Atomic Absorption Spectra

About 1800, the English physicist William Wollaston (1766–1828) noted, while observing the spectrum of the sun, that a number of dark lines appeared to be superimposed on the continuous bright (blackbody) spectrum. These lines were studied in 1814 by the German instrument maker and physicist Joseph Fraunhofer (1787–1826); they are called **Fraunhofer lines** in his honor. A few of the solar Fraunhofer lines are shown in Figure 41.3. Fraunhofer observed about 600 of them, and thousands are known today.

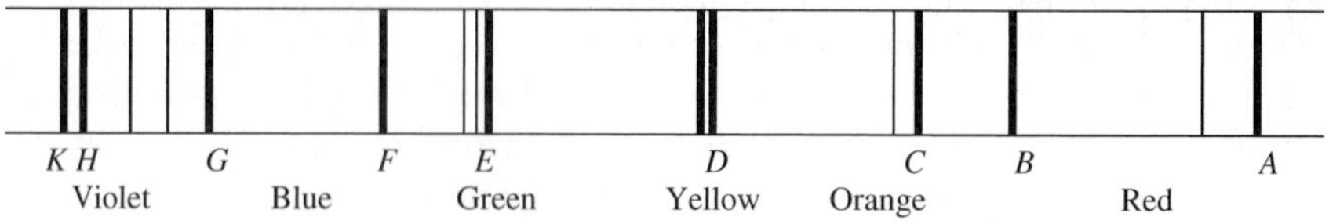

**FIGURE 41.3** The spectrum of the sun, showing Fraunhofer lines. The letters beneath the spectrum are conventional labels for the Fraunhofer lines.

Fraunhofer lines are dark only by contrast with the brighter background. When observed with a high-resolution spectrometer in such a way that the brighter light at adjacent wavelengths is excluded, the Fraunhofer lines are themselves quite bright. Moreover, each Fraunhofer line corresponds to a bright line in a known atomic spectrum. This latter point was demonstrated dramatically in the 1850s by the English chemist Edward Frankland (1825–1899). Frankland produced artificial Fraunhofer lines by passing bright light from a hot blackbody through a container filled with relatively cool sodium vapor. Each line in the resulting dark-line spectrum—called an *absorption spectrum*—corresponded precisely to a known bright line in the ordinary *emission spectrum* of sodium (Figure 41.4).

What is happening in Frankland's experiment? The spectrum of the light incident on the sodium vapor is characteristic of the hot blackbody where the light originated. The cooler sodium vapor absorbs energy from the light that passes through it. Whatever rules restrict the absorption of light to specific wavelengths appear to be the *same* rules that restrict the emission of light to the same specific wavelengths.

The Fraunhofer lines are readily accounted for on this basis. Light emitted by the sun's *photosphere*—the hot, incandescent layer that we "see"—must pass through the sun's less dense and relatively cool outer atmosphere. Atoms in this outer atmosphere absorb radiation at the wavelengths characteristic of their spectra. (Fraunhofer absorption occurs in the earth's atmosphere as well, though these lines tend to be less prominent than the solar lines.)

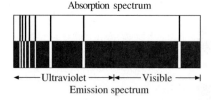

**FIGURE 41.4** Absorption lines, produced by passing a continuous spectrum through relatively cool sodium vapor, compared with the emission lines in the spectrum of sodium.

## Balmer's Rule

Among the many empirical rules devised by spectroscopists to impose some order on the multitude of observed lines was one proposed in 1885 by the Swiss high-school teacher Johann Jakob Balmer (1825–1898). As already noted, the spectrum of hydrogen is especially simple; there are only four lines in the visible region. Working with precise measurements reported by the Swedish physicist Anders J. Ångström (1814–1874), Balmer made the empirical observation that the wavelengths of all four lines could be expressed very accurately (within 0.1 nm in modern units, or about 2 parts in $10^3$) by the rule

$$\lambda_j = b \, \frac{j^2}{j^2 - 2^2}, \quad j = 3, 4, 5, 6. \tag{41.1}$$

In this equation, $b$ is an empirical constant whose value is 364.6 nm. The four lines described by this rule are the first four (counting from the right) in Figure 41.2a. The fourth line, at $\lambda_6 = 410.17$ nm, is at the limit of vision in the extreme violet. Balmer knew that any lines corresponding to $j \geq 7$ would lie in the ultraviolet region. He was delighted to learn that these lines had not only been observed (in the spectra of very hot stars), but agreed accurately in wavelength with the values predicted by his rule. Figure 41.5, in which wavelength increases to the left, shows individual lines corre-

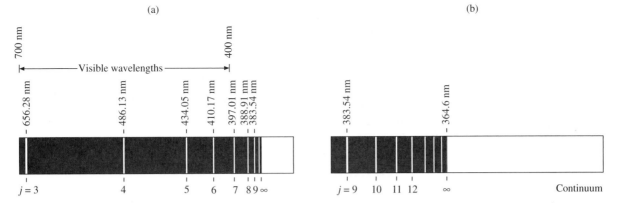

700 nm

400 nm

Visible wavelengths

656.28 nm

486.13 nm

434.05 nm

410.17 nm

397.01 nm

388.91 nm

383.54 nm

383.54 nm

364.6 nm

$j = 3$    4    5    6    7  8 9 $\infty$

$j = 9$    10    11    12    $\infty$    Continuum

**FIGURE 41.5** (*a*) The Balmer series. (*b*) Expanded view of the region near the series limit. Drawings from spectrograms.

sponding to values of $j$ up to 15. Beyond that, the lines are too close together to reproduce here. But a good original spectrum shows many more. As the value of $j$ becomes large, the lines converge to a value $\lambda_\infty = 364.6$ nm, called the **series limit**. The entire series of lines is called the **Balmer series**. The lines are conventionally labeled with Greek letters, starting with the longest-wavelength line. The corresponding wavelengths are $\lambda_\alpha$, $\lambda_\beta$, $\lambda_\gamma$, and so forth.

It is gratifying that Balmer's rule works so well, but many empirical rules work well. What distinguishes Balmer's rule is that it seems to account for *all* the lines of hydrogen—at least, all those that had been observed until Balmer's time. But there was better to come. In 1890, the Swedish physicist and mathematician Johannes R. Rydberg (1854–1919) rewrote Balmer's rule in the neater and more general form

$$\frac{1}{\lambda} = R_X\left(\frac{1}{n^2} - \frac{1}{j^2}\right), \quad \begin{array}{l} n = 1, 2, 3, \ldots, \\ j = n + 1, n + 2, n + 3, \ldots. \end{array} \tag{41.2}$$

When we set $R_X = R_H = 4/b$ and $n = 2$, this **Rydberg formula** reduces to Balmer's rule. But Rydberg found that kindred series of lines in other elements (rule 2, page 1121) could be fitted to the formula if the proper empirical choice of the **Rydberg constant** $R_X$ was made for each chemical element. Moreover, the variation in $R_X$ from element to element is small.

Rydberg's formula contains a strong motivation to search for further series of lines in the hydrogen spectrum, corresponding to values of $n$ other than the Balmer-series value $n = 2$. Between 1906 and 1914, the American physicist Theodore Lyman (1874–1954), a pioneer in ultraviolet spectroscopy, discovered the series of lines corresponding to the value $n = 1$, now called the **Lyman series**. Since then, five more series have been discovered in the spectrum of hydrogen, so there are observations corresponding to all values of $n$ up to 7 (Table 41.1). The dates of discovery are in order of the technical difficulty presented by the series. Of particular interest is the longest-wavelength line in the Lyman series ($\lambda = 121.566$ nm), called the Lyman $\alpha$ line. It is by far the most intense line in the spectrum emitted by the sun. However, it

**TABLE 41.1 Series Observed in the Hydrogen Spectrum**

| Series Name | Year Discovered | $n$ in Equation 41.2 | Series Limit (nm) | Spectral Region |
|---|---|---|---|---|
| Lyman | 1906–1914 | 1 | 91.126 | far ultraviolet |
| Balmer | (1885) | 2 | 364.506 | visible-ultraviolet |
| Paschen | 1908 | 3 | 820.14 | infrared |
| Brackett | 1922 | 4 | 1458.03 | infrared |
| Pfund | 1924 | 5 | 2278.17 | infrared |
| Humphreys | 1953 | 6 | 3280.56 | infrared |
| Hansen-Strong | 1973 | 7 | 4465.21 | infrared |

can be seen only from high-altitude balloons and rockets or from spacecraft because it is completely absorbed by the earth's atmosphere, which is opaque to far-ultraviolet radiation.

So successful is the Rydberg formula in predicting the spectrum of hydrogen that any model of the hydrogen atom must explain why the formula works. We will consider this problem in Section 41.4. But first, we consider another line of evidence leading to a model of the hydrogen atom.

## Rutherford Scattering and the Nucleus

Once J. J. Thomson had discovered the electron in 1897 (see Section 28.8), it became clear that electrons constitute an important part of matter. If matter is made up of atoms and if electrons are part of matter, then electrons are part of atoms. One can make three inferences:

1. Atoms, whose very name comes from the Greek word meaning "indivisible," are *not* indivisible.

2. Because atoms are electrically neutral, they must contain positive charge as well as electrons.

3. The mass of an electron is 0.02% or less of the mass of a typical atom. Consequently, either atoms contain thousands of electrons or else most of the mass of an atom is due to atomic constituents other than the electrons.

On the basis of these inferences, various atomic models were suggested at about the turn of the century. In most of them, the positive charge was smeared out over the volume of the atom and the electrons orbited in the electric field. From the classical point of view, a major difficulty with this class of models is that an electron in orbit must be accelerated and must therefore radiate electromagnetic energy continuously (Section 34.5). As the electron radiates, it quickly loses its energy and comes to rest. Moreover, the frequency of revolution of the electron changes continuously as it loses energy, and so the emission cannot be monochromatic. This conclusion is contradicted by the observation of line spectra.

Shortly after Thomson discovered the electron (see Section 28.8), he suggested a static model of the atom, in which the electrons do not orbit but distribute themselves throughout the smeared-out positive charge until an equilibrium exists between the attraction of the electrons by the positive charge and the mutual repulsion of the electrons. This model was called the *plum-pudding atom.** Thomson suggested that, when disturbed, the electrons would oscillate harmonically about their equilibrium positions, radiating monochromatic light. Unfortunately, detailed calculations showed that (1) the electrons would oscillate anharmonically, with varying frequency, in any such electric field and (2) such a static atom would be unstable. Also, the most prominent form of the plum-pudding model postulated that each atom contain thousands of electrons, in order to account for the fact that atomic masses were known to be thousands of times as great as the electron mass. In spite of these implausibilities, the plum-pudding atom in its various forms was for a time widely favored because nothing better suggested itself.

With the intent of acquiring more experimental information about atomic structure, Rutherford decided in 1909 to probe the atom by means of the energetic $\alpha$ *particles* produced in the radioactive decay of certain heavy elements, notably radium. Alpha particles have mass close to 4 u and carry charge $+2e$. Rutherford had himself established in the previous year that $\alpha$ **particles** are doubly ionized helium atoms; the details are discussed in Section 43.2.

Rutherford's apparatus is shown in Figure 41.6a. An evacuated case consists of the

(a)

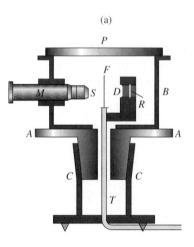

(b)

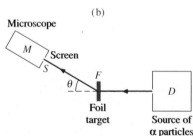

**FIGURE 41.6** (*a*) The scattering apparatus used in the Rutherford-Geiger-Marsden experiment. (*b*) Schematic top view, showing the geometry of the experiment.

*Plum pudding is a British delicacy in which raisins and bits of dried or candied fruit are embedded in a flour-suet "cake." In this analogy, the electrons are the raisins, and the smeared-out positive charge is the cake.

cover $P$, cylindrical chamber $B$, and base $AC$. Inside this case is a radioactive $\alpha$-particle source, labeled $R$. It is surrounded by a lead shield, and $\alpha$ particles can emerge only through a small hole at $D$. A thin metal foil target is shown edge on at $F$. (Gold was the first metal to be used because it can readily be beaten into *gold leaf*, of the order of 100 atoms thick.) The source and target are mounted on a vertical rod that passes through a vacuum seal at $T$, around which the outer case of the apparatus can be turned while the source and target remain stationary. Mounted in a hole in the wall of the case is the low-power microscope $M$, whose objective is focused on a small screen $S$. The screen is impregnated with tiny zinc sulfide crystals. When an energetic particle strikes one of the crystals, the crystal *scintillates*; that is, it emits a flash of light. Though dim, the flashes can be seen by an observer whose eyes are dark adapted.

In an experimental run, the observer turns the case and the microscope so as to observe $\alpha$ particles that are scattered by the foil through any desired angle $\theta$ (Figure 41.6b). With the scattering angle $\theta$ set, the scintillations are counted over a fixed time interval. By repeating the observation at various scattering angles, the observer can plot the number of $\alpha$ particles scattered through each angle $\theta$ as a function of $\theta$.

The plum-pudding model of the atom predicts very little scattering. As the $\alpha$ particle passes through the foil, it is not scattered significantly by the electric field of the smeared-out positive charge. This is because the positive charge is distributed uniformly throughout the foil and does not exert a net sideward force on the $\alpha$ particle. If the $\alpha$ particle passes near an electron, it does experience an asymmetric electric force. But the electron mass is only about $10^{-4}$ that of the $\alpha$ particle. Consequently, it is the electron, not the $\alpha$ particle, that experiences substantial scattering. The scattering process is analogous to what would happen if you rolled a bowling ball through a stack of Ping-Pong balls.

Rutherford and his associate Geiger made such $\alpha$-scattering measurements and observed precisely what they expected: Most incident particles were scattered at negligibly small angles; as $\theta$ was increased from zero, the number of particles observed fell off very rapidly, nearly disappearing at angles between 1° and 2°.

What happened then is best described in Rutherford's own words:

*One day Geiger came to me and said, ''Don't you think that young Marsden, whom I am training in radioactive methods, ought to begin a small research?'' Now I had thought that, too, so I said, ''Why not let him see if any $\alpha$-particles can be scattered through a large angle?'' I may tell you in confidence that I did not believe that they would be, since we knew that the $\alpha$-particle was a very fast, massive particle, with a* *great deal of energy, and you could show that if the scattering was due to the accumulated effect of a number of small scatterings the chance of an α-particle's being scattered backwards was very small. Then I remember two or three days later Geiger coming to me in great excitement and saying, "We have been able to get some of the α-particles coming backwards. . . ." It was quite the most incredible event that has ever happened to me in my life. It was almost as incredible as if you fired a 15-inch shell at a piece of tissue paper and it came back and hit you.*

Rutherford's point is that, though the number of scattered α particles does indeed decrease rapidly with increasing θ, the surprise is that *any α particles at all are scattered through large angles—even 180°.* Rutherford concluded that an α particle can be scattered through a large angle—even backscattered—only if it is repelled by an object much more massive than itself, carrying a very concentrated positive charge. An atomic model consistent with this conclusion is one in which nearly all of the mass of the atom (about 197 u for gold) is concentrated in a volume very much smaller than the volume of the entire atom. This highly concentrated, positively charged bit of matter is called the **nucleus** of the atom (from a Latin word meaning "kernel"). The rest of the volume of the atom—by far the greater part—must be occupied by the negatively charged electrons, which we can picture as forming a cloud around the nucleus.

The nuclear model of the atom accounts for the scattering observations of Rutherford, Geiger, and Marsden in the following way:

1. Because a nucleus presents such a small "target," most α particles never come near a nucleus as they pass through the metal foil.* The deflecting impulse exerted on an α particle by Coulomb repulsion between the (relatively distant) nucleus and the α particle is usually quite small. Although the electron clouds surrounding the nuclei of the gold foil more or less fill the space through which the α particle passes, the passage of the α particle through the cloud also deflects it but little, as already noted. Hence the overall deflection angle is quite small for most α particles.

2. An incident α particle belonging to the small fraction that do happen to come close to nuclei is strongly deflected by the Coulomb repulsion between the nucleus and the α particle and is thus scattered through a large angle.

Rutherford developed a quantitative expression that describes the dependence on the scattering angle θ of $N(\theta)$, the number of monoenergetic α particles counted per unit time in the small angular range between θ and $\theta + \Delta\theta$. Suppose that the atomic mass $M$ is very much greater than the α-particle mass and that the repulsive force between the nucleus and the α particle is given by Coulomb's law. Suppose also that $Ze$ is the (positive) electric charge on the nucleus. The overall electrical neutrality of the atom assures us that the electron cloud surrounding the nucleus contains $Z$ electrons, with a total charge $-Ze$. Then the result, which we will not derive here, is

$$N(\theta) \propto \left( \frac{Ze^2}{2\pi\epsilon_0 M} \right)^2 \frac{\cos\frac{\theta}{2}}{\sin^3\frac{\theta}{2}}. \tag{41.3}$$

This is one form of the *Rutherford scattering formula.*

For small θ, Equation 41.3 simplifies to $N(\theta) \propto 1/\theta^3$. This small-angle form displays somewhat more simply than the general form what is true for both: For small scattering angles, the number of scattered α particles decreases rapidly with increasing angle, just as the earlier (pre-Marsden) observations had indicated. But the observations of Rutherford, Geiger, and Marsden agreed quite well with the general form of Equation 41.3 over the entire range of θ, as did much more precise measurements made later by others. Thus the experiment rules out the plum-pudding model of the atom, for which no large-angle scattering is possible because there is nothing in the "tissue paper" to deflect the "15-inch shell."

---

*The term "near" must be interpreted on the scale of the dimensions of the α particle and the nucleus, which, as we will soon see, have diameters of order $10^{-15}$ m—about $10^{-5}$ times the diameter of the atom.

More important is the fact that the angular behavior expressed in the Rutherford scattering formula cannot be explained on any basis other than a nuclear model of the atom. Thus the evidence for a nuclear atom is strong. Moreover, the Rutherford atom contains $Z$ electrons—a much smaller number than the thousands required to account for the total mass of the original version of the Thomson atom.

EXAMPLE  41.1

Find the ratio $N(1°)/N(179°)$, and thus get a quantitative idea of the frequency with which $\alpha$ particles are backscattered through large angles.

**SOLUTION:** For $\alpha$ particles scattering from a given metal foil, the quantity in parentheses in Equation 41.3 is the same for all angles. So you have

$$\frac{N(1°)}{N(179°)} = \frac{\cos 0.5° \sin^3 89.5°}{\cos 89.5° \sin^3 0.5°} = 1.7 \times 10^8.$$

That is, almost 200 million $\alpha$ particles are scattered through a narrow angular window centered on $\theta = 1°$ for every particle scattered through the same narrow angular window centered on $\theta = 179°$. The backscattered $\alpha$ particles Marsden saw were rarities, but they revolutionized our picture of atomic structure nevertheless.

The scattering experiment makes it possible to set an upper limit on the radius of the nucleus. When $\alpha$ particles penetrate into the nucleus, the scattering deviates from Equation 41.3. If such penetration occurs at all, it must certainly occur for the backscattered $\alpha$ particles (for which $\theta \simeq 180°$) because these are the particles that approach the nucleus most closely. On the basis of observations of backscattering from targets having values of $Z$ much less than that of gold, Rutherford concluded in 1911 that the nuclear radius must be smaller than $10^{-14}$ m. (See Problems 41.6, 41.7, and 41.25.) Because nuclei are not tiny billiard balls with sharply defined surfaces, there is no hard-and-fast value for the nuclear radius. Nevertheless, experiments show that most nuclei can be characterized by radii that conform fairly closely to the relation

$$r = r_0 A^{1/3}, \quad \text{where } r_0 = 1.37 \times 10^{-15} \text{ m.} \tag{41.4}$$

In this expression, $A$ is the atomic mass of the substance in question, expressed in atomic mass units (Section 18.2). For gold ($A = 197$ u), the expressions yields $r = 8.0 \times 10^{-15}$ m; for aluminum ($A = 27$ u), it yields $r = 4.1 \times 10^{-15}$ m. Owing to the cube-root relation, the range of nuclear radii is not great; it lies between $2 \times 10^{-15}$ m and $9 \times 10^{-15}$ m for all elements.

To get a feeling for the size of the nucleus, you should compare these values with the radius of a typical atom—about $10^{-10}$ m. The ratio of the cross-sectional (target) area of the nucleus to that of the entire atom of which it is a part is something like $(10^{-15} \text{ m})^2/(10^{-10} \text{ m})^2 = 10^{-10}$. No wonder large-angle Rutherford scattering is an uncommon event!

# The Bohr Atom

Although Rutherford's work clearly establishes the existence of an atomic nucleus containing nearly all the mass of the atom and having a positive charge equal to $Ze$, it leaves us with three serious questions:

1. What is the nucleus made of?
2. What keeps the nucleus from disintegrating under the influence of its positive charge?
3. How do the electrons and the nucleus fit together to form the entire atom?

Niels Henrik David Bohr (1885–1962) was born into a prominent academic and banking family in Copenhagen. Showing promise as a physics student, he went to the Cavendish Laboratory after receiving his doctorate in 1911, with the intent of working with J. J. Thomson. But he and Thomson did not hit it off, and he went on to Manchester to work with Rutherford. It was there that Bohr developed his revolutionary picture of the hydrogen atom, for which he won the Nobel Prize in 1922. In 1921, a foundation established by the Carlsberg brewery furnished funds for the establishment of the Bohr Institute in Copenhagen, which became a major center for scientific research, first in physics and later in biology as well. Subsequently, Bohr was awarded life occupancy of the Carlsberg mansion, where the seeds of many important scientific advances were planted at dinner conversations. During the Second World War, Bohr made a daring escape from occupied Denmark in a fishing boat. He spent the rest of the war at Los Alamos participating in the Manhattan Project, which led to the development of the atomic bomb. Although he was a poor lecturer, Bohr was a brilliant teacher, famous for his ability to get to the heart of a matter and pose tough Socratic questions that set his listeners off on fruitful courses. Through apparently casual conversations, Bohr contributed to the solution of countless knotty problems. As a scientist, administrator, and elder statesman, he influenced most of the leading physicists whose careers flourished from the 1920s through the period immediately following World War II. Bohr's profound philosophical insights affect the course of physics to this day. His personal attitude toward achievement is summarized in a German verse he liked to quote:

Gifts? who has them not?
Talent? a children's toy!
Foremost, seriousness makes the man;
Foremost, diligence the genius.

The first two questions we must defer to later chapters—Chapter 43 for the first and Chapter 45 for the second. The third question we consider now.

An incomplete but very useful answer to the third question was set forth in 1913 by the Danish physicist Niels Bohr.

Remember the salient facts for which any satisfactory theory of the atom must account:

1. the discrete (line) form of atomic spectra in general and the Rydberg series of hydrogen in particular;

2. the existence of a tiny, massive nucleus.

Although these properties are common to all atoms, it makes sense to consider the hydrogen atom first, in view of its relatively simple spectrum and its presumably relatively simple structure.

A nuclear atom, dictated by the result of the Rutherford-Geiger-Marsden experiment, requires orbiting electrons. From the classical point of view, a stationary electron will crash into the nucleus as a result of the strong Coulomb force exerted on the electron by the nucleus; an orbiting electron will not do so. The picture is reminiscent of the solar system, with "planetary" electrons revolving around the "solar" nucleus—an analogy we will find very helpful. In particular, imagine the hydrogen atom to consist of a single electron circling a nucleus, as shown in Figure 41.7. By 1919, Rutherford and others had shown that the hydrogen nucleus is itself a basic constituent of matter (as the electron is). For this reason, the hydrogen nucleus is called a **proton** (from the Greek word meaning "first"). The proton contains nearly all of the mass of the hydrogen atom and has charge $+e$. We know that the ultimate picture of the atom cannot be so simple, but we must start somewhere.

In this picture, the electron undergoes continuous acceleration. What about the problem of energy loss through electromagnetic radiation, which Maxwell's equations predict for any accelerated charged particle? Just as Planck had done, Bohr made an argument that is completely classical *except* for a single bold hypothesis. Bohr's hypothesis has two parts. The first part is as follows:

1. *Electrons are restricted to certain particular stable orbits.* That is to say, the electron orbits are *quantized.* Just as Einstein had done in 1905, in his explanation of the photoelectric effect, Bohr took the quantum hypothesis and extended it to yet a new realm.

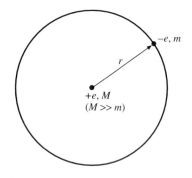

**FIGURE 41.7** The Bohr hydrogen atom.

From the hypothesis of stability of the allowed orbits, it follows that *an electron in an allowed orbit does not radiate energy*. Bohr called these allowed orbits **stationary states**.

Before we consider the second part of Bohr's hypothesis, let us explore some of the consequences of the first part, using the solar-system analogy explicitly. For a planet in a circular orbit, Equation 14.21 relates the potential energy $U$ to the total energy $E$:

$$E = \tfrac{1}{2}U. \tag{41.5}$$

The relation between kinetic energy $K$ and total energy $E$ of the planet also will be useful. We have $E = K + U$, and therefore $K = E - U$. Using Equation 41.5, we obtain

$$K = -E, \tag{41.6}$$

which agrees with Equation 14.20. Equations 41.5 and 41.6 depend only on the fact that the force law is an inverse-square law, and not on any other details. Consequently, these equations apply just as well to a circular orbit followed by a planetary electron under the influence of the inverse-square Coulomb's-law force, $F = -e^2/(4\pi\epsilon_0\, r^2)$, as they do to a circular orbit followed by an actual planet, under the influence of the inverse-square Newton's-law force, $F = -GMm/r^2$.

The electric potential in the vicinity of the proton is given by Equation 25.14b, $V = q/4\pi\epsilon_0 r = e/4\pi\epsilon_0 r$. Consequently, the electric potential energy of the electron, whose charge is $-e$, is

$$U = -eV = -\frac{e^2}{4\pi\epsilon_0}\frac{1}{r}. \tag{41.7a}$$

According to Equation 41.5, the total energy of the electron is one-half this value, or

$$E = -\frac{e^2}{4\pi\epsilon_0}\frac{1}{2r}. \tag{41.7b}$$

Similarly, using Equation 41.6, we can write the kinetic energy of an electron in a circular orbit of radius $r$ in the form

$$K = \frac{e^2}{4\pi\epsilon_0}\frac{1}{2r}. \tag{41.7c}$$

Consider two possible allowed orbits for an electron. For simplicity, assume that both orbits are circular. If the electron is to pass from one orbit to another of different radius, it must gain or lose energy. Thus we come to the second part of Bohr's hypothesis:

**2.** *The energy gain or loss is accomplished by the absorption or emission of a photon whose energy $h\nu$ is exactly equal to the energy difference between the two orbits.*

If we call the initial electron energy $E_i$ and the final energy $E_f$, we have

$$\pm(E_f - E_i) = h\nu, \tag{41.8}$$

where the $+$ sign applies when the photon is absorbed and the $-$ sign applies when the photon is emitted. Because only certain electron energies $E_i$ and $E_f$ are allowed in stationary states, only photons of certain energies can be absorbed or emitted when the electron makes a transition from one stationary state to another. This restriction on photon energy is consistent with the observed line spectrum of hydrogen, as well as those of other elements.

Which orbits are allowed? In other words, what is the *quantization condition* for the stationary states? We use a dimensional argument to guide us toward a plausible answer, which we then test against experimental observation. Because all the quantization conditions we have encountered so far contain Planck's constant, we presume that this one does, too. The simplest possible condition is

$$\text{some dynamical quantity} = nh, \quad n = 1, 2, 3, \ldots. \tag{41.9}$$

What is the dynamical quantity we want? Its dimensions must be the same as those of

Planck's constant, [energy · time]. We can manipulate these dimensions as follows:

$$[E][t] = \left[\frac{ml^2}{t^2}\right][t] = [mv][vt] = [\text{momentum}] \cdot [\text{length}].$$

What particular momentum and length does it make sense to insert into Equation 41.9? The momentum is presumably that of the electron, $mv$. For the length, it seems plausible to insert the circumference of the orbit, which is $2\pi r$. So we have

$$mv\, 2\pi r = nh,$$

or
$$mvr = n\frac{h}{2\pi}. \tag{41.10a}$$

The left side of this equation is the **orbital angular momentum** $L$ of the electron, which has a particular value $L_n$ for each value of $n$. We can thus write the **Bohr quantization condition** in the form

$$L_n = n\frac{h}{2\pi}. \tag{41.10b}$$

*When the electron is in a stationary state (an allowed orbit), its angular momentum has a value equal to an integer times the constant $h/2\pi$.*

The integer $n$ in Equations 41.10a and 41.10b is called a **quantum number**. You will see in Chapter 42 that a detailed description of the stationary states of atoms requires several quantum numbers, of which $n$ is the most important. For this reason, $n$ is called the **principal quantum number**.

If the angular momentum of the electron in a stationary state is quantized, other quantities of interest must be quantized as well, because they are related to the angular momentum. Let us begin with the allowed orbit radii. One way to evaluate the radii is to relate the kinetic energy to the angular momentum. By manipulating the general definitions $K = \frac{1}{2}mv^2$ and $L = mvr$, we find

$$K = \tfrac{1}{2}mv^2 = \frac{(mvr)^2}{2mr^2} = \frac{L^2}{2mr^2}. \tag{41.11}$$

Using Equation 41.10b, we find that the Bohr quantization condition restricts the kinetic energy of the electron to the values

$$K_n = \frac{n^2h^2}{8\pi^2mr_n{}^2}. \tag{41.12}$$

The subscript $n$ is used because $K_n$ and $r_n$ have particular values for every value $n = 1, 2, 3, \ldots$.

Although the kinetic energy of an electron in a stationary state is restricted by the Bohr quantization condition to the particular values $K_n$ given by Equation 41.12, any of these values may be expressed by means of the purely classical Equation 41.7c, $K = (e^2/4\pi\epsilon_0)(1/2r)$. For the particular value $K_n$, this equation can be written

$$K_n = \frac{e^2}{4\pi\epsilon_0}\frac{1}{2r_n}.$$

Combining this value with Equation 48.12, we have

$$\frac{e^2}{4\pi\epsilon_0}\frac{1}{2r_n} = \frac{n^2h^2}{8\pi^2mr_n{}^2}.$$

We solve for $r_n$ to obtain

$$r_n = \frac{\epsilon_0h^2}{\pi me^2}n^2. \tag{41.13}$$

The fraction on the right side of this equation contains only constants and so is itself a constant. The radii of the stationary-state, circular orbits are thus in the ratio $1:4:9:\cdots$.

The smallest possible orbit corresponds to the value $n = 1$. This is the orbit of lowest energy and is called the **ground-state orbit**. Its radius is called the **Bohr radius** $a_0$ of the hydrogen atom. Its value is

$$a_0 = \frac{\epsilon_0 h^2}{\pi m e^2} \tag{41.14a}$$

$$= \frac{8.854 \times 10^{-12} \text{ C}^2/\text{N·m}^2 \times (6.626 \times 10^{-34} \text{ J·s})^2}{\pi \times 9.109 \times 10^{-31} \text{ kg} \times (1.602 \times 10^{-19} \text{ C})^2},$$

which yields $a_0 = 5.29 \times 10^{-11}$ m. $\tag{41.14b}$

The diameter of the hydrogen atom predicted by the Bohr theory is $2a_0$, or about $10^{-10}$ m.

How does the value of $a_0$ compare with experiment? The radius of an atom is not a fixed quantity; its value depends on the particular experiment used to measure it. Thus, we cannot count on precise agreement between experiment and the calculated Bohr radius to substantiate the Bohr theory. Nevertheless, all experimental values agree with one another—and with the calculated Bohr radius—within about 20%.

We are now ready to evaluate $E_n$, the energy of the electron in the stationary state corresponding to $n$. We write the expression for the electron energy given by Equation 41.7b for the special case $E = E_n$:

$$E_n = -\frac{e^2}{4\pi\epsilon_0} \frac{1}{2r_n}.$$

Into this expression, we substitute the value of $r_n$ given by Equation 41.13:

$$E_n = -\frac{e^2}{4\pi\epsilon_0} \frac{\pi m e^2}{2\epsilon_0 h^2} \frac{1}{n^2},$$

which simplifies to

$$E_n = -\frac{me^4}{8\epsilon_0^2 h^2} \frac{1}{n^2}. \tag{41.15}$$

The smallest possible value of $E_n$ is the value corresponding to $n = 1$:

$$E_1 = -\frac{me^4}{8\epsilon_0^2 h^2}. \tag{41.16}$$

This is called the **ground-state energy** of the hydrogen atom.

The energy of an electron at rest at an infinite distance from the proton is the energy corresponding to $n = \infty$. We choose $E_\infty = 0$, consistent with our conventional choice of the zero point for electrostatic potential energy. The energy required to *ionize* a hydrogen atom that is originally in the ground state is $E_\infty - E_1 = me^4/8\epsilon_0^2 h^2$. In Problem 41.30, you will see that the value calculated from this expression agrees very well with the experimentally determined *ionization energy* of atomic hydrogen, $\Delta E = 13.6$ eV.

We are now ready for a crucial check against experiment: Do transitions between any two energy levels expressed by Equation 41.15 correspond to the emission or absorption of photons whose wavelengths conform to the Rydberg formula? According to Equation 41.8, emission of a photon occurs when the electron makes a transition from some stationary state $E_j$ to a lower energy state $E_n$:

$$-(E_f - E_i) = -(E_n - E_j) = E_j - E_n = h\nu.$$

We make the substitution $h\nu = hc/\lambda$ and write

$$\frac{1}{\lambda} = \frac{1}{hc}(E_j - E_n).$$

Using Equation 41.15 twice to express $E_j$ and $E_n$, we have

$$\frac{1}{\lambda} = \frac{me^4}{8\epsilon_0^2 h^3 c}\left(\frac{1}{n^2} - \frac{1}{j^2}\right). \tag{41.17}$$

Compare this with Equation 41.2, the Rydberg formula. For hydrogen, the formula is

$$\frac{1}{\lambda} = R_H\left(\frac{1}{n^2} - \frac{1}{j^2}\right).$$

The quantity in parentheses is certainly the same for the empirically determined Rydberg series and the Bohr theory. That is, the Bohr theory explains the relative values of the wavelengths in the Rydberg series. Does the Bohr theory account for the value of the Rydberg constant $R_H$ as well? To find out, let us compare the empirical Rydberg constant with the constant factor in Equation 41.17.

The Rydberg constant for hydrogen is $4/b$, where $b$ is the Balmer constant $364.6 \times 10^{-9}$ m. Thus the empirical value of $R_H$ is

$$R_H = \frac{4}{364.6 \times 10^{-9} \text{ m}} = 1.097 \times 10^7 \text{ m}^{-1}.$$

Using the best current values listed inside the front cover of this book, we find that the constant in Equation 41.17 is

$$\frac{me^4}{8\epsilon_0^2 h^3 c} = \frac{9.109\ 3897 \times 10^{-31} \text{ kg} \times (1.602\ 177\ 33 \times 10^{-19} \text{ C})^4}{8 \times (8.854\ 187\ 817 \times 10^{-12} \text{ C}^2/\text{N·m}^2)^2 \times (6.626\ 0755 \times 10^{-34} \text{ J·s})^3 \times 2.997\ 924\ 58 \times 10^8 \text{ m/s}}$$
$$= 1.097\ 3731 \times 10^7 \text{ m}^{-1}.$$

The two values are in agreement.

Values of $\lambda$ calculated by means of Equation 41.17 agree very well with the corresponding measured wavelengths. For example, Equation 41.17 gives the calculated wavelength of the Balmer $\alpha$ line ($n = 2, j = 3$) as 656.11 nm, and the observed value is 656.48 nm. The two values differ by less than 6 parts in $10^4$.* Most of the difference can be accounted for by taking into consideration the fact that the proton mass, though much greater than the electron mass, is not infinite; see Problem 41.54.

The connection between the quantization of the energy of the hydrogen atom and the Rydberg series is shown graphically in the **energy-level diagram** of Figure 41.8.

Energy-level diagrams are used very frequently in spectroscopy. Although there are few sets of spectral lines as simple as the Rydberg series, every line represents a transition from some atomic energy level to some other level. By careful spectrographic analysis, it is usually possible to construct the energy-level diagram of the system under study.

## The Correspondence Principle

You have seen how the lines in any of the Rydberg series are packed closer and closer together as they approach the series limit. As you can see from Figure 41.7, this is because the energy levels $E_n$ lie closer and closer together with increasing $n$. For $E_n$ less (but not much less) than $E_\infty = 0$, the energy difference between adjacent lines becomes so small that it cannot be observed experimentally. For energies greater than $E_\infty$, there are no longer closed electron orbits, and therefore there is no quantization condition. All energies $E > 0$ are allowed; this energy region is labeled "continuum" in Figure 41.7.

Thus, the quantized properties of the atom gradually merge into classical, continuous properties as the electron orbits become bigger and bigger. In retrospect, it is not surprising that this happens. Our everyday experience provides no evidence of quantum behavior. Although atoms do exhibit quantum behavior, that behavior does not conflict with everyday experience because everyday experience does not extend to so small a

---

*When Bohr made this comparison in 1913, the available values of $h$ and $e$ were much less accurate than the modern values. Nevertheless, he found agreement within 8%. He must have been far from displeased with the result.

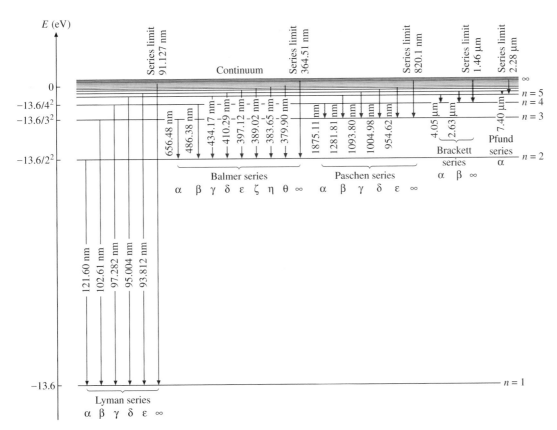

**FIGURE 41.8** Energy-level diagram for the Bohr atom. The first five series—Lyman ($n = 1$), Balmer ($n = 2$), Paschen ($n = 3$), Brackett ($n = 4$), and Pfund ($n = 5$)—are shown. The horizontal lines represent the energies of the stationary states. The vertical lines represent transitions from the upper energy level $E_j$ to the lower level $E_n$; each transition is labeled with the observed wavelength of the corresponding spectral line corrected to vacuum. The series limits are shown.

scale. There is a gentle merger of the quantum and classical worlds at the boundary between them.

This merger has been observed in many contexts, some of which we will consider in Chapters 42 and 43. In 1919, Bohr asserted that gentle transition from the quantum world to the continuous classical world is a universal principle, called the **correspondence principle**. The correspondence principle has useful scientific as well as philosophical applications.

## Moseley's Law and the Periodic Table

The Bohr model is remarkably successful in accounting for the major features of the hydrogen atom, including its spectrum, ionization potential, and size. Can we extend the model to account for more complicated atoms? In the final analysis, the answer is No, as you will see in Chapter 42. Nevertheless, qualitative atomic models, based in part on the Bohr picture, can be useful sources of insight into atomic structure and behavior. We begin by reviewing two salient properties of atoms. The first property is the tendency of the chemical elements to fall into families with respect to their physical and chemical properties. The second property is the way in which atoms produce *X-ray emission spectra* when they are bombarded with energetic electrons. After reviewing these not obviously related properties, we will see how they fall together in the light of an extension of the Bohr theory and yield new insight into the structure of the atom.

### The Periodic Law

It has been known for a long time that chemical elements fall into "families," the members of each family having similar chemical and physical properties. For example, the *alkali metals* lithium, sodium, potassium, rubidium, and cesium are highly active metals of valence +1 that form strongly basic hydroxides; all are soft, and all have low density. They have many other properties in common as well. Copper, silver, and gold are soft, lustrous, chemically rather inert metals of valence +1 or +2; all have high density. The *halogens* fluorine, chlorine, bromine, and iodine are active nonmetals of valence −1; all are colored vapors in the gaseous state, and all combine strongly to form such salts as NaCl.

When the chemical elements are arranged in order of their atomic masses, properties such as those considered in the preceding paragraph appear to repeat themselves in a roughly cyclic way. This observation is called the **periodic law**. In 1872, the Russian chemist Dmitri I. Mendeleyev (1834–1907) showed that myriad chemical resemblances fell into order quite neatly if the elements were listed in a **periodic table**, a modern version of which is shown in Appendix 7. With a few exceptions, the listing is in order of the atomic masses of the elements. Successive rows are arranged so that chemical families lie in vertical columns.

The order in the periodic table has two complementary aspects. As we proceed across a row from left to right, we see a transition from highly active, unambiguously metallic elements through less active elements of multiple valence to highly active, unambiguously nonmetallic elements and finally to the inert gases in the last column. As we proceed down a column, we see a steady variation in properties within each group. The Group IA alkali metals, for example, become more active with increasing atomic mass; the Group VIIA halogens become less active with increasing atomic mass.

The exceptions to atomic-mass order are few. Nevertheless, we are obliged to violate atomic-mass order where that order clearly conflicts with the order of chemical properties. For example, tellurium (Te, number 52) and iodine (I, number 53) are very different from one another. But tellurium is very much like the Group VIA elements selenium (Se, number 34) and sulfur (S, number 16), and iodine is very much like the Group VIIA elements bromine (Br, number 35) and chlorine (Cl, number 17). These irregularities are puzzling, even though the overall regularity of the periodic table is clear.

Mendeleyev's first periodic table had numerous gaps because only about sixty elements had been identified at the time. Indeed, one of the triumphs of Mendeleyev's table was its power in predicting the properties of undiscovered elements. With the help of such predictions, other chemists soon discovered the elements germanium (Ge, number 32), gallium (Ga, number 31), scandium (Sc, number 21), and argon (Ar, number 18). With one exception (technetium, Tc, number 43), all of the elements from hydrogen through uranium (U, number 92) occur naturally, and all had been discovered by the 1920s.

It seems natural to number the elements in the order in which they appear in the periodic table. This number Z is called the **atomic number** of the element. Very roughly speaking, the atomic number of an element (always an integer because of the way in which it is defined) is one-half the atomic mass.

Each horizontal row in the periodic table is called a **period**. Note that the periods get longer and longer with increasing atomic number. (Some of the elements in the last two periods are pulled out of their places in the table to keep the table from becoming awkwardly wide.)

## X Rays

The German physicist Wilhelm Conrad Roentgen (1845–1923) was the first to generate X rays, in 1895. His discovery was an accidental by-product of a study of cathode rays (Section 28.8), but he quickly exploited the fortunate accident and determined many of the properties of X rays. X rays are produced when electrons that have been accelerated through a large potential difference (usually upward of 30 kV) lose their energy on striking a target. The target is most conveniently made of metal.

As we noted in Section 34.5, X radiation is electromagnetic radiation of wavelength less than about 10 nm. Unlike visible light, X rays penetrate most substances quite readily. But, because the wavelength of X rays is roughly comparable to the regular spacing between atoms in crystalline solids, crystals can be used as diffraction gratings. By using crystal diffraction techniques, it is possible to determine X-ray wavelengths with considerable accuracy.

Figure 41.9 shows a typical X-ray spectrum. There is a continuous spectrum that extends indefinitely to long wavelengths but has a sharp **cutoff wavelength** $\lambda_{min}$ at the short-wavelength end. Superimposed on the continuous spectrum is a line spectrum comprising several discrete wavelengths at which intense radiation is emitted. As a general rule, X-ray line spectra are much simpler than visible-light spectra.

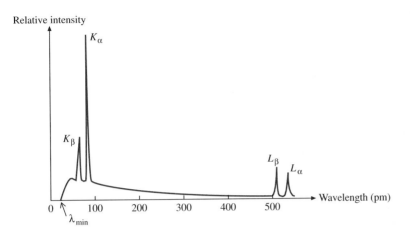

**FIGURE 41.9** X-ray spectrum of zirconium. Shown are the background radiation (bremsstrahlung), the cutoff wavelength $\lambda_{min}$, and four lines of the discrete X-ray spectrum.

The existence of a cutoff wavelength is easy to explain. The wavelength of any photon is related to the energy of the photon by the expression $\lambda = hc/E$. The energy of X-ray photons must be provided by the energetic electrons that bombard the target. Whenever it strikes the target, an electron having initial energy $E$ loses its energy in collisions with the atoms that make up the target. Such a collision is shown schematically in Figure 41.10. An electron passes near a nucleus having relatively large mass and positive charge $Ze$. The main result of this encounter is a (negative) acceleration of the electron, which emits an X-ray photon as it is accelerated.

Most electrons undergo a series of collisions in the target, losing a part of their initial energy in each. Some electrons lose substantially all of their energy in the first collision. But in any case, the greatest amount of energy that can be removed from the electron in a single collision is $E$. The cutoff wavelength for the X-ray photons is therefore

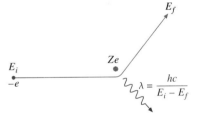

**FIGURE 41.10** An energetic electron encounters an atomic nucleus in the target of an X-ray tube. The trajectory of the electron is bent by the strong Coulomb attraction of the positively charged, massive nucleus. As it is accelerated, the electron emits a photon.

$$\lambda_{min} = \frac{hc}{E}. \tag{41.18}$$

The cutoff wavelength is determined by the energy of the electrons and is completely independent of the target material.

In order to understand the continuous part of the X-ray spectrum, let us consider the multiple collisions that are the more common fate of the electrons. No two electrons make identical series of collisions. All photon energies smaller than $E$ are possible, and this continuous range of possibilities accounts for the fact that the spectrum is continuous. It is possible to calculate the probability that an electron will lose a certain amount of energy $\Delta E$ in a collision. We do not need to make this calculation here, but the result must depend on the target material. For example, nuclei with large charge are likely to accelerate electrons more than do nuclei with small charge. The detailed shape of the continuous spectrum thus depends both on the energy of the electrons striking the target and on the target material. Because the electrons are "braked" to a stop by their multiple collisions, the continuous radiation is also called **bremsstrahlung**. The name comes from the German words *brems*, meaning "brake," and *strahlung*, meaning "radiation." In English, the loose and redundant usage "bremsstrahlung radiation" is often encountered, but you should avoid it.

## The X-Ray Line Spectrum

We now consider the X-ray line spectrum, first clearly identified in 1913 by the British physicist William H. Bragg (1862–1942). For any target material, the lines fall into groups. For historical reasons, the groups are called the *K, L, M, . . . series*. The *K* series lies at the shortest wavelengths. In analogy with the lines in the various series of the hydrogen spectrum, the individual lines in the series are called $K_\alpha$, $K_\beta$, $K_\gamma$, . . . , $M_\alpha$, $M_\beta$, $M_\gamma$, and so on, beginning with the line of longest wavelength in each series. The wavelength of each spectral line is independent of the energy of the electrons. However, the intensity is not. Indeed, a given line will not appear at all if the electron energy is not great enough to excite it.

Unlike optical spectra, X-ray line spectra are similar from element to element. With increasing atomic mass, the entire line spectrum shifts to shorter wavelengths. This behavior was investigated in detail for many elements by the British physicist H. G. J. Moseley. Moseley found empirically that, if he plotted the square root of the frequency of a given X-ray line for each element—say, the $K_\alpha$ line—versus the atomic mass $A$ of the element, the points fell fairly close to a straight line; see the red curve in Figure 41.11. But he obtained a *much better* fit by plotting $\sqrt{\nu}$ versus the *atomic number Z*, as in the blue curve of Figure 41.11. Later, Moseley found that the fit was improved still further by plotting $\sqrt{\nu}$ versus $Z - 1$ rather than $Z$. A plot of $\sqrt{\nu}$ versus $Z$ (or $Z - 1$) is called a **Moseley plot**, and the fact that $\sqrt{\nu}$ is proportional to $Z$ (or $Z - 1$) is called **Moseley's law**.

**FIGURE 41.11** Plots of $\sqrt{\nu}$ versus $A$ (red curve) and $\sqrt{\nu}$ versus $Z$ (blue curve). The fit of points to the blue curve is significantly better. A still better fit is obtained by plotting $\sqrt{\nu}$ versus $Z - 1$.

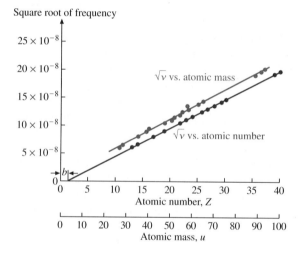

The Moseley plot implies that *atomic number Z is a more fundamental property of atoms than atomic mass A*, at least as far as X rays are concerned. Most dramatic is the way the anomalous cases in the mass sequence become perfectly regular in the atomic-number sequence. Iodine, for example, has a smaller mass than tellurium but falls beneath bromine, and not selenium, in the periodic table. Similarly, tellurium belongs beneath selenium, not bromine. But their atomic-number order (Te = 52, I = 53) coincides with their chemical (periodic-law) order. We must find a *physical* basis for the significance of the atomic number.

## Moseley's Law and the Periodic Law

We can account quantitatively for Moseley's law and qualitatively for the periodic law on the basis of an extension of the Bohr model called the *shell model* of the atom. We argue that an atom consists of a nucleus with charge $Ze$ and mass just a little smaller than the total atomic mass $A$, surrounded by a cloud of $Z$ electrons. (The difference between the nuclear mass and $A$ is just the mass of $Z$ electrons.) We have already noted that, as a rough general rule, $A \simeq 2Z$.

The electrons do not behave independently of one another. Rather, they are accommodated in shells. In this extension of the Bohr picture (which should not be taken too literally), the shells may be visualized as generalizations of the Bohr orbits of hydrogen. Each shell has a characteristic range of Coulomb energies. The shells are numbered, from the inside out, by a quantum number $n = 1, 2, 3, \ldots$ that corresponds to the principal quantum number in the Bohr hydrogen atom. Each shell has a certain capacity for electrons; the capacity increases with increasing $n$.

In atoms of increasing atomic number, the stationary states are occupied by electrons from the lowest-energy state upward. In hydrogen, a single electron lies in a stationary ground state in the shell corresponding to the quantum number $n = 1$. This shell is called the $K$ shell. In helium, the second electron occupies the second state, filling the $K$ shell. In lithium, the first shell is full and the third electron lies in the second shell, called the $L$ shell, for which $n = 2$, and so on.

As a step toward understanding the periodic law in reference to the shell model of the atom, let us look more closely at helium and lithium. Helium is chemically inert, suggesting that a filled shell implies chemical inertness. This suggestion is substantiated by the observation that all Group VIIIA elements are chemically inert (or nearly so). In lithium, the $K$ shell—filled with the first two electrons—lies closer to the nucleus than does the third electron in the $L$ shell. The charge $-2e$ of the electrons in the $K$ shell partly shields the third electron from the electric field of the nucleus. We argue that the outer electron lies in an electric field whose source is a net charge $3e - 2e = e$. Consequently, the lithium atom acts somewhat like a hydrogen atom, with a single outer electron available for chemical reaction and thus a valence $+1$. Moreover, we may expect the ionization potential of the outer electron in lithium to be less than that of hydrogen because the outer electron of lithium lies farther from the nucleus than the single electron of hydrogen. Indeed, the ionization potentials of the alkali metals decrease systematically with increasing atomic number, as shown in Table 41.2.

### TABLE 41.2 Ionization Potentials of Hydrogen and the Alkali Metals

| Element | Atomic Number Z | Ionization Potential (V) |
|---|---|---|
| Hydrogen, H | 1 | 13.6 |
| Lithium, Li | 3 | 5.4 |
| Sodium, Na | 11 | 5.1 |
| Potassium, K | 19 | 4.3 |
| Rubidium, Rb | 37 | 4.2 |
| Cesium, Cs | 55 | 3.9 |
| Francium, Fr | 87 | <4 |

As a general rule, the outer electrons of atoms are largely shielded from the nuclear charge $Ze$ by the inner electrons. As the atomic number increases, the ionization potentials of the outermost electrons remain of the same order of magnitude as that of hydrogen—that is, within an order of magnitude of 10 V. But the inner electrons are shielded less; the innermost electrons "see" the field of the full nuclear charge $Ze$. We expect that these inner electrons are difficult to remove from the atom and do not participate in chemical reactions.

We have thus sketched the outline of an explanation of the periodic law. As electrons are added to the outermost shell, the first atom in a period is monovalent because there is only one electron that is easy to remove. The second atom in the period is either monovalent or divalent because one or two electrons can be removed, and so on. (You should not be surprised that the details are more complicated than this.)

You have probably encountered a more elaborate version of this qualitative picture in studying chemistry. For our purposes, however, the present sketch is adequate, and we now use it to account for Moseley's law.

Neglecting the relatively small energy differences among the stationary states of the electrons within the same shell, we again assume that the energy levels described by the principal quantum number $n$ are analogous to the levels in the hydrogen atom. Suppose that, in striking the target of an X-ray tube, an energetic electron collides with one of the two electrons in the $K$ (innermost) shell of an atom, knocking the electron out of the atom. An electron from one of the outer shells of the atom can now occupy the stationary state left vacant. If the electron happens to come from the $L$ shell, it must dispose of the energy difference $\Delta E$ between its initial and final stationary states by emitting a photon. Because the initial energy $E_L$ and the final energy $E_K$ are defined within fairly narrow bounds, all the photons in such transitions have nearly the same energy. We thus have an explanation for the general properties of the X-ray line spectrum.

## EXAMPLE 41.2

(a) Find the energy $E_K$ of an electron in the $K$ shell of a tungsten (W) atom, for which $Z = 74$. (b) Find the energy $E_L$ of an electron in the $L$ shell of a tungsten atom, at an instant when one of the $K$ electrons has been knocked out of the atom but no outer electron has fallen into the $K$ shell to replace it. (c) Find the energy and wavelength of the photon emitted when the empty $K$ state is filled by an $L$ electron.

### SOLUTION:

(a) Calculate $E_K$ for tungsten.

You can write Equation 41.15, which gives the energy of the $n$th stationary state of hydrogen, in the form

$$E_n = -\frac{me^2e^2}{8\epsilon_0^2h^2}\frac{1}{n^2}.$$

However, you must modify this equation to account for the fact that the tungsten nucleus has charge $Z = 74e$. In addition, you must allow for the fact that a $K$ electron is shielded from the nucleus (at least in part) by the other $K$ electron. Although you cannot account for this shielding exactly, you cannot be very far wrong if you substitute the factor $[(Z - 1)e]^2$ for one of the factors $e^2$ in the preceding equation. Setting $n = 1$ for a $K$ electron, you have

$$E_K = -\frac{m(Z - 1)^2e^4}{8\epsilon_0^2h^2} \qquad \textbf{(41.19)}$$

$$= -\frac{9.11 \times 10^{-31} \text{ kg} \times (73)^2 \times (1.60 \times 10^{-19} \text{ C})^4}{8 \times (8.85 \times 10^{-12} \text{ C}^2/\text{N·m}^2)^2 \times (6.63 \times 10^{-34} \text{ J·s})^2}$$

$$= -1.16 \times 10^{-14} \text{ J},$$

or $\qquad E_K = -72.2$ keV.

(b) Find $E_L$ at the instant before the $L$ electron falls into the $K$ shell.

With one $K$ electron missing, the $L$ electron is shielded from the nucleus by just one $K$ electron. Thus the effective charge on the nucleus is again $Z - 1$. Setting $n = 2$ for an $L$ electron, you have

$$E_L = -\frac{m(Z - 1)^2e^4}{8\epsilon_0^2h^2}\frac{1}{2^2}. \qquad \textbf{(41.20)}$$

By comparing this equation with Equation 41.19, you can see that $E_L = E_K/4$, so

$$E_L = -2.90 \times 10^{-15} \text{ J} = -18.0 \text{ keV}.$$

(c) Find the energy and wavelength of the emitted photon.

The energy is equal to the difference:

$$\Delta E = E_L - E_K = -\frac{3E_K}{4} = 8.7 \times 10^{-15} \text{ J} = 54.2 \text{ keV}.$$

The corresponding photon wavelength is

$$\lambda = \frac{hc}{\Delta E}$$

$$= \frac{6.63 \times 10^{-34} \text{ J·s} \times 3.00 \times 10^8 \text{ m/s}}{8.7 \times 10^{-15} \text{ J}}$$

$$= 22.9 \text{ pm}.$$

This wavelength lies in the X-ray region of the spectrum. The experimental value is $\lambda = 21.4$ pm; not bad for a crude approximation.

If you reflect on the implications of Example 41.2, you will see that the $K$-series lines in the X-ray spectrum correspond to the ultraviolet Lyman series of the hydrogen atom, with the energy shifted to larger values (and the wavelength to smaller values) by the presence of greater charge on the nucleus. $K$-series lines correspond to the transitions $L \rightarrow K$, $M \rightarrow K$, $N \rightarrow K$, and so on. Similarly, the $L$-series lines correspond to the Balmer series, with transitions $M \rightarrow L$, $N \rightarrow L$, and so on. Indeed, we can characterize all of the lines in the X-ray spectrum by means of the generalized Rydberg formula

$$\frac{1}{\lambda} = \frac{m(Z - 1)^2e^4}{8\epsilon_0^2h^3c}\left(\frac{1}{n^2} - \frac{1}{j^2}\right); \qquad \textbf{(41.21)}$$

compare with Equation 41.17.

Moseley's law follows directly from Equation 41.21. Because $1/\lambda = \nu/c$, we have $\nu \propto (Z-1)^2$, or

$$\sqrt{\nu} \propto Z - 1.$$

(41.22)

# The Laser

The Bohr model of the atom provides a quantitative (though incomplete) picture only of the hydrogen atom and the related ions $He^{2+}$, $Li^{3+}$, and so forth. Nevertheless, as you have just learned, an extension of the Bohr model can be used to account with reasonable accuracy for the gross features of the X-ray spectra of all atoms. Underlying all the explanations we have made so far is a single principle of spectroscopy:

> *An atom can exist only in certain specific stationary states, each of which is characterized by a specific energy, called an* energy level. *A photon can be absorbed or emitted by an atom only if the energy of the photon is equal to the difference between two energy levels of the atom.*

As a general rule, the energy-level structure of atoms is much more complicated than that of hydrogen, and we cannot expect to describe the details of such structures in terms of the simple Bohr model. Nevertheless, the very existence of energy levels leads to the possibility of an important optical device called the *laser*.

## Absorption, Spontaneous Emission, and Stimulated Emission of Photons

Figure 41.12 shows a small part of the energy-level diagram of an atom. The energy levels $E_1$ and $E_2$ are only two among many levels that make up the complete diagram. The atom can be *excited* from $E_1$ to $E_2$ by the absorption of a photon of energy $h\nu = E_2 - E_1$, as shown in part $a$ of the diagram. The process is reversible; if the atom is already in the excited state $E_2$, it can *decay* to the state $E_1$ by **spontaneous emission** of a photon of the same energy $h\nu = E_2 - E_1$ (Figure 41.12b).

There is no way to predict when spontaneous emission will take place for any particular excited atom. But there is a mean decay time $\tau$ for the transition. That is, if many atoms are in the excited state $E_2$ at $t = 0$ and no further excitation takes place, only $1/e$ of the atoms will still be in the excited state at time $\tau$; the rest will have decayed to state $E_1$. A typical mean decay time for such transitions is $\tau \simeq 10^{-8}$ s, but the range of mean decay times is tremendous. In some cases, $\tau$ is as small as $10^{-15}$ s; for phosphorescent materials (materials that "glow in the dark" after being exposed to light), however, $\tau$ can be of the order of minutes or hours. We are particularly interested here in transitions for which $\tau$ is considerably greater than $10^{-8}$ s; say, $\tau \simeq 10^{-3}$ s.

In 1916, Einstein showed that a third transition process, called **stimulated emission**, must exist as well. This process is shown in Figure 41.12c. Suppose that a photon of energy $h\nu$ is incident on an atom that happens to be in the state $E_2$. The incident photon cannot be absorbed because the atom is already excited. It can be shown that the oscillating electric field associated with the incident photon *stimulates* the transition of the atom back to the state $E_1$, with the emission of a photon. The emitted photon has exactly the same energy $h\nu$ as the incident photon, and *the two photons are in phase with one another*. On the average, a collection of excited atoms will spend less time in the excited state if photons of energy $h\nu$ are present than if they are not. The more intense the incident radiation field (of which the incident photons are part), the less time the atoms spend in the excited state on the average.

If the collection of atoms is in thermal equilibrium, more atoms are in the lower state $E_1$ than in the upper state $E_2$. Consequently, most of the incident photons are absorbed by atoms in the lower state, and only a few serve to stimulate emission by atoms in the upper state. But suppose we somehow produce a **population inversion**; that is, we arrange matters so that most of the atoms are in the excited state $E_2$. Under

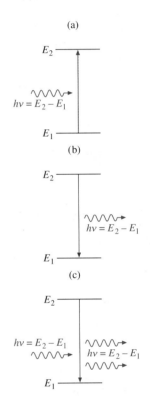

**FIGURE 41.12** Part of the energy-level diagram of an atom. As in the Bohr model of the hydrogen atom, the energies $E_1$ and $E_2$ are expressed with respect to a common reference level $E_\infty = 0$, whose greater value lies beyond the scale of the diagram. (*a*) Absorption. A photon of energy $h\nu = E_2 - E_1$ is absorbed by an atom initially in the energy state $E_1$. The atom makes a transition to the higher energy level $E_2$. (*b*) An atom initially in the excited state $E_2$ decays to the state $E_1$ with the spontaneous emission of a photon of the energy $h\nu = E_2 - E_1$. (*c*) A photon of energy $h\nu = E_2 - E_1$ is incident on an atom in state $E_1$ and stimulates the emission of an identical photon.

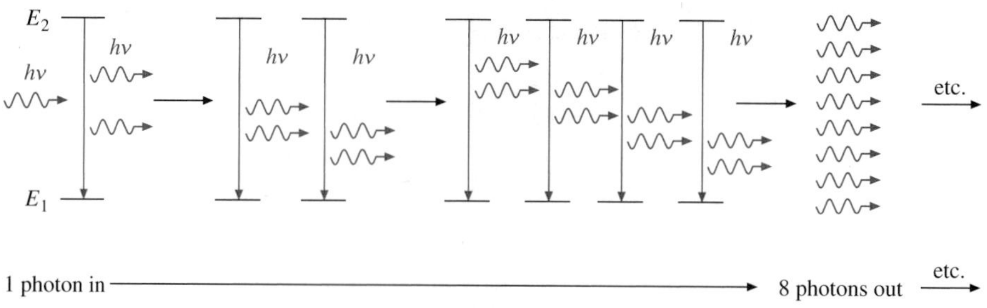

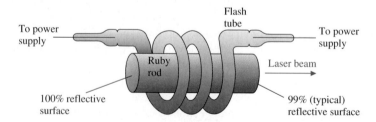

**FIGURE 41.13** Stimulated emission of photons by a collection of atoms, all of which are initially in the excited state $E_2$. All of the emerging photons are of equal energy $h\nu$, and all are in phase.

see Problem 36.8.

### Origins of the Maser and the Laser

The maser came first; it was invented in 1954 by the American physicist Charles H. Townes (b. 1915) and independently by the Russian physicists Nikolai G. Basov (b. 1922) and Aleksandr M. Prokhorov (b. 1916). In 1958, Townes and Arthur L. Schawlow (b. 1921) extended the theory of the maser into the visible region of the electromagnetic spectrum and set forth the theoretical requirements for a laser. All four scientists won Nobel Prizes, the first three in 1964 and Schawlow in 1981. A working laser was first realized in 1960 by the American physicist Theodore H. Maiman (b. 1927). For a good account of the early history of the maser and the laser, see Bela A. Lengyel, "Evolution of Masers and Lasers," *Amer. J. Phys.* **34**, 903 (1966).

these conditions, a few incident photons of the proper energy $h\nu$ can produce a chain reaction, in which photons produced by stimulated emission themselves produce further stimulated emission. This process is shown in Figure 41.13. All of the photons emerging from the process are in phase, and they constitute a beam of *coherent light* (Section 36.4). Because a single incident photon begets a large number of identical photons, we say that *the incident light is amplified*. A device in which this process occurs is called a **laser**, an acronym for **l**ight **a**mplification by **s**timulated **e**mission of **r**adiation. If the photon frequency $\nu = (E_2 - E_1)/h$ happens to lie in the microwave range, the device is called a **maser**, for **m**icrowave **a**mplification by **s**timulated **e**mission of **r**adiation.

How can we produce the necessary population inversion? There are many possible ways, and we consider two of them here.

### The Ruby Laser

Maiman's ruby laser is shown in Figure 41.14. The critical component is a single-crystal rod of synthetic ruby. Ruby is crystalline aluminum oxide (sapphire), $Al_2O_3$, containing in solid solution a small amount of chromium (typically 0.05%) in the form of $Cr^{3+}$ ions. The $Cr^{3+}$ ions take part directly in the action of the laser; the $Al_2O_3$ acts as a transparent medium that contains the $Cr^{3+}$.

The ruby rod is ground into a precise right cylinder. The length of the cylinder is typically 5 cm, and its diameter is typically 0.5 cm. The ends of the rod are coated with reflective surfaces. A multilayer interference coating, designed for the wavelength of the laser, is usually used.* At one end, the reflectivity is made as close to 100% as possible; at the other end, the coating is designed to allow passage of something like 1% of the light striking it. The ruby with its end coatings is thus a Fabry-Perot interferometer; see Problem 36.8.

Wrapped around the ruby is a xenon-filled flash tube much like those used in flash photography. When a high-voltage pulse is applied to it, the tube produces a brief but very bright flash of bluish white light. The entire assembly is usually surrounded by reflective material such as aluminum foil (not shown in Figure 41.13) in order to maximize the amount of light that illuminates the ruby.

**FIGURE 41.14** A ruby laser. The ends of the ruby rod are very precisely parallel. One end is coated with a highly reflective surface; the other is coated so as to reflect most, but not all, of the light striking it. Surrounding the ruby rod is a xenon flash tube similar to the flash tubes used in photography.

The three energy levels of $Cr^{3+}$ relevant to the laser action are shown in Figure 41.15. The upper level, $E_3$, is in fact a collection of many closely spaced levels called an *absorption band*. When the xenon light flashes, many of the photons are of proper

---

*The principle of operation of the coating is similar to that of the antireflection coatings described in Section 36.3. In the present case, however, the thickness of the layers is designed to promote reflection rather than to prevent it.

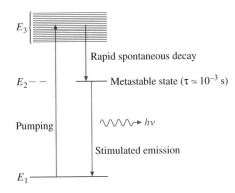

Rapid spontaneous decay

Metastable state ($\tau \approx 10^{-3}$ s)

$h\nu$

Stimulated emission

Pumping

**FIGURE 41.15** Energy-level diagram for a ruby laser. The lowest energy state $E_1$ is the ground state of the $Cr^{3+}$ ion. The upper level $E_3$ is a collection of many closely spaced levels. Stimulated emission occurs between $E_2$ and $E_1$.

energy to excite ions from the lowest energy state $E_1$ to some part of the absorption band. This process is called **pumping**.

The mean decay time $\tau_{32}$ for transition from the absorption band levels to $E_2$ is quite short. In ruby, the transition is **nonradiative**; that is, the energy $E_3 - E_2$ is lost not by emission of a photon but by transfer of mechanical energy to the crystal, whose internal vibrations are made more vigorous. The spontaneous decay time $\tau_{21}$ of level $E_2$ is of order $10^{-3}$ s—orders of magnitude greater than $\tau_{32}$. Consequently, ions ''pile up'' in state $E_2$.* If the pumping is sufficiently energetic, state $E_1$ is largely depleted and contains many fewer ions than state $E_2$. That is, population inversion is achieved.

A few ions in state $E_2$ emit photons spontaneously. The resulting photons stimulate emission from other ions, as shown in Figure 41.13. Because the ends of the ruby rod are reflective, most photons make many trips back and forth through the crystal before escaping through the partially transmitting end. Because trapping of the photons inside the crystal greatly enhances the chance that a photon will interact with an excited ion and produce stimulated emission, the state $E_2$ is emptied quite rapidly. This rapid stimulated emission, called **laser action**, ceases when the population is no longer inverted. The laser beam—the light emerging from the laser—is highly coherent and highly directional, and its wavelength is quite sharply defined. For ruby, the characteristic wavelength is 694.3 nm, in the red region of the visible spectrum. Because the laser process involves three energy levels, the ruby laser is called a *three-level laser*.

### The Helium-Neon Laser

The helium-neon laser, invented in 1961, is widely used in applications requiring a continuous source of coherent light of low intensity. The most familiar applications are in video-disk and compact-disk players and in supermarket checkout scanners. The He-Ne laser is sketched in Figure 41.16. A gaseous-discharge tube is filled with a mixture of He and Ne gases, usually in the proportion 5:1. The pressure is about $10^{-3}$ atm. Quartz windows are cemented onto the ends of the tube, oriented at Brewster's angle (Section 35.8) with respect to the tube axis. (We will consider the reason for this orientation shortly.) Carefully aligned mirrors, one totally and the other partially reflective, serve the same purpose as the reflective end coatings in the ruby laser.

**FIGURE 41.16** The helium-neon laser. The glass discharge tube is filled with a mixture of He and Ne gas at low pressure. The end windows are of quartz, oriented at Brewster's angle with respect to the tube axis. The 100% reflective mirror at the left and the partially reflective mirror at the right define the laser cavity. Laser light emerges through the latter mirror, parallel to the tube axis.

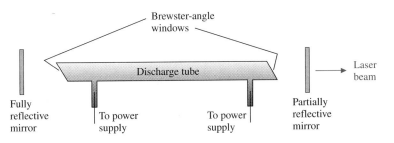

Brewster-angle windows

Discharge tube

Laser beam

Fully reflective mirror

To power supply

To power supply

Partially reflective mirror

*A state such as $E_2$, having a relatively long but not infinite decay time, is called a *metastable state*.

Helium-neon lasers can be made to emit light at several different wavelengths in the red and the infrared. We consider here the process that leads to the emission of red light at 632.8 nm. The relevant part of the energy-level diagram of the system is shown in simplified form in Figure 41.17. The electrical discharge excites helium atoms from the ground state $E_0$ into various excited states. Two of these states are of interest. They are close together in energy and are both represented in the diagram by the energy level $E_2$.

**FIGURE 41.17** Energy-level diagram for the He-Ne laser. The excited helium atom gives up its energy to a ground-state neon atom by inelastic collision, represented by the dashed arrow.

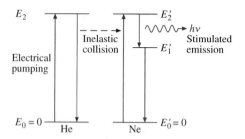

It happens that neon has an energy level $E_2'$ whose value is almost the same as that of the helium level $E_2$. When excited He atoms collide with Ne atoms, they often collide inelastically, transferring their excitation energy to the Ne atoms. The decay time for the excited state $E_2$ of helium is long enough that the necessary collisions take place frequently.

As a result of such inelastic collisions, neon atoms are raised to the excited state $E_2'$, which is metastable. Stimulated emission results in the transition $E_2' \rightarrow E_1'$. Subsequent rapid spontaneous emission returns the atom to the ground state.

The *Brewster-angle windows* lead to the production of polarized light. Consider light traveling in either direction that strikes a window. The component polarized with its electric vector normal to the plane of incidence of the windows is largely reflected, and it leaves the laser. Only light with its electric vector parallel to the plane of incidence can pass the window without loss by reflection. Thus only light of that polarization can pass repeatedly through the laser, and laser action involves only light of that polarization.

In addition to their role in producing polarized light, which is often useful in itself, the Brewster-angle windows reduce the *threshold*—the amount of pumping energy required to maintain a population inversion and thus produce continuous laser action. Because of the restriction of the laser transition to only one of the two possibilities that would exist with unpolarized light, the lifetime of the excited state $E_2'$ is essentially doubled.

*Shopping as a quantum phenomenon: The operation of the laser that reads the bar codes depends on the fact that light is quantized.*

## Symbols Used in Chapter 41

| | | | | |
|---|---|---|---|---|
| $a_0$ | Bohr radius | | $M$ | mass of an atom |
| $A$ | atomic mass | | $n$ | principal quantum number |
| $b$ | Balmer's constant | | $q$ | electric charge |
| $c$ | speed of light | | $R_X, R_H$ | Rydberg constant for element X, for hydrogen |
| $e$ | elementary charge | | $U$ | potential energy |
| $E$ | total energy | | $V$ | electric potential |
| $h$ | Planck's constant | | $Z$ | atomic number |
| $K$ | kinetic energy | | $\epsilon_0$ | permittivity of free space |
| $L$ | angular momentum | | $\lambda$ | wavelength |
| $m$ | electron mass | | $\nu$ | frequency |

Atoms acting individually emit and absorb light at sharply defined wavelengths. The **line spectrum** of the atoms of a chemical element is characteristic of that element. The lines in the **absorption spectrum** correspond to the lines in the **emission spectrum**.

All of the lines of the hydrogen spectrum fall into series that can be accounted for by means of the **Rydberg formula** (Equation 41.2):

$$\frac{1}{\lambda} = R_X \left( \frac{1}{n^2} - \frac{1}{j^2} \right), \quad \begin{array}{l} n = 1, 2, 3, \ldots, \\ j = n + 1, n + 2, n + 3, \ldots. \end{array}$$

In this equation, $R_X$ is an empirical constant whose value for hydrogen is $R_H = 1.0972 \times 10^7$ m$^{-1}$.

In the Rutherford-Geiger-Marsden experiment, energetic $\alpha$ particles are scattered by the atoms in a thin metal foil. The observed dependence on the scattering angle $\theta$ of the number of particles scattered per unit time can be accounted for only by a *nuclear atom*—that is, an atom in which nearly all of the mass and all of the positive charge are concentrated in a tiny **nucleus**, whose radius is given approximately by Equation 41.4.

The Rydberg formula can be derived on the basis of the **Bohr model** of the hydrogen atom, in which an electron follows a "planetary" orbit around a nucleus consisting of a single proton. The Bohr model combines a classical picture of a system bound by the Coulomb force, together with a two-part *ad hoc* assumption:

1. The electron is restricted to certain stable orbits, called **stationary states**, in which it does not radiate electromagnetic energy.

2. The electron can accomplish the energy gain or loss required to make a transition from one stationary state to another by absorbing or emitting a photon whose frequency is given by Equation 41.8,

$$\pm(E_f - E_i) = h\nu.$$

The stationary states are determined by the **Bohr quantization condition** (Equation 41.10b):

$$L_n = n \frac{h}{2\pi}, \quad n = 1, 2, 3, \ldots.$$

The integer $n$ is the **principal quantum number**. It follows from this condition that the atomic radius is given by Equation 41.13,

$$r_n = \frac{\epsilon_0 h^2}{\pi m e^2} n^2,$$

and that, most important, the energy of the stationary states is given by Equation 41.15,

$$E_n = -\frac{m e^4}{8 \epsilon_0^2 h^2} \frac{1}{n^2}.$$

Although the quantum nature of the Bohr atom may conflict with "common sense" acquired in familiar surround-

ings where all transitions are smooth, we ought not to expect that common-sense inferences will be valid in a realm where we have no familiar experience. Bohr's **correspondence principle** asserts that quantum systems merge into continuous ones as their scale makes the transition from microscopic (roughly atom-sized) to macroscopic. This is exemplified by the behavior of Equations 41.15 and 41.17 as the quantum number $n$ becomes large.

The Bohr model of the hydrogen atom can be generalized to provide a qualitative picture of other atoms as well. This picture can be used to account for **Moseley's law**, according to which the frequencies of corresponding X-ray lines observed in the spectra of atoms of various atomic numbers $Z$ are related by Equation 41.22, $\sqrt{\nu} \propto Z - 1$. Moreover, the Bohr picture accounts for the fact that the X-ray line spectrum of a chemical element conforms to the modified Rydberg formula given by Equation 41.21. The atomic number $Z$ assumes physical significance as the number of elementary charges in the nucleus.

The transition of an atom from one stationary state to another can take place in any of three ways: **absorption**, **spontaneous emission**, and **stimulated emission**. Any particular spontaneous emission process has a characteristic mean decay time $\tau$. Stimulated emission is more prompt than spontaneous emission, and it is induced by the presence of photons whose energy conforms to the Bohr condition of Equation 41.8. The photons that emerge from a stimulated emission process are in phase.

In a laser, a **population inversion** is produced by a **pumping** process. Stimulated emission is promoted by the geometry of the laser system, and the emitted light is highly coherent, monochromatic, and directional.

## KEY TERMS

**Section 41.2  Atomic Spectra**
gaseous discharge tube ▪ discrete spectrum ▪ Fraunhofer lines ▪ Balmer series, series limit ▪ Rydberg formula, Rydberg constant

**Section 41.3  Rutherford Scattering and the Nucleus**
α particle

**Section 41.4  The Bohr Atom**
proton ▪ stationary state ▪ orbital angular momentum ▪ Bohr quantization condition ▪ principal quantum number ▪ ground-state orbit, Bohr radius ▪ ground-state energy ▪ energy-level diagram ▪ correspondence principle

**Section 41.5  Moseley's Law and the Periodic Table**
periodic law, period ▪ atomic number ▪ cutoff wavelength

**Section 41.6  The Laser**
spontaneous emission, stimulated emission ▪ population inversion ▪ maser ▪ pumping

# QUERIES

**41.1** *(2) Flame test.* In performing qualitative analyses for metal ions, chemists often use a simple test. A small amount of the unknown substance is put on a thin wire made of an inert, refractory metal (such as platinum) and placed in the not-very-luminous flame of a Bunsen burner. A skilled chemist can make fairly quick preliminary determinations by observing the color given to the flame by the unknown sample. Explain how a flame test works.

**41.2** *(2) Mix-up.* You can see from Figure 41.7 that some of the series of lines that make up the hydrogen spectrum overlap other series. Is there any series that lies completely within the lines of another series? Explain.

**41.3** *(2) Enlightenment.* Even before the era of interplanetary probes, it was possible to learn something about the chemical composition of planetary atmospheres by spectroscopic means. Explain how this can be done, in view of the fact that the atmospheres are not luminous.

**41.4** *(2) Remote thermometry.* How can you estimate the surface temperature of a star by looking at its line spectrum?

**41.5** *(3) Earnest speaking.* Explain why Rutherford's remark about a 15-inch shell bouncing off a piece of tissue paper is so appropriate.

**41.6** *(3) Whoosh!* What is the form of the trajectory of an α particle that passes near a nucleus?

**41.7** *(4) Pickering series.* Consider the spectrum of a singly ionized helium atom ($He^+$) from the viewpoint of the Bohr model. A single electron circles a nucleus containing two protons. The emission spectrum resembles that of neutral hydrogen, but with the wavelength of each line shifted to a smaller value—that is, with the photon energies shifted toward higher values. Explain. [The $He^+$ spectrum lines are named after Edward C. Pickering (1846–1919), the American astronomer who first saw them in 1896 in the spectra of hot stars. The wavelengths are given by the Rydberg formula if the Rydberg constant is multiplied by 4; see Problem 41.31.]

**41.8** *(5) All roads lead to Planck.* One way to evaluate Planck's constant is to measure the cutoff wavelength of the X-ray spectrum as a function of the potential difference that drives the electrons into the target of the X-ray tube. Explain how this experiment yields the desired result.

**41.9** *(5) Shades of Rydberg!* Define the series limits for X-ray lines.

**41.10** *(5) You can't get one without the other.* When the K lines of the X-ray spectrum are excited, why would you expect to see L lines as well?

**41.11** *(6) Four levels are better than three?* Three-level lasers such as the ruby laser require much pumping power to produce the necessary population inversion because the energy state $E_1$ is the ground state and contains many atoms. This difficulty is ameliorated in *four-level lasers* such as the *neodymium laser*, in which the active $Nd^{3+}$ ions are dissolved in glass or in a transparent crystal (usually yttrium aluminum garnet, called YAG for short). The energy-level diagram is shown below. Describe the action of the laser, and explain why it takes much less power than a three-level laser to produce population inversion of the laser levels $E_2$ and $E_1$.

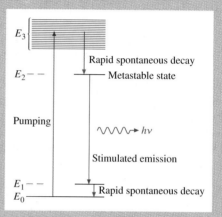

**41.12** *(6) Giant pulses.* In an ordinary pulsed laser (such as a ruby laser), stimulated emission begins shortly after population inversion is achieved and before the pumping flash has ended. The depletion of the upper energy level by laser action limits the inversion that can be achieved by the flash lamp. In the *giant-pulse laser*, a much greater degree of population inversion can be achieved by inserting a very fast-acting electro-optic shutter into the laser system. Laser action is delayed until the end of the flash by keeping the shutter closed and then opening it when the flash has ended and the population inversion has reached a maximum value. Describe the characteristics of the pulse, and suggest some advantages of the giant-pulse laser.

# PROBLEMS

## GROUP A

**41.1** *(2) Rydberg covers Balmer.* Show that Equation 41.2 reduces to Equation 41.1 if you set $n = 2$ and $R_H = 4/b$.

**41.2** *(2) Calculating with a Paschen.* Use the Rydberg formula to calculate the values $\lambda_\infty$ for the seven series given in Table 41.1.

**41.3** *(2) Bracketting the values.* Use the Rydberg formula to calculate the values of $\lambda_\alpha$ for the seven series given in Table 41.1.

**41.4** *(2) Balmer stands alone.* Calculate **(a)** the longest wavelength in the Lyman series and **(b)** the shortest wavelength in the Paschen series. By doing so, show that only Balmer-series lines lie in the visible part of the spectrum.

**41.5** *(2) Pounding it in.* Does the Pfund series overlap the Brackett series? Base your answer on explicit calculation.

**41.6** *(3) Getting up close, I.* You repeat the Rutherford-Geiger-Marsden experiment, using $\alpha$ particles of energy 4.77 MeV. **(a)** What is the distance of closest approach of the $\alpha$ particles to the target nuclei if the target is made of lead? **(b)** At this distance, do the $\alpha$ particles penetrate into the nucleus?

**41.7** *(3) Getting up close, II.* You repeat the Rutherford-Geiger-Marsden experiment, using $\alpha$ particles of energy 4.77 MeV. **(a)** What is the distance of closest approach of the $\alpha$ particles to the target nuclei if the target is made of lithium? **(b)** Show that, at this distance, the $\alpha$ particles penetrate into the nucleus. Rutherford's scattering observations under such circumstances showed deviation from the predictions of Equation 41.3, and he concluded that he was observing the effects of the *strong nuclear force*—the force that holds the nucleus together against the mutual Coulomb repulsion of the positive charges it contains.

**41.8** *(4) Growing a little.* Find the radii of the second and third Bohr orbits for the hydrogen atom.

**41.9** *(4) Is relativity relevant? I.* **(a)** Find the speed of the electron in a Bohr hydrogen atom as the electron circles the nucleus in the ground state. **(b)** Is a relativistic treatment of the atom necessary? Explain. **(c)** Is the speed of the electron in excited states greater than or less than that in the ground state? Is a relativistic treatment necessary?

**41.10** *(4) Packing a lot in.* When hydrogen atoms are excited to the first excited state ($n = 2$), they remain in that state on the average for about 10 ns before returning to the ground state. How many revolutions does the electron make around the nucleus in that time?

**41.11** *(4) Growing, but how much?* What is the diameter of a hydrogen atom in an excited state in which the electron energy is just 0.1 eV less than that required for ionization?

**41.12** *(4) Still growing.* What is the diameter of a hydrogen atom in the stationary state for which $n = 1000$?

**41.13** *(5) First cut.* What is the cutoff wavelength for X rays generated by electrons accelerated across a potential difference of 60 kV?

**41.14** *(5) Second cut.* What is the cutoff wavelength for X rays generated by electrons accelerated across a potential difference of 90 kV?

**41.15** *(5) Watching too much TV?* The electrons that strike the face of a typical color television tube and "write" the image on the phosphorescent screen have been accelerated through a potential difference of about 20 kV. **(a)** What is the cutoff wavelength of the X rays generated when the electrons hit the screen? X rays in this wavelength range are called "soft." **(b)** Some persons have expressed concern that children can be exposed to excessive soft X rays, though these X rays do not penetrate glass readily and there is no solid evidence of harm from such exposure. Nevertheless, to be on the safe side, why should you encourage small children not to sit immediately in front of the screen, as they seem to love to do?

**41.16** *(5) Cutting it fine.* What minimum potential difference is required if an X-ray tube is to produce radiation of wavelength 10.6 pm?

**41.17** *(5) Line voltage, I.* **(a)** Show that the wavelength of the $K_\alpha$ line for molybdenum is 72.0 pm. **(b)** What is the minimum voltage across the X-ray tube that will produce this line?

**41.18** *(5) Line voltage, II.* **(a)** Find the wavelength of the $K_\alpha$ line produced by an X-ray tube with an iron target. **(b)** What is the minimum voltage across the X-ray tube that will produce this line?

**41.19** *(5) Line voltage, III.* **(a)** Find the wavelength of the $K_\alpha$ line produced by an X-ray tube with a copper target. **(b)** What is the minimum voltage across the X-ray tube that will produce this line?

**41.20** *(5) Gold and hydrogen.* A gold atom emits a $K_\alpha$ photon. The wavelength of the $K_\alpha$ line of gold ($Z = 79$) is 18.5 pm. **(a)** What is the energy difference between the initial and final stationary states of the atom? **(b)** What are the energies of the corresponding states in hydrogen? **(c)** Calculate the ratio of the answers to parts **a** and **b**. Does this ratio conform to the prediction of the generalized Bohr theory?

**41.21** *(6) Stimulating diversity, I.* What is the energy difference $\Delta E$ between the two laser levels in a ruby laser? Take $\lambda = 694.3$ nm.

**41.22** *(6) Stimulating diversity, II.* In a He-Ne laser, what is the energy difference $\Delta E$ between the laser levels that are involved in the emission of light of wavelength 632.8 nm?

**41.23** *(6) Double duty.* The He-Ne laser can also operate between a pair of levels for which $\Delta E = 1.08$ eV. Find the wavelength of the light emitted.

**41.24** *(6) Eye surgery.* Lasers are used to treat patients suffering from detached retinas. A pulse of light is directed to a tiny spot on the retina and "welds" it to its substrate. For this purpose, a neodymium laser is usually used. A typical pulse length is 1 ns, and the total pulse energy is about 1 mJ. If the weld is to be a circular spot of diameter 25 $\mu$m, what is the power per unit area delivered to the spot?

**41.25** *(3) Close, but not too close.* Rutherford concluded, on the basis of α-particle backscattering experiments with aluminum foils, that the sum of the radii of the α particle and the Al nucleus was less than $7 \times 10^{-15}$ m. (Problems 41.6 and 41.7 suggest Rutherford's line of argument.) What must the energy of the α particle be if it is to come that close to the Al nucleus? Take $Z = 13$.

**41.26** *(3) Tough nut to crack?* What fraction of the total volume of an atom is occupied by the nucleus? An order-of-magnitude answer is sufficient.

**41.27** *(4) Photoelectrons from hydrogen.* If you performed a photoelectric experiment on a sample of hydrogen, at what wavelength would you find the photoelectric threshold?

**41.28** *(4) Twanging all the strings.* In a gaseous discharge tube (and in other circumstances as well), atoms acquire the energy that they later radiate as photons by means of inelastic collision with energetic electrons. **(a)** What is the smallest possible maximum energy of the electrons if all the lines of the hydrogen spectrum are to be excited? **(b)** Set a lower limit on the voltage across a gaseous discharge tube for which this can happen when the pressure in the tube is $10^{-3}$ atm and the tube is 15 cm long. (Hint: Assume that intermolecular collisions are randomizing, so a molecule must acquire the necessary energy in a distance equal to the mean free path. The mean free path of the molecules is $l = n^{-1/3}$, where $n$ is the number of molecules per unit volume.)

**41.29** *(4) Going round in circles.* Starting with Coulomb's law, derive Equation 41.5 explicitly for an electron in a circular orbit around a proton. Assume that the mass of the proton is very much larger than that of the electron.

**41.30** *(4) Ionization potential of atomic hydrogen.* Experiment shows that the amount of energy required to ionize atomic hydrogen—that is, to remove the electron entirely from the proton—is 13.6 eV. **(a)** Show that the Bohr model predicts a value in close agreement with the experimental value. **(b)** What is the maximum wavelength of a photon capable of ionizing a hydrogen atom? In what region of the spectrum does this wavelength lie?

**41.31** *(4) Was Pickering Bohred? I.* Show that the spectrum of singly ionized helium (He$^+$) is described by the modified Rydberg formula

$$\frac{1}{\lambda} = \frac{4me^4}{8\epsilon_0 h^3 c} \left( \frac{1}{n^2} - \frac{1}{j^2} \right).$$

(See Query 41.7.)

**41.32** *(4) Beyond Pickering.* Write an equation that describes the wavelengths of the lines in the emission spectrum of doubly ionized lithium (Li$^{2+}$), in which a single electron orbits a nucleus containing three protons. (Hint: Refer to Problem 41.31.)

**41.33** *(4) Was Pickering Bohred? II.* Find $\Delta E$ and $\lambda$ for the α line and for the series limit of the Pickering series that corresponds to the Balmer series, for which $n = 2$. In what part of the spectrum do these lines lie? (See Problem 41.31.)

**41.34** *(4) Star light, star bright.* Consider the series of lines in the spectrum of singly ionized helium that corresponds to the Lyman series of hydrogen. **(a)** What is the range of wavelengths of this series? **(b)** What is the photon energy for the longest wavelength? (See Problem 41.31.)

**41.35** *(4) Was Pickering Bohred? III.* In searching for the series of spectral lines named after him (Query 41.7), Pickering was guided by the Balmer formula. In the spectra he studied, he saw what appeared to be the Balmer series, except that between every two adjacent Balmer lines there was an extra line. Pickering found that he could account for all of the lines with a modified Rydberg formula:

$$\frac{1}{\lambda} = R \left[ \frac{1}{(n/2)^2} - \frac{1}{(j/2)^2} \right].$$

Explain why this formula works. (See Problem 41.31.)

**41.36** *(4) More than meets the eye.* In the Bohr hydrogen atom, a ground-state electron lies at the distance $a_0$ from the nucleus. **(a)** Calculate the Coulomb force exerted on the electron by the nucleus. **(b)** Calculate the acceleration of the electron. **(c)** If the electron is in a circular orbit, find its frequency of rotation. Compare this frequency with the range of frequencies of the lines in the Lyman series. (This is the range in which the electron ought to emit light according to classical physics.) Is the agreement good or bad?

**41.37** *(4) Count 'em!* A collection of hydrogen atoms is excited to the state corresponding to the principal quantum number $n$. Show that the number of lines in the spectrum is $n(n - 1)/2$.

**41.38** *(4) The Franck-Hertz experiment.* In 1914, James Franck (1882–1964) and Gustav Hertz (1887–1975) performed the experiment illustrated below. Although they were not aware of Bohr's work at the time, their experiment served to substantiate Bohr's view that atoms can make transitions between definite stationary states by emitting or absorbing photons. The central part of the apparatus is a glass tube with two electrodes, filled with mercury vapor at low pressure, as shown in part *a*. As the voltage between the electrodes is in-

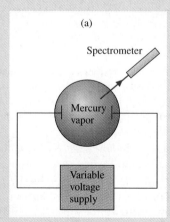

(a)

Spectrometer

Mercury vapor

Variable voltage supply

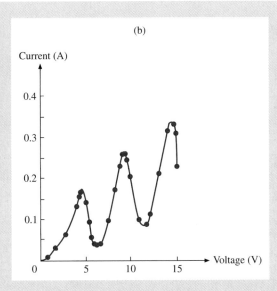

(b)

creased, the electric current through the tube varies, as shown in part b. The first maximum occurs at $V = 4.88$ V. At that point, the tube begins to glow. The spectrometer is used to measure the wavelength of the light. What is that wavelength?

**41.39** *(4) Muonium.* A *muon* is a particle having the same charge as an electron but mass 207 times as great (Section 38.4). A muon does not live very long, but its lifetime is sufficient to form a hydrogenlike atom, called a *muonium*, consisting of a muon in orbit around a proton. Although the proton mass $M$ is only about 9 times the muon mass $m_\mu$ (compared with 1836 times the electron mass $m$), make the approximation that $M \gg m_\mu$. **(a)** Derive an expression for the radius of the muonium atom, and calculate the Bohr radius. **(b)** Derive an expression for the energy levels of the muonium atom. **(c)** Calculate the wavelength of the Lyman $\alpha$ line of the muonium spectrum. In what region of the electromagnetic spectrum does this line lie?

**41.40** *(5) Rip-off.* Use the generalized Bohr model to calculate **(a)** the first ionization energy of lithium—that is, the energy required to remove one electron—and **(b)** the third ionization energy of lithium—that is, the energy required to remove the third and last electron.

**41.41** *(5) Is relativity relevant? II.* Consider one of the $K$ electrons in an atom of uranium ($Z = 92$). **(a)** Assume that you can ignore electric fields associated with the other electrons (because they lie farther away from the nucleus), and

calculate the ground-state energy of the electron. **(b)** Calculate the "Bohr radius" of the electron, and use Newtonian physics to calculate its speed. Is a relativistic correction needed for massive atoms such as uranium?

**41.42** *(5) Getting it all, I.* What is the minimum voltage required if an X-ray tube with a copper target is to produce the entire $K$ series?

**41.43** *(5) Getting it all, II.* What is the minimum voltage required if an X-ray tube with a platinum target is to produce the entire $K$ series?

**41.44** *(5) Moseley plot.* **(a)** To check the validity of Moseley's law, make a Moseley plot, using the following values for the wavelengths of the $K_\alpha$ lines of various elements:

| Element | Z | λ (pm) |
|---------|-----|--------|
| Li | 3 | 22 800 |
| Al | 13 | 834 |
| Mn | 25 | 251 |
| Cu | 29 | 154 |
| Mo | 42 | 71.4 |
| Ba | 55 | 39.0 |
| W | 74 | 21.4 |
| Bi | 83 | 16.6 |
| U | 92 | 13.1 |

(Hint: You can save yourself a little trouble by plotting $1/\sqrt{\lambda}$ instead of $\sqrt{\nu}$.) **(b)** Measure the slope of your plot, and use it to calculate Planck's constant. How good is the agreement with the accepted value?

**41.45** *(5) X-ray crystal ball.* The atomic number of indium (In) is 49. Beginning with information provided in the table of Problem 41.44, calculate the wavelength of the $K_\alpha$ line for indium. (Hint: What choice makes your calculation easiest?)

**41.46** *(6) An advantage of coherence.* An important property of laser light is its coherence. **(a)** $N$ light waves, each of amplitude $A$, are superposed. If the phases of the individual waves are random, what is the ratio $I/I_0$ of the intensity of the superposed wave to that of one of the component waves? **(b)** $N$ light waves, each of amplitude $A$, are superposed *in phase.* What is the ratio $I/I_0$ in this case? **(c)** A typical gas laser involves simultaneous emission by $10^{16}$ atoms. What is the ratio of the intensity of the coherent laser beam to that which is obtained when the tube is operated with $10^{16}$ atoms emitting in a nonlasing mode?

## GROUP C

**41.47** *(3) Single event versus a random series.* Rutherford's theory of $\alpha$-particle scattering depends on deflection of the $\alpha$ particle by a single tiny nucleus. The plum-pudding model of the atom requires that the observed scattering angles be the result of many small deflections as the $\alpha$ particle passes through the foil. Show that the Rutherford picture predicts that, other things being equal, $N(\theta)$, the number of particles per unit time observed at a particular angle $\theta$, will be directly proportional to the foil thickness $t$, and show that the plum-

pudding picture predicts that $N(\theta)$ will be proportional to $\sqrt{t}$. [Actual measurement shows that $N(\theta) \propto t$.] (Hint: Here is a standard result of statistics. Suppose a man is so drunk that, as he walks along the sidewalk, he cannot remember in which direction he made his last step. Consequently, some of his steps are directed up the street and others are directed down the street. If each step on the average is of length $l$, his distance from the starting point after $n$ steps is $l\sqrt{n}$.)

**41.48** *(3) Classical failure.* According to classical electromagnetic theory, an electron that experiences acceleration $a$ will radiate energy at the instantaneous rate

$$\frac{dE}{dt} = \frac{e^2}{6\pi\epsilon_0 c^3}\, a^2.$$

The theory also predicts that an oscillating electron will emit radiation whose frequency is the same as the oscillation frequency of the electron. **(a)** What is the acceleration of an electron in a classical planetary hydrogen atom, in which it revolves around a stationary proton in an orbit of radius $R$? (See part **b** of Problem 41.36.) **(b)** What is the initial rate of energy loss if the electron starts out in an orbit whose radius $r_0$ is the Bohr radius $5 \times 10^{-11}$ m? **(c)** As an order-of-magnitude approximation, assume that this energy-loss rate remains constant, and calculate the order of magnitude of the time required for the electron to crash into the proton.

**41.49** *(4) Twanging Lyman's lyre.* The hydrogen atoms in the sun that emit Lyman $\alpha$ radiation are mainly excited by inelastic collisions with energetic electrons. **(a)** What minimum energy must an electron have to produce such a collision? **(b)** Make the assumption (which is not accurate) that the root-mean-square energy of the electrons is equal to the minimum energy you found in part **a**. On this assumption, what is the temperature of the layer of the sun where the Lyman $\alpha$ radiation originates? **(c)** As a matter of fact, the Lyman $\beta$ radiation from the sun is considerably less intense than the Lyman $\alpha$ radiation. What does this tell you about the actual temperature of the relevant solar layer, compared with the value you calculated in part **b**? (Hint: Consider the shape of the Maxwell-Boltzmann energy distribution curve and the fractional energy difference between the Lyman $\alpha$ and $\beta$ photons.)

**41.50** *(4) Correspondence principle at work.* **(a)** Find the frequency of revolution $\nu_n$ of an electron in a hydrogen Bohr orbit whose quantum number is $n$. **(b)** Write the frequency $\nu_{n-1}$ for the next lower orbit, corresponding to the quantum number $n - 1$. **(c)** What is the frequency $\nu'$ of the photon emitted when the electron makes the transition $n \to n - 1$? **(d)** Show that $\nu_n > \nu' > \nu_{n-1}$; that is, the frequency of the emitted photon is intermediate between the initial and final revolution frequencies of the electron. **(e)** Show that, as $n \to \infty$, the frequency of the emitted photon approaches the frequency of revolution of the electron—$\nu' \to \nu$—in accordance with classical electromagnetic theory, as the correspondence principle requires.

**41.51** *(4) Little atom, big field.* **(a)** Show that, according to the Bohr model, the magnetic field at the nucleus of a hydrogen atom due to the orbital motion of the electron in the ground state is

$$\mathscr{B} = \frac{\pi\mu_0 m^2 e^7}{8\epsilon_0^{\,3} h^5} = \frac{\mu_0 e h}{8\pi^2 m a_0^{\,3}}.$$

**(b)** Evaluate this quantity.

**41.52** *(4) Reduced-mass correction to the Rydberg constant.* The mass of the proton is 1836 times the mass of the electron. **(a)** Show that, in order to take into account the fact that the proton mass is much larger than the electron mass but not infinite, the electron mass $m$ in Equation 41.17 should be replaced by the reduced mass $\mu$, given by

$$\frac{1}{\mu} = \frac{1}{1} + \frac{1}{1836}.$$

(Hint: See Problem 9.67.) **(b)** Recalculate the Rydberg constant $R_H$ for hydrogen, using the reduced mass in place of the electron mass. **(c)** Does this correction by itself account for the discrepancy between the Balmer $\alpha$ wavelength given by Equation 41.17 and the observed value 656.28 nm?

**41.53** *(4) Rydberg constant for singly ionized helium.* Following the argument of Problem 41.52, calculate the Rydberg constant $R_{He}$ for He$^+$ ions. (See Problem 41.31.)

**41.54** *(4) Proton-electron mass ratio.* Mass spectrographic data show that the mass of the proton is 1.007 27 u and the mass of the $\alpha$ particle is 4.001 51 u. Spectrographic analysis shows that the corresponding Rydberg constants are $R_H = 1.096\ 775\ 76 \times 10^7$ m$^{-1}$ and $R_{He} = 1.097\ 222\ 67 \times 10^7$ m$^{-1}$. Use this information to find the ratio of the mass of the proton to that of the electron. (Hint: Consider the reduced masses of the two atoms; see Problem 41.52.)

**41.55** *(4) Deuterium.* In addition to the common form of hydrogen, whose nucleus consists of a single proton of mass $M$, there is an isotope called deuterium or "heavy hydrogen." The nucleus of deuterium is formed by the fusion of a proton and a neutron, and its mass is roughly twice that of a proton, though its charge is exactly the same. **(a)** Make an argument to show that the presence of deuterium in a sample of hydrogen results in a splitting of each of the lines of the spectrum into a *doublet*, consisting of a pair of closely spaced lines, the difference in wavelength being $\Delta\lambda = (m/2M)\lambda$. **(b)** Find the wavelength difference for the Balmer $\alpha$ line, for which $\lambda = 656.3$ nm. (Hint: Consider the reduced masses of the two atoms; see Problem 41.52.)

**41.56** *(4) Positronium.* A short-lived hydrogenlike atom can be made from an electron and a *positron*. A positron is the *antiparticle* of the electron; it has the same mass as an electron but charge $+e$. Positronium does not last very long because the atom radiates photons as the electron and positron come closer and closer together, finally annihilating one another with the production of a photon pair. Because their masses are equal, the electron and the positron revolve about their common center of mass, which always lies halfway between them. **(a)** Derive an expression for the radius of a positronium atom, and find its Bohr radius. **(b)** Derive an expression for the energy levels of the positronium atom. **(c)** Find the wavelength of the Lyman $\alpha$ line. In what region of the electromagnetic spectrum does it lie? (Hint: Consider the reduced mass of the system; see Problem 41.52.)

**41.57** *(5) Screening constant.* In Equation 41.21, the factor $(Z - 1)^2$ is used in place of the factor $Z^2$ to take account, in an approximate way, of the partial screening (shielding) of the nuclear charge by the "other" electron in the $K$ shell. Even as an approximation, the argument works only for the $K$ series.

But more accurate account can be taken of screening by modifying Equation 41.21 to the form

$$\frac{1}{\lambda} = \frac{m(Z - b)^2 e^4}{8\epsilon_0^2 h^3 c} \left( \frac{1}{n^2} - \frac{1}{j^2} \right),$$

where the *screening constant b* is to be determined experimentally. Measurement of the wavelength of the $L_\alpha$ line emitted by tungsten yields $\lambda = 143$ pm. **(a)** Evaluate $b$. **(b)** What is the wavelength of the $L_\alpha$ line for copper $(Z = 29)$?

# Wavelike Properties of Matter

**L O O K I N G   A H E A D**

———— The "old" quantum mechanics, based on the Bohr model of the atom, leaves more questions unanswered than it answers. Beginning in 1923, the "new" quantum mechanics dealt with the problems of microscopic physics at a deeper level.

———— The de Broglie relation associates a wavelength $\lambda$ with every particle of momentum $p$ and justifies the Bohr quantization condition.

———— The amplitude of the square of the wave function at any location is a measure of the probability of finding the associated particle at that location.

———— The uncertainty principle sets a limit to the accuracy with which we can know the position and momentum of a particle simultaneously.

———— Schrödinger's equation is the quantum-mechanical equivalent of Newton's second law of motion; it describes the evolution of the wave function of a particle.

———— The spectrum of the hydrogen atom can be described completely by application of quantum-mechanical techniques. This is one of the first great triumphs of the theory.

*Left:* A quantum corral. A scanning tunneling microscope has been used to place 48 iron atoms in a ring on a copper surface and to form an image of the wave pattern produced by the electrons confined in the ring. The ring diameter is 14.3 nm, but the wave pattern is pretty much the same as one you will see in a circular container of water (such as a coffee maker) if you drip water into the center. If you remain skeptical of the idea that particles exhibit wavelike behavior, this photo should be a convincer.

*God doesn't shoot craps.* (Der liebe Gott würfelt nicht.)

—ALBERT EINSTEIN

*Don't tell God what to do.*

—Attributed to NIELS BOHR

## Introduction

The Bohr model of the atom has afforded considerable insight into the structure of atoms and their interactions with electromagnetic radiation. But the picture is far from complete. In the decade (1913–1923) following Bohr's pioneering work, much effort was devoted to expanding the scope of the Bohr theory, but with limited success. The developments based on the theory came later to be called the *"old" quantum mechanics*. Many physicists were persuaded that the limitations of the old quantum mechanics were connected with the lack of any theoretical justification for the existence of stationary states in atoms. Section 42.2 examines *de Broglie's postulate*—the breakthrough in 1923 that accounted for the stationary states and laid the foundation for a much deeper understanding of the nature of matter. In Sections 42.3 through 42.6, we follow the developments that came to be called the *"new" quantum mechanics*. In Section 42.3, we apply de Broglie's picture to the simple case of a particle in a one-dimensional box. Section 42.4 deals with the *Heisenberg uncertainty principle*. In Section 42.5, we consider the *Schrödinger equation*, which is the quantum-mechanical counterpart of Newton's second law in classical systems. Most of the section is a discussion of the description of atomic structure made possible by the Schrödinger equation. Section 42.6 presents experimental evidence for the quantization of the angular momentum of electron orbits that underlies the success of the Bohr model.

"TRY TO IMAGINE MY NEW GAME AS A WAY OF CAPTURING THE BEST QUALITIES OF BOTH."

# de Broglie's Postulate

We have made much of the fact that light exhibits particle-like properties as well as wave-like properties. What about matter? Can it be that matter exhibits wavelike properties as well as particle-like properties? If so, we can unify wavelike light and particle-like matter—distinct domains in classical physics—into a single domain. This unifying idea was first set forth in 1923 by Louis de Broglie, while he was a graduate student at the Sorbonne. The quantum properties of light *and* matter fall naturally out of de Broglie's synthesis.

As a prelude to this unification, let us review what we know about the photon as a particle. The energy and wavelength of a photon are connected by Equation 40.10,

$$E = \frac{hc}{\lambda}. \qquad (42.1)$$

Because it is an extreme relativistic particle, the photon has zero rest mass. Consequently, its energy and momentum are related by Equation 40.14,

$$E = pc. \qquad (42.2)$$

From the purely mathematical point of view, it is child's play to eliminate $E$ from this pair of equations and write $hc/\lambda = pc$, or

$$\lambda = \frac{h}{p}. \qquad (42.3)$$

This is the **de Broglie relation**. The quantity $\lambda$ is called the **de Broglie wavelength**.

de Broglie asserted that *this equation (42.3) is valid for all particles*, including nonrelativistic ones—particles whose speed $v$ is much less than $c$. This assertion is called **de Broglie's postulate**. Its role in quantum mechanics is analogous to the role of Newton's first law in classical mechanics. For nonrelativistic particles, we can write the momentum in the classical form $p = mv$. Equation 42.3 then becomes

$$\lambda = \frac{h}{mv} \quad \text{for } v \ll c. \qquad (42.4)$$

Although de Broglie's mathematics is simple, the physical implications are very deep. *Matter has wavelike properties* because the wavelength $\lambda$ implies that there is a wave associated with every particle that moves. Why, then, have you never noticed any wavelike behavior in familiar, macroscopic matter?

**Louis de Broglie**

Louis Victor Pierre Raymond, prince de Broglie (pronounced roughly as de Brull-**yee**′) (1892–1987), French physicist, belonged to an aristocratic family long engaged in military, political, and diplomatic affairs. Not long after he shifted his interest from medieval history to physics, de Broglie did the work for which he was awarded the Nobel Prize in 1929.

## EXAMPLE 42.1

Find the de Broglie wavelength of **(a)** a 450-g billiard ball moving at 8.5 m/s, **(b)** an electron that has been accelerated from rest through a potential difference of 10 V, **(c)** a 10-MeV electron of energy, and **(d)** a 3.5-eV photon.

**SOLUTION:**
**(a)** Find $\lambda$ for the billiard ball.
  Equation 42.4 gives you

$$\lambda = \frac{6.63 \times 10^{-34} \text{ J·s}}{0.450 \text{ kg} \times 8.5 \text{ m/s}} = 1.7 \times 10^{-34} \text{ m}.$$

This is a tiny wavelength indeed; something like $10^{20}$ waves would fit into the diameter of an atomic nucleus. No wonder the wavelike properties of a billiard ball are not apparent!

**(b)** Find $\lambda$ for a 10-eV electron.
  This kinetic energy lies well within the nonrelativistic range

for electrons. Consequently, you can use the classical relation between momentum and kinetic energy: $p = \sqrt{2mE}$. Equation 42.3 thus becomes

$$\lambda = \frac{h}{p} = \frac{h}{\sqrt{2mE}}$$

$$= \frac{6.63 \times 10^{-34} \text{ J·s}}{\sqrt{2 \times 9.11 \times 10^{-31} \text{ kg} \times 10 \text{ eV} \times 1.60 \times 10^{-19} \text{ J/eV}}}$$

$$= 3.9 \times 10^{-10} \text{ m} = 0.39 \text{ nm}.$$

This wavelength is small on the everyday scale of things. But it is not small on the atomic scale; it is about the same as a typical atomic diameter. (Remember that twice the Bohr radius of hydrogen is about 0.1 nm.) More significantly, this wavelength is at least as large as the diameter of the electron considered a classical particle. (You know this because you know that many electrons "fit" into the typical atom.)

**(c)** Find $\lambda$ for a 10-MeV electron.

The rest mass of the electron is about 0.5 MeV, and so a 10-MeV electron is highly relativistic. You can therefore use the extreme relativistic relation $p = E/c$, with $E$ the total relativistic energy $E_0 + K = 10.5$ MeV (Equation 39.15). Equation 42.3 becomes

$$\lambda = \frac{h}{p} = \frac{hc}{E},$$

which is the same as Equation 42.1. You thus have

$$\lambda = \frac{6.63 \times 10^{-34}~\text{J·s} \times 3.00 \times 10^8~\text{m/s}}{1.05 \times 10^7~\text{eV} \times 1.60 \times 10^{-19}~\text{J/MeV}}$$
$$= 1.2 \times 10^{-13}~\text{m}.$$

This wavelength is intermediate between a typical atomic diameter ($\approx 10^{-10}$ m) and a typical nuclear diameter ($\approx 10^{-15}$ m).

**(d)** Find $\lambda$ for a 3.5-eV photon.

The photon is an extreme relativistic particle, and so you use Equation 42.1 to write

$$\lambda = \frac{h}{p} = \frac{hc}{E} = \frac{6.63 \times 10^{-34}~\text{J·s} \times 3.00 \times 10^8~\text{m/s}}{3.5~\text{eV} \times 1.6 \times 10^{-19}~\text{J/eV}}$$
$$= 3.6 \times 10^{-7}~\text{m} = 360~\text{nm}.$$

There is really nothing new in this; you have just calculated the wavelength of the photon, using the Einstein relation $E = hc/\lambda$, which does not involve de Broglie's postulate. But note the close connection between the calculation for a photon and that for a highly relativistic electron—or, for that matter, any highly relativistic particle.

## de Broglie's Postulate and the Bohr Quantization Condition

Example 42.1 shows why we do not notice the wavelike properties of macroscopic bodies. What is the use, then, of de Broglie's postulate? Consider the Bohr model of the hydrogen atom. You learned in Section 41.4 that the stationary states can be specified by Equation 41.12, which restricts the kinetic energy of the orbital electron to the values

$$K_n = \frac{n^2 h^2}{8\pi^2 m r_n^2}. \tag{42.5}$$

Because the electron in the Bohr atom is nonrelativistic, we have (as in part **b** of Example 42.1) the relation $K = p^2/2m$. Using this relation, we write $p^2 = n^2 h^2/4\pi^2 r_n^2$. The de Broglie relation in the form of Equation 42.3 then gives us

$$p^2 = \frac{h^2}{\lambda^2} = \frac{n^2 h^2}{4\pi^2 r_n^2}.$$

Solving for $\lambda$, we obtain

$$\lambda = \frac{2\pi r_n}{n}, \tag{42.6a}$$

or

$$n\lambda = 2\pi r_n. \tag{42.6b}$$

The right side of Equation 42.6b is just the circumference of the circular electron orbit corresponding to the principal quantum number $n$. Thus the equation tells us that *the electron wavelength must fit into the orbit an integral number of times*. This is a restatement of the Bohr quantization condition in the form of a standing-wave condition! In the ground state ($n = 1$), the wavelength fits once around the orbit and forms the standing wave shown in Figure 42.1a. In the first excited state ($n = 2$), the wavelength fits twice around the orbit, forming the standing wave shown in Figure 42.1b, and so on.

What if the Bohr quantization condition is not met, as shown in Figure 42.1e? In this case, the electron wave does not superpose on itself to form a standing wave. Rather, there is a superposition of waves having the same wavelength but arbitrary phase relations. Figure 42.1f shows the mean amplitude for 100 successive orbits. You can see how the waves tend to cancel each other; the amplitude of the electron wave is zero.

What does this zero mean amplitude imply? We argue, along with de Broglie, that we must take the wave properties of matter seriously; there is a wave associated with every moving particle. But in an orbit that does not satisfy the Bohr quantization condition, we have *no* wave. It follows that *there can be no electron* in such an orbit—at least, not for a time long enough to allow the electron to execute more than a few orbits.

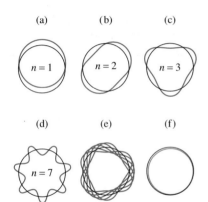

(a)   (b)   (c)

$n = 1$   $n = 2$   $n = 3$

(d)   (e)   (f)

$n = 7$

**FIGURE 42.1** The Bohr quantization condition as a standing-wave condition for the de Broglie waves of an electron. The waves (red lines) are shown as sinusoidal for simplicity, though this is not essential to the argument. For clarity, the waves are shown together with circles centered on the nucleus. According to the Bohr theory, the orbit radii increase as $n^2$. For convenience in comparing the various parts of this figure, all orbit radii are shown to be the same; that is, the actual radii have been divided by $n^2$. (*a*) The ground state, $n = 1$. (*b*) The first excited state, $n = 2$. (*c*) $n = 3$. (*d*) $n = 7$. (*e*) Electron wave for an orbit that does not satisfy the Bohr condition. Five successive orbits are shown. (*f*) Mean amplitude of the waves for 100 successive revolutions like those shown in part *e*.

EXAMPLE **42.2**

Force an electron into a forbidden orbit by setting $n$ equal to $\pi$—a notorious noninteger! Assume that 100 revolutions around the nucleus yield sufficient cancellation to give a wave of negligible amplitude. How long can the electron remain in this orbit?

**SOLUTION:** For the circular orbit corresponding to $n$, the orbit period $\tau_n$ is just the orbit circumference divided by the electron speed $v_n$:

$$\tau_n = \frac{2\pi r_n}{v_n}.$$

There are several ways to express $v_n$ in terms of $r_n$, and here is one. Begin with the Bohr quantization condition stated in terms of the angular momentum, $L_n = mv_n r_n$. From Equation 41.10a, you have

$$mv_n r_n = \frac{nh}{2\pi},$$

or

$$v_n = \frac{nh}{2\pi m r_n}.$$

Using this value of $v_n$ to express the period $\tau_n$, you obtain

$$\tau_n = 2\pi r_n \frac{2\pi m r_n}{nh} = \frac{4\pi^2 m r_n^2}{nh}.$$

Into this equation you substitute the value of $r_n$ given by

Equation 41.13, $r_n = \epsilon_0 h^2 n^2 / \pi m e^2$:

$$\tau_n = \frac{4\pi^2 m}{nh} \frac{\epsilon_0^2 h^4}{\pi^2 m^2 e^4} n^4 = \frac{4\epsilon_0^2 h^3}{me^4} n^3.$$

The time required for a single revolution of the electron around the nucleus is thus

$$t_n = \frac{4 \times (8.85 \times 10^{-12}\,\text{C}^2/\text{N·m}^2)^2 \times (6.63 \times 10^{-34}\,\text{J·s})^3}{9.11 \times 10^{-31}\,\text{kg} \times (1.60 \times 10^{-19}\,\text{C})^4} n^3$$
$$= (1.53 \times 10^{-16}) n^3\,\text{s}.$$

For the ground state, $n = 1$, the period is

$$\tau_1 = 1.53 \times 10^{-16}\,\text{s}.$$

Here, however, you are interested in the time required for 100 revolutions in the forbidden state $n = \pi$. You have

$$100\tau_\pi = 100 \times 1.53 \times 10^{-16}\,\text{s} \times \pi^3 = 4.74 \times 10^{-13}\,\text{s}.$$

We have argued that this is an upper limit on the time the electron can spend in the state $n = \pi$, because the waves associated with successive revolutions superpose to yield zero amplitude. Compare the time $100\tau_\pi$ with the typical mean lifetime of stationary states, $10^{-8}$ s, quoted in Section 41.6. You will see in Section 42.3 that a short lifetime implies that a photon of sharply defined energy cannot be emitted at all.

## Electron Diffraction

de Broglie's assignment of wavelike properties to objects that had always been treated as ideal particles was greeted with some skepticism. Indeed, the Sorbonne physics faculty were in disagreement as to whether to award a doctorate for what many regarded as clever but purely speculative work. Fortunately for de Broglie, Einstein visited the Sorbonne and immediately understood the far-reaching possibilities of de Broglie's work.

As always, however, the final appeal must be to experiment. Given that de Broglie had proposed wavelike properties, it was immediately evident that one should look for a uniquely wavelike feature, such as diffraction, in an experiment involving electrons or other particles. Davisson and Germer were the first to observe electron diffraction, in 1927, followed within months by G. P. Thomson.

EXAMPLE **42.3**

When 54-eV electrons are normally incident on nickel crystal (Figure 42.2), a strong diffraction maximum is observed at $\theta = 50°$. X-ray diffraction measurements (see Problem 36.53) show that the spacing between nickel atoms is $D = 215$ pm. **(a)** Find the de Broglie wavelength of the electrons. **(b)** Predict the first-order diffraction angle $\theta$, and compare with the observed angle.

**SOLUTION:**
**(a)** What is the electron wavelength?

You could use the procedure of part **b** of Example 42.1 to calculate $\lambda$. But you can save a little trouble by noting the result of that example: A 10-eV electron has wavelength 0.39 nm $=$ 390 pm. You are now interested in a 54-eV electron. From the

same example, you know that $\lambda \propto 1/\sqrt{E}$. So you can write

$$\frac{\lambda}{390\,\text{pm}} = \sqrt{\frac{10\,\text{eV}}{54\,\text{eV}}}.$$

Solving for $\lambda$, you have

$$\lambda = 168\,\text{pm}.$$

**(b)** At what angle should the de Broglie waves be diffracted in first order?

You use the diffraction-grating equation (Equation 36.30), which you write in the form

$$j\lambda = D \sin \theta.$$

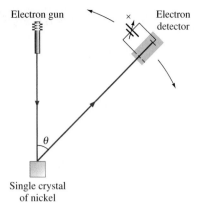

Electron gun    ✕    Electron
detector

$\theta$

Single crystal
of nickel

FIGURE 42.2 Schematic view of the Davisson-Germer experiment. The entire apparatus is inside an evacuated chamber, not shown. The electron gun provides a beam of electrons whose energy can be adjusted in the range from 25 eV to a few hundred eV. The electrons are normally incident on a single crystal of nickel. The detector, aimed at the crystal, can be swung to any angle $\theta$ in an arc centered on the crystal. The detector collects only those electrons whose energy is essentially equal to that of the incident electrons—that is, electrons that have been scattered elastically from the crystal.

Here you are interested in the first order, $j = 1$. So you have

$$\theta = \sin^{-1} \frac{\lambda}{D} = \sin^{-1} \frac{168 \text{ pm}}{215 \text{ pm}} = 51.4°.$$

This compares well with the measured value, 50°.

---

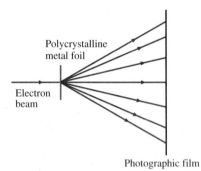

Polycrystalline
metal foil

Electron
beam

Photographic film

FIGURE 42.3 Thomson's electron diffraction experiment. A beam of electrons is incident on a thin, polycrystalline metal foil. The electrons are diffracted and fall on a photographic film.

Thomson's experimental results display the diffraction effect in a dramatic way. The geometry of his experiment is shown in Figure 42.3. The situation is more complicated than the diffraction of visible light from a two-dimensional grating because here the metal-foil "grating" consists of tiny three-dimensional crystals having all possible orientations with respect to the incident beam. We will not carry out a detailed analysis. However, the situation is completely analogous to X-ray diffraction. Figure 42.4 shows how close the analogy is. By 1930, diffraction had been demonstrated for hydrogen atoms and helium atoms as well as for electrons. Today, *neutron diffraction* is a powerful tool for the investigation of the properties of solids.

FIGURE 42.4 (*a*) X-ray and (*b*) electron diffraction patterns made using the same polycrystalline aluminum foil. The energy of the electrons was adjusted so that their wavelength would be the same as that of the X rays.

(a)

(b)

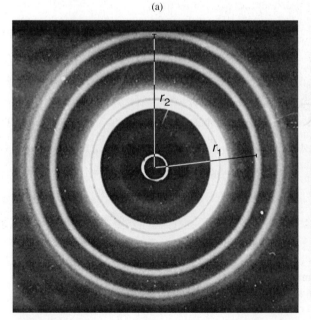

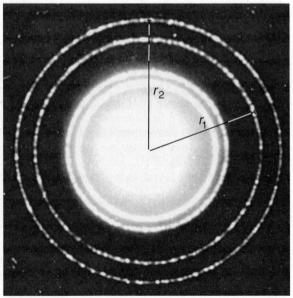

EXAMPLE 42.4

Find the momentum, speed, and energy of an X-ray photon, an electron, and a neutron—all of wavelength 100 pm. Take the neutron mass to be $m_n = 1.68 \times 10^{-27}$ kg.

**SOLUTION:** For all three cases, you can write

$$p = \frac{h}{\lambda} = \frac{6.63 \times 10^{-34} \text{ J} \cdot \text{s}}{1 \times 10^{-10} \text{ m}} = 6.63 \times 10^{-24} \text{ kg} \cdot \text{m/s}.$$

The photon has speed $c$. For the relativistic energy you have

$$
\begin{aligned}
E &= pc \\
&= 6.63 \times 10^{-24} \text{ kg} \cdot \text{m/s} \times 3.00 \times 10^8 \text{ m/s} \\
&= 1.99 \times 10^{-15} \text{ J} \\
&= 12.4 \text{ keV}.
\end{aligned}
$$

Is the electron nonrelativistic? Assume that it is, and write

$$v = \frac{p}{m} = \frac{6.63 \times 10^{-24} \text{ kg} \cdot \text{m/s}}{9.11 \times 10^{-31} \text{ kg}} = 7.28 \times 10^6 \text{ m/s}.$$

This value is only about 2% of $c$, and the electron is indeed nonrelativistic. You can thus ignore the mass-energy and write $E = K = p^2/2m$. This gives you

$$
\begin{aligned}
E &= \frac{(6.63 \times 10^{-24} \text{ kg} \cdot \text{m/s})^2}{2 \times 9.11 \times 10^{-31} \text{ kg}} = 2.41 \times 10^{-17} \text{ J} \\
&= 151 \text{ eV}.
\end{aligned}
$$

For the neutron, you have

$$v = \frac{p}{m} = \frac{6.63 \times 10^{-24} \text{ kg} \cdot \text{m/s}}{1.68 \times 10^{-27} \text{ kg}} = 3950 \text{ m/s}.$$

Again, the energy is $E = p^2/2m$. You have

$$
\begin{aligned}
E &= \frac{(6.63 \times 10^{-24} \text{ kg} \cdot \text{m/s})^2}{2 \times 1.68 \times 10^{-27} \text{ kg}} = 1.31 \times 10^{-20} \text{ J} \\
&= 0.0818 \text{ eV}.
\end{aligned}
$$

Because the neutron mass is much greater than the electron mass, the neutron energy is much smaller than the electron energy corresponding to the same wavelength. This fact has a very useful application, as you will see in the following text.

---

X rays, electrons, and neutrons can all produce diffraction patterns that yield information about the materials that diffract them. For delicate materials, neutron diffraction is often desirable because the small energy required makes it possible to investigate samples without damaging them. In addition, neutrons interact mainly with atomic nuclei, whereas X rays and electrons interact mainly with electrons. Thus neutrons can reveal properties of a sample other than those made evident by X-ray or electron diffraction.

As Example 42.4 shows, 150-eV electrons have the same wavelength—and hence the same theoretical resolving power—as 12-keV photons. In addition, electrons are not nearly as penetrating as X rays; thus electrons can often be used to form an image having much more contrast than is possible with X rays. These possibilities are exploited in a wide variety of instruments called *electron microscopes*.

A final illustration of the wavelike properties of electrons is given in Figure 42.5.

(a)            (b)

 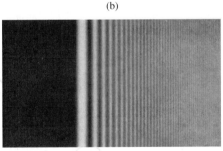

**FIGURE 42.5** Straight-edge diffraction patterns produced (*a*) by 38-keV (6.3-pm) electrons streaming past the edge of a small crystal of magnesium oxide and (*b*) by visible light illuminating a razor blade.

# The Meaning of the de Broglie Wave: Particle in a Box

In Section 42.2, we justified Bohr's quantization principle for the hydrogen atom in terms of the wavelike properties of the electron. But we must not forget about the particle-like properties of the electron. What is the connection between the two?

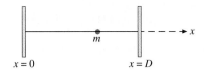

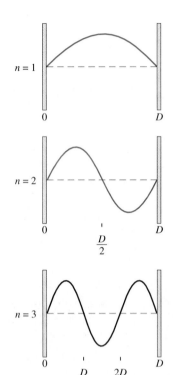

**FIGURE 42.7** The first three standing-wave modes for a particle in a one-dimensional box.

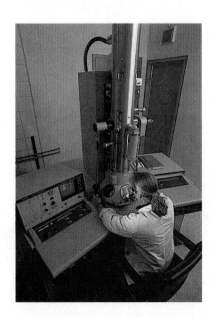

*Electron microscope*

Figure 42.6 shows a nonrelativistic particle (say, an electron) in a one-dimensional box. The geometry of this system is simpler than that of the three-dimensional hydrogen atom. This makes it much easier to follow the physical principles as we work through the mathematical details. The particle is free to move along the $x$ axis between the walls at $x = 0$ and $x = D$. The particle experiences no force except when it collides elastically with the walls and reverses its motion. Consequently, its speed $v$ and the magnitude of its momentum $|p| = mv$ have fixed values. Because $|p|$ is precisely defined, so is the de Broglie wavelength $\lambda = h/p$ (Equation 42.3). [The wave must be a sinusoid, because no other wave has just one Fourier component (Section 21.6).]

When the particle moves in the positive $x$ direction, the momentum is $+p$ and the wave is a sinusoid of wavelength $\lambda$ running in the positive $x$ direction. When the particle moves in the negative $x$ direction, the momentum is $-p$ and the wave is a sinusoid of wavelength $\lambda$ running in the negative $x$ direction. The two waves superpose to form a standing wave (Section 21.7).

We describe the standing wave by means of the *wave function* $\psi(x, t)$. With a slight change in notation, the wave function has the form of Equation 21.51,

$$\psi(x, t) = \psi_0 \sin kx \sin \omega t, \quad 0 \le x \le D, \tag{42.7a}$$

and
$$\psi(x, t) = 0, \qquad\qquad x < 0 \text{ and } x > D, \tag{42.7b}$$

where $k = 2\pi/\lambda$ is the wave number and $\omega$ the angular frequency. In this chapter we will be concerned only with the time-independent part of the wave, which we can write

$$\psi(x) = \psi_0 \sin \frac{2\pi x}{\lambda}. \tag{42.8}$$

Now we must pay attention to the conditions imposed on the wave by the ends of the box. For the hydrogen atom, we argued that the particle cannot be where the wave is not. We must maintain the converse as well: The wave cannot be where the particle is not. Because the particle can never be outside the box, the amplitude of the wave must fall to zero at $x = 0$ and $x = D$. That is, the standing wave has nodes at the ends of the box. Consequently, just as with the vibrating string considered in Section 21.7, the wavelength is restricted to the values given by Equation 21.56b, which we write

$$\lambda_n = \frac{2D}{n}, \quad n = 1, 2, 3, \dots. \tag{42.9}$$

The wave equation that describes the particle in the box thus has the form

$$\psi = \psi_0 \sin \frac{n\pi x}{D}, \quad n = 1, 2, 3, \dots. \tag{42.10}$$

Figure 42.7 shows the first three standing-wave modes.

Because the wavelength is restricted to the values given by Equation 42.9, there is a corresponding restriction on $|p|$: *The momentum of the particle is quantized* and is restricted to the values

$$p_n = \frac{h}{\lambda_n} = \frac{nh}{2D}, \quad n = 1, 2, 3, \dots. \tag{42.11}$$

Compare this equation with the quantization condition for the hydrogen atom given by Equation 41.10b,

$$L_n = \frac{nh}{2\pi}, \quad n = 1, 2, 3, \dots.$$

In the present one-dimensional, linear case, the linear momentum is quantized. For the electron revolving around the hydrogen nucleus, the angular momentum is quantized instead. But the two conditions are very similar. In both cases, *quantization is a direct result of the wavelike properties of the particle, taken together with the restriction of the particle to a certain region of space.*

Because the momentum of the particle in the box is quantized, its kinetic energy is quantized as well. Using Equation 42.11, we have

$$K_n = \frac{p_n^2}{2m} = \frac{n^2 h^2}{8mD^2}, \quad n = 1, 2, 3, \dots. \tag{42.12}$$

The allowed values of $n$ do not include zero. Thus the particle cannot be at rest! The smallest kinetic energy it can have is the **zero-point energy** $K_1 = h^2/8mD^2$. Nonzero values of the zero-point energy are the rule for quantized systems.

What is the physical reason for this nonzero value? If the kinetic energy is to be zero, the momentum $p$ must be zero as well. But the de Broglie relation $\lambda = h/p$ implies that the corresponding wavelength must be $\lambda = \infty$. This can be true only for an unbounded particle. But our particle is restricted to the space $0 \le x \le D$. As a general rule, the more closely a particle is restricted, the greater its zero-point energy.

## *The Born Interpretation*

We have argued that the wave and the particle go together—that the presence of either implies the presence of the other. We now make this statement quantitative.

Given a specific wave function $\psi$ like that of Equation 42.10, where is the particle associated with the wave? For a classical wave, $\psi^2$ is a measure of the energy density of the wave (Section 21.5). If the wave is a standing wave—say, in a stretched string—$\psi^2(x)\,dx$ describes how the total energy is distributed over the medium. A string element of length $dx$ located at an antinode (where $\psi^2$ has its maximum value) contains much energy. On the other hand, an element $dx$ located at a node (where $\psi^2 = 0$) contains no energy. In 1925, the Austrian physicist Erwin Schrödinger (1887–1961) argued that waves associated with matter behave analogously. He suggested that the particle associated with a wave is "smeared out" over the region where the wave function has nonzero values and that $\psi^2(x)\,dx$ is a measure of the fraction of the particle present between $x$ and $x + dx$. In 1926, however, the German (later Scottish) physicist Max Born (1882–1970) showed that this interpretation was not self-consistent. He made an alternative suggestion that has assumed a central position in quantum physics: The particle is not smeared out but is in a specific place. According to the **Born interpretation** of the wave function, *the quantity $\psi^2(x)\,dx$ is a measure of the probability that the particle lies between $x$ and $x + dx$.* The term $\psi^2(x)$ is called the **probability density**. Figure 42.8 shows the probability densities for the first three quantum states of a particle in a one-dimensional box.

Of course, the particle has to be *somewhere*. The probability that it is anywhere at all is therefore 1. That is to say, the wave function is subject to the **normalization condition**

$$\int_{-\infty}^{\infty} \psi^2(x)\,dx = 1. \qquad (42.13)$$

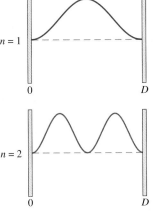

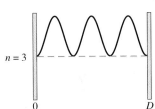

**FIGURE 42.8** Probability densities $\psi^2$ for the first three standing-wave modes of a particle in a one-dimensional box.

EXAMPLE 42.5

**(a)** For a particle in a one-dimensional box, find the probability $P_{mid}$ that the particle is located in the middle half of the box—that is, between $x = D/4$ and $x = 3D/4$. **(b)** Evaluate $P_{mid}$ for the ground state.

**SOLUTION:**
**(a)** Evaluate $P_{mid}$.

First, you must normalize according to Equation 42.13. Using the value of $\psi$ given by Equation 42.10, you have for the normalization condition

$$\int_0^D \psi_0^2 \sin^2 \frac{n\pi x}{D}\,dx = 1.$$

The limits of integration reflect the fact that $\psi$ is zero everywhere outside the box. You can either work out the necessary indefinite integral or else find it in a table. The integral is

$$\int \sin^2 \frac{n\pi}{D} x\,dx = \frac{x}{2} - \frac{D}{4n\pi} \sin \frac{2n\pi}{D} x.$$

You can therefore write the normalization condition in the form

$$\psi_0^2 \left[ \frac{x}{2} - \frac{D}{4n\pi} \sin \frac{2n\pi}{D} x \right]_0^D = 1,$$

which simplifies to

$$\psi_0^2 = \frac{2}{D}. \qquad (42.14)$$

Note that this result does not depend on $n$.

According to the Born postulate, the probability $dP(x)$ of finding the particle in any infinitesimal region between $x$ and $x + dx$ is

$$dP(x) = \psi^2(x)\,dx.$$

Using the value of $\psi^2(x)$ given by Equation 42.10, together with Equation 42.14, you have

$$dP(x) = \frac{2}{D} \sin^2 \frac{n\pi x}{D}\,dx.$$

You can find the probability $P_{mid}$ of finding the particle some-where—anywhere—in the middle half of the box by integrating this quantity between the limits $x = D/4$ and $x = 3D/4$:

$$P_{mid} = \int_{D/4}^{3D/4} dP(x) = \frac{2}{D} \int_{D/4}^{3D/4} \sin^2 \frac{n\pi x}{D}\, dx.$$

The necessary indefinite integral is the same as the one you used in normalization, and you find

$$P_{mid} = \frac{2}{D}\left( \frac{3D}{8} - \frac{D}{4n\pi} \sin \frac{3n\pi}{2} - \frac{D}{8} + \frac{D}{4n\pi} \sin \frac{n\pi}{2} \right)$$

$$= \frac{1}{2} + \frac{1}{2\pi n}\left( \sin \frac{n\pi}{2} - \sin \frac{3n\pi}{2} \right).$$

Note that the value of $P_{mid}$ does depend on $n$. For even values, both $\sin n\pi/2$ and $\sin 3n\pi/2$ are zero, and the term in parentheses is therefore zero. For $n = 1, 5, 9, 13, \ldots$, you have $\sin n\pi/2 = 1$ and $\sin 3n\pi/2 = -1$; thus the term in parentheses is 2. For $n = 3, 7, 11, 15, \ldots$, you have $\sin n\pi/2 = -1$ and $\sin 3n\pi/2 = 1$; thus the term in parentheses is $-2$. Hence there

are three sets of solutions for $P_{mid}$:

$$P_{mid} = \frac{1}{2} + \frac{1}{n\pi} \quad \text{for } n = 1, 5, 9, 13, \ldots, \quad \textbf{(42.15a)}$$

$$P_{mid} = \frac{1}{2} \quad \quad \text{for } n \text{ even,} \quad \textbf{(42.15b)}$$

and $\quad P_{mid} = \frac{1}{2} - \frac{1}{n\pi} \quad \text{for } n = 3, 7, 11, 15, \ldots. \quad \textbf{(42.15c)}$

**(b)** Find $P_{mid}$ for the ground state.
  Setting $n = 1$, you have

$$P_{mid} = \frac{1}{2} + \frac{1}{\pi} = 0.818.$$

That is, at any instant, you have about an 80% chance of finding the particle in the middle half of the box if the particle is in the ground state. Alternatively, if you look repeatedly, you will find the particle in the middle half of the box about four times out of five. What is the probability of finding the particle in the rest of the box?

---

### The Correspondence Principle Again

The result of Example 42.5 for $n = 1$ is quite different from what you would expect for a classical particle. A classical particle moving at constant speed and bouncing back and forth between the walls will spend the same amount of time in every segment $dx$ of the box. Consequently, the classical value of $P_{mid}$ is $\frac{1}{2}$. But you have already seen that the energy of macroscopic particles is always far above the ground-state energy; that is, $n$ is very large (see Problem 42.37 for an explicit consideration of this point). Note what happens to Equations 42.15a and 42.15c as $n$ becomes large: The value of $P_{mid}$ approaches the classical value $\frac{1}{2}$. This is another example of the Bohr correspondence principle.

---

**SECTION 42.4**

## The Heisenberg Uncertainty Principle

The wavelike aspect of matter has still more consequences for the behavior of matter on the atomic scale. We now consider the *uncertainty principle*, first formulated by the German physicist Werner Heisenberg (1901–1976) in 1927.

  Suppose, for simplicity, that the motion of a particle is restricted to the $x$ axis. The motion is completely described if we have *simultaneous* measurements of its position $x$ and its momentum $p$. In practice, we cannot make error-free measurements of $x$ and $p$; our measurements always leave us with uncertainties $\Delta x$ and $\Delta p$. However, there is nothing in classical physics that prevents us from improving our measurements. If we are dissatisfied with our precision, we are free to redesign our apparatus and make a measurement that yields smaller values of $\Delta x$ and $\Delta p$. Although no measurement can ever yield $\Delta x = 0$ or $\Delta p = 0$, we can in principle approach both of these ideals as closely as we wish.

  The wavelike aspect of matter imposes a limit short of the classical ideal. This limit is intrinsic in the properties of matter and has nothing to do with instrumental limitations. No matter how error-free our instruments, *we can never make simultaneous measurements of x and p whose results are more precise than the limit*

$$\Delta x\, \Delta p \gtrsim \frac{h}{2\pi}. \quad \textbf{(42.16a)}$$

The quantities $x$ and $p$ are called *conjugate quantities*. A similar principle applies to any pair of conjugate quantities. For example, when we measure the energy $E$ of a particle and specify the time $t$ at which the measurement was made, the results can never be more precise than the limit

$$\Delta E \, \Delta t \gtrsim \frac{h}{2\pi}. \qquad \text{(42.16b)}$$

Either of Equations 39.16a and 39.16b is called the **Heisenberg uncertainty principle**. The same name is applied to similar limiting statements for other pairs of conjugate quantities.

## *Bohr's Ideal Microscope*

As an initial, physical justification of the uncertainty principle, we consider a thought experiment first suggested by Bohr. In this experiment, we use the ideal microscope of Figure 42.9 to find the location $x$ of a particle. We are free to illuminate the particle with light of any wavelength. However, our choice of wavelength will affect the precision with which we can locate the particle. According to Equation 36.24, the limit of resolution of a microscope is roughly equal to the wavelength of the light used:

$$\Delta x \simeq \lambda. \qquad \text{(42.17)}$$

If we are to improve the resolution of the observation, we must reduce the wavelength of the light used.

If a photon is to be useful for observing the particle, it must collide with the particle and be scattered into the microscope. But in colliding with the particle, the photon gives up some of its momentum $p = h/\lambda$ to the particle and thus disturbs the momentum of the particle. We can minimize this disturbance (in our ideal experiment) by using just one photon.

In itself, the fact that the photon transfers momentum to the particle does not introduce an uncertainty into our knowledge of the momentum of the particle, because we could always make a correction to account for the momentum transfer. Unfortunately, we do not know exactly how much momentum the photon has transferred to the particle. This is because we do not know where the photon entered the microscope; it might have done so anywhere within the angle $\alpha$ in Figure 42.9. The $x$ component of the photon momentum may therefore have any value between $p \sin(\alpha/2)$ and $-p \sin(\alpha/2)$. If we optimize our microscope, we can make $\alpha/2$ approach $90°$; thus we have

$$\Delta p = 2p \sin(\alpha/2) \simeq 2p. \qquad \text{(42.18)}$$

We multiply Equation 42.18 by Equation 42.17 to obtain

$$\Delta x \, \Delta p \simeq 2p\lambda.$$

Using the de Broglie relation, $p = h/\lambda$, we have

$$\Delta x \, \Delta p \simeq 2h. \qquad \text{(42.19)}$$

This result is consistent with the one stated in Equation 42.16a, $\Delta x \, \Delta p \gtrsim h/2\pi$.

Here is the trade-off in which Equation 42.19 involves us: We can increase the resolution of the microscope and thus reduce $\Delta x$ by decreasing $\lambda$. But doing this increases the photon momentum $p$ and therefore increases $\Delta p$ as well, in accordance with Equation 42.18. Alternatively, we can reduce $\Delta p$ by using photons of longer wavelength. But doing this increases the position uncertainty $\Delta x$.

In deriving Equation 42.19, we have assumed an ideal experiment in which the equipment does not introduce special measurement errors of its own. Thus the uncertainty expressed in Equation 42.19 is fundamental, and there is no way to overcome it by refinement of experimental apparatus. This situation is quite different from the classical one, in which (in principle) refinement can always produce a more precise result.

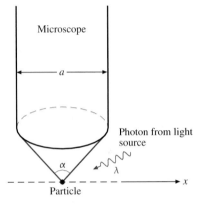

**FIGURE 42.9** In Bohr's thought experiment, an ideal microscope is used to observe the $x$ coordinate of a particle. A single photon of wavelength $\lambda$ is used to illuminate the particle.

## Nonexistence of Electrons in the Nucleus

An important consequence of the uncertainty principle is that electrons cannot exist inside the nucleus. We have seen that the nucleus of an atom of atomic number $Z$ has charge $Ze$ and mass $A \simeq 2Z$. It is tempting to argue that the nucleus contains $A$ protons and $A - Z$ electrons. The net nuclear charge is then $Z$, as observed. When we take into account the $Z$ orbital electrons, the atom consists of $A$ protons and $A$ electrons and no other kinds of particles. Example 42.6 shows that this model of the nucleus cannot be correct.

## EXAMPLE 42.6

Assume that the proton-electron model of the nucleus is correct, and use the uncertainty relation to find (a) the uncertainty $\Delta p$ in the momentum of an electron in a silver nucleus ($Z = 47$, $A \simeq 108$ u), (b) the corresponding uncertainty in the energy of the electron, and (c) the electrostatic energy $E$ that binds the electron to the nucleus.

**SOLUTION:**

(a) Find $\Delta p$ for an electron in a silver nucleus.

According to the model, each electron is somewhere in the nucleus. The uncertainty $\Delta x$ in the position of an electron is just the diameter of the nucleus. You determine the diameter of the nucleus by using Equation 41.4, $r = 1.37 \times 10^{-15}$ m $\times A^{1/3}$. This gives you

$$\Delta x = 2r = 2 \times 1.37 \times 10^{-15} \text{ m} \times (108)^{1/3}$$
$$= 1.3 \times 10^{-14} \text{ m.}$$

You use Equation 42.16a to determine the uncertainty in momentum:

$$\Delta p = \frac{h}{2\pi \, \Delta x} = \frac{6.63 \times 10^{-34} \text{ J·s}}{2\pi \times 1.3 \times 10^{-14} \text{ m}}$$
$$= 8.1 \times 10^{-21} \text{ kg·m/s.}$$

(b) What is $\Delta E$ for the electron?

To a crude approximation, each electron moves freely within the nucleus but collides repeatedly with the "walls" of the nucleus like a particle in a box. Thus the energy is purely kinetic; $E = K$. You need to relate $\Delta E$ to $\Delta p$, but the proper relation between the two quantities depends on whether the electron is relativistic or not. Assume that it is relativistic; you then use Equation 39.22 to write $E = pc$. You differentiate both sides

of this equation to obtain

$$\Delta E = c \, \Delta p \quad \text{for an extreme relativistic particle.}$$

The numerical values give you

$$\Delta E = 3 \times 10^8 \text{ m/s} \times 8.1 \times 10^{-21} \text{ kg·m/s}$$
$$= 2.4 \times 10^{-12} \text{ J.}$$

This energy is equivalent to 15 MeV. The rest mass of an electron is about 0.5 MeV; indeed, the electron is extremely relativistic.

(c) Find the electrostatic energy $U$ that binds an electron to the nucleus.

According to the model, there are $A - Z = 108 - 47 = 61$ electrons in the nucleus. The net charge on the nucleus is $Ze = 47e$. Imagine that one of the electrons happens to be at the periphery of the nucleus, a distance $r = \Delta x/2 = 0.65 \times 10^{-14}$ m from the center. The electron "sees" a charge $(Z + 1)e = +48e$ and therefore has a potential energy

$$U = \frac{1}{4\pi\epsilon_0} \frac{(48e)(-e)}{r}$$
$$= -\frac{(8.99 \times 10^9 \text{ N·m/C}^2) \times 48 \times (1.6 \times 10^{-19} \text{ C})^2}{0.65 \times 10^{-14} \text{ m}}$$
$$= -1.7 \times 10^{-12} \text{ J.}$$

Compare this with the uncertainty $\Delta E = 2.4 \times 10^{-12}$ J that you found in part **b**. The uncertainty in the energy of the bound electron is greater than the magnitude of the binding energy, and the electron cannot long remain in the nucleus. Thus the model is inconsistent with the uncertainty principle and must be rejected.

## THINKING LIKE A PHYSICIST

The proof sketched in Example 42.6, that electrons cannot exist inside the nucleus, swept away a neat, simple model of nuclear structure. It led to at least one other puzzle as well: Many nuclei emit electrons in the radioactive decay process called β⁻ emission. If the electrons do not exist before they are emitted, they must come into existence at the moment of emission. Conversely, electrons must be annihilated in the radioactive process called electron capture, in which a nucleus absorbs one of its inner orbital electrons. (We will

consider β⁻ emission and electron capture in more detail in Chapter 43.) If we insist on thinking of electrons as miniature billiard balls, we find it hard to accept the idea that they can be created and annihilated. But creation and annihilation of electrons are entirely consistent with relativistic mass-energy conservation, a principle in which we have great confidence.

The proton-electron picture of atomic structure has another subtler fault—asymmetry. Why should the fundamen-

## The Uncertainty Principle as an Intrinsic Property of Waves

We now show that the uncertainty principle is a consequence of properties common to all waves. That is, the uncertainty principle is "grafted" onto objects such as electrons, which we ordinarily think of as particles, as a consequence of their wavelike aspect.

Consider again the de Broglie relation, $p = h/\lambda$. If the momentum of a particle is to be precisely defined ($\Delta p = 0$), the wavelength must be precisely defined as well ($\Delta \lambda = 0$). It is not difficult to write a time-independent wave function having a precisely defined wavelength. Indeed, such a wave function is the most elementary kind of all. We can write it in the form

$$\psi = \psi_0 \cos kx, \tag{42.20}$$

where $k = 2\pi/\lambda$ is the *wave number*. (Compare with Equation 21.21.)

This wave function represents a sinusoidal wave *of infinite extent*, stretching from $x = -\infty$ to $x = \infty$. The time-independent probability density is

$$\psi^2 = \psi_0^2 \cos^2 kx. \tag{42.21}$$

A particle having this probability density is not localized at all; that is, $\Delta x = \infty$.

The uncertainty in $x$ will have a finite value if the wave function describes a localized **wave packet**—a wave whose amplitude is nonzero only over a certain distance $\Delta x$, as in Figure 42.10. Such a wave packet can be constructed by superposing sinusoidal wave functions of slightly different wave numbers. By superposing wave functions having an increasing range $\Delta k$ of wave numbers, we decrease the length $\Delta x$ of the wave packet—that is, of the region in space for which $\psi^2$ is not negligible. In doing this, we localize the particle. (The superposition is mathematically akin to the production of beats by the superposition of sound waves of slightly different frequency; see Section 22.6, and especially Figure 22.9.) A detailed calculation shows that, as $\Delta k$ increases and $\Delta x$ decreases, their product $\Delta x \, \Delta k$ retains a constant value. By judicious choice of the amplitudes of the superposed components, that constant value can be minimized; the minimal value is

$$\Delta x \, \Delta k = 1. \tag{42.22}$$

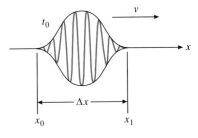

FIGURE 42.10 A wave packet associated with an electron having velocity $v$. The position uncertainty of the electron is $\Delta x$.

(Indeed, if we go to the extreme and superpose an infinite number of sinusoids having all possible values of $k$, the result is a single, infinitely narrow pulse. This pulse is the opposite extreme from the wave function of Equation 42.20; it represents perfect localization and complete uncertainty of momentum: $\Delta x = 0$ and $\Delta p = \infty$.)

Equation 42.22 is the result of a *mathematical* calculation. Let us now consider its *physical* consequences. Using the de Broglie relation, we write

$$\Delta k = \Delta \left( \frac{2\pi}{\lambda} \right) = \Delta \left( \frac{2\pi p}{h} \right) = \frac{2\pi}{h} \Delta p.$$

Equation 42.22 can thus be written

$$\frac{2\pi}{h} \Delta x \, \Delta p = 1,$$

or
$$\Delta x \, \Delta p = \frac{h}{2\pi}. \tag{42.23}$$

This is the Heisenberg uncertainty principle (Equation 42.16a).

## Philosophical Interlude: Does God Shoot Craps?

If you feel uncomfortable with the Born interpretation and the Heisenberg uncertainty principle, you can console yourself with the knowledge that you are in very good company. You learned in Chapters 17 through 20 that it is often necessary—even desirable—to deal with complicated systems statistically. But no fundamental problem arises from this statistical treatment. For example, it would be difficult or impossible, for practical reasons, to follow the history of every molecule in the cylinder of a steam engine. But statistical methods make it possible to learn a great deal about the behavior of the steam as a whole and thus about the operation of the engine.

In the present case, however, we deal with *individual* particles in statistical terms. According to the Born interpretation, determining the wave function of a particle enables us not to pinpoint the position of the particle but only to make a statement about the probability of finding it in any specified region. The uncertainty principle sets a fundamental limit on our knowledge of the dynamical quantities that determine the future behavior of the system. From the classical point of view, if you know exactly the position and momentum of a particle at any instant, and the forces that act on it, you can predict its future behavior without error. But the uncertainty principle guarantees that you cannot have the knowledge you need.

Einstein was the most notable of those physicists who did not think that quantum mechanics was the ultimate description of physical reality on the atomic level. That is not to say that Einstein thought quantum mechanics was false. To the contrary, it had spectacular success in describing the systems to which it was applied—as far as it went. But, in spite of the fact that no one contributed more than he to the development of quantum mechanics, Einstein was never convinced that nature operated on a statistical basis at its most fundamental level. As one alternative possibility, he suggested that there exists a deterministic "subquantic world"; in this view, the statistical behavior of particles on the atomic level merely reflects the fact that they are composed of large numbers of much smaller particles.

Bohr, an equally powerful philosopher, felt much more comfortable with the quantum world. At the risk of oversimplifying it, his view of the world was "What you see is what you get." From the 1920s on, Einstein and Bohr engaged in an intense though friendly debate as to the meaning of the statistical properties of the quantum world. The quotations at the beginning of this chapter summarize their views. Many of the arguments on both sides are elegant, and many are very subtle.

With the passage of time, quantum mechanics has been applied successfully to the solution of a vast variety of physical problems. But no evidence of a subquantic world has surfaced. To the contrary, other fundamental pictures of the basic structure of matter have emerged—all consistent with a statistical interpretation. Three generations of physicists have been nurtured from the start on a diet of quantum mechanics—including the Born interpretation and the uncertainty principle—and they have thrived. To a large extent, the Bohr-Einstein debate has cooled as most young physicists have adopted Bohr's position as a matter of course in the process of learning quantum mechanics. Nevertheless, as you have seen, old ideas sometimes do reemerge in new clothes. In the meantime, fortunately, it is possible to do a lot of good physics without worrying about the ultimate disposition (if any) of the Bohr-Einstein debate.

# Schrödinger's Equation and Atomic Structure

In studying waves on a string in Chapter 21, we began by considering the kinematics of waves. We focused our attention on simple sinusoidal waves and did not worry about *why* such waves might propagate on the string. Only later did we turn to the dynamics of waves. We derived the *wave equation* that governs the propagation of waves. We found that the wave equation, taken together with the boundary conditions, imposes specific conditions on the waves.

We have now completed the kinematic part of an analogous development for de Broglie waves. We studied these waves, focusing our attention on their properties without considering the wave equation that governs their propagation. The next logical step is to consider the wave equation for the propagation of de Broglie waves. That equation is the **Schrödinger equation**, which dates from 1926.

Unfortunately, the mathematical and physical ideas embodied in the Schrödinger equation are considerably more complicated than those required to deal with the classical wave equation. Although we cannot treat the Schrödinger equation in detail in this book, we can use analogy to understand its role in quantum physics.

In classical mechanics, we use Newton's second law of motion, together with the specific conditions imposed by a physical system, to determine the motion of a body. In classical wave theory, the wave equation, taken together with the boundary conditions of a system, determines the form and evolution of a wave excited in the system. In classical electrodynamics, the same role is played by Maxwell's equations. An analogous role is taken in quantum mechanics by Schrödinger's equation. Together with the boundary conditions imposed by a particular system, it can be used to determine the form and evolution of the wave function that describes the system.

Even without solving the Schrödinger equation explicitly, we can consider some of the many physical phenomena accurately described by its solutions. The remainder of this chapter deals with matters of this kind.

## The Pauli Exclusion Principle

To study the structure of atoms, we must apply the Schrödinger equation to systems containing two or more particles. In doing so, we must take into account another fundamental law of nature, called the **Pauli exclusion principle**:* *No two electrons can simultaneously occupy the same quantum state—a state described by a single set of quantum numbers.*

The structure of atoms more complex than hydrogen, with electrons occupying shells of increasing energy from the inside out, depends directly on the exclusion principle. In Section 41.5, we merely advanced this structure as a plausible hypothesis for explaining the periodic table and Moseley's law. Here, we assert that the structure is a consequence of an intrinsic property of particles that can be described by Schrödinger's equation.

## The Hydrogen Atom Again

The scientific world view that began in 1923 with de Broglie's postulate is often called the *"new" quantum mechanics*. The new quantum mechanics superseded the *"old" quantum mechanics*, based on the Bohr model of the atom, which prevailed from 1913 to 1924. You have seen how the new quantum mechanics incorporates—and justifies—all the basic principles of the old quantum mechanics. But the new quantum mechanics goes much further; it accounts for many phenomena beyond the explanatory powers of

---

*After the Austrian-Swiss physicist Wolfgang Pauli (1900–1958), who first proposed the principle in 1925.

the old quantum mechanics. (Nevertheless, you will see that the Bohr model can yield simple insights into situations whose complete interpretation requires complicated calculation.)

The problem of interpreting atomic spectra in terms of atomic structure has engaged physicists and chemists for the better part of a century. The greatest early triumph of the new quantum mechanics was the demonstration that the theory could account, with exquisite accuracy, for the observed spectra of atoms and molecules. To this day, the problem has not been worked out in every detail. Atomic spectra—and atoms—are so complicated that work still goes on in this field. But there is little doubt that every spectral line can be, and ultimately will be, accounted for in terms of the theory as it stands. The situation is much like that in celestial mechanics. Work, often very challenging, still goes on to account for the motion of various asteroids. But no one imagines that new fundamental ideas will be needed.

The Bohr theory accounts for the gross features of the hydrogen atom and hints at the structure of other atoms as well. As we have just noted, the Schrödinger equation can be used to provide detailed solutions, though carrying out the solutions is sometimes quite difficult. Nevertheless, the spectra of all atoms can be accounted for in terms of four quantum numbers that arise from the solutions.

### The Principal Quantum Number n

You are already familiar with the *principal quantum number n*, which we considered in the context of the hydrogen atom. We restate the *quantization condition* of Equation 41.15 here for completeness:

$$E_n = -\frac{me^4}{32\pi^2\epsilon_0^2\hbar^2}\frac{1}{n^2}. \tag{42.24}$$

We have rewritten this equation slightly. The symbol $\hbar$ (called "*h*-bar") represents the quantity

$$\hbar \equiv \frac{h}{2\pi}. \tag{42.25}$$

In concord with the uncertainty principle (Equation 42.16b), the energies $E_n$ are no longer the only allowed energies. However, they are the *most probable* energies. Neighboring values of energy $E_n \pm \Delta E$ are possible, though their probability drops off quickly with increasing $\Delta E$.

### The Orbital Angular Momentum Quantum Number l

Equation 42.24 does not completely describe the hydrogen atom. Even in the overly simple terms of the Bohr theory, it is possible to consider noncircular orbits. As in the planetary-orbit case, the angular momentum of an electron depends on the shape of its orbit. But the angular momentum is quantized; this is a basic assumption of the Bohr theory. This quantization of angular momentum "falls out automatically" in the solution of the Schrödinger equation for any atom. The quantization of angular momentum **L** is described by the **angular momentum quantum number** *l*, which arises from the solution of the Schrödinger equation as naturally as does the principal quantum number *n*. The *quantization rule* for angular momentum is

$$L = |\mathbf{L}| = \sqrt{l(l+1)}\,\hbar, \quad l = 0, 1, 2, \ldots, n-1. \tag{42.26}$$

An important feature of the angular momentum is that the value $L = 0$ is allowed. This cannot be accounted for either classically or on the basis of the Bohr model. The classical analogy is a planet that oscillates along a straight line passing through the sun! From the Bohr-theory point of view, states of increasing *l* correspond to elliptical orbits, all with the same semimajor axis but of decreasing eccentricity. Even for the maximum value of *l*, there is quantitative disagreement between Equation 42.26 and the Bohr theory. The maximum value of $l$ corresponds to angular momentum $L = \sqrt{(n-1)n}\,\hbar$, which differs from the Bohr-theory value $L = n\hbar$.

The energies corresponding to various values of $n$ and $l$ are conveniently plotted on an *energy-level diagram* like that shown for hydrogen in Figure 42.11. For the special case of the hydrogen atom, the additional quantum number $l$ does not alter the observed spectrum. This is because varying $l$ does not affect the energy $E_n$. When different quantum states correspond to the same energy, the states are called **degenerate**. In atoms other than hydrogen, states corresponding to the same value of $n$ but different values of $l$ have different energies and are not degenerate.

Using the Bohr model, we can develop a qualitative insight into why energy states corresponding to different values of $l$ are not necessarily degenerate in atoms more complex than hydrogen. As a consequence of the Pauli exclusion principle, the electrons are distributed around the nucleus in a series of shells, each shell corresponding to a particular value of $n$.

Now, consider electrons in the outermost shell. Their Bohr orbits have the same semimajor axis, given by the value of $r_n$ that arises from an equation like Equation 41.13. As you learned in Chapter 14, the energy of a planet bound in an elliptical or circular orbit by the inverse-square gravitational force depends only on the semimajor axis. The same statement applies to "classical" electrons in an inverse-square electric field. The semimajor axis, in turn, depends only on the value of $n$.

We might conclude, wrongly, that the electron energy does not depend on $l$, which determines only the eccentricity of the orbit, and thus that the energy levels for different values of $l$ are degenerate. But let us look more closely at the shapes of the orbits. The orbit corresponding to $l = n - 1$, the maximum possible value, is roughly circular. But the more eccentric orbits, corresponding to smaller values of $l$, must approach the nucleus more closely. That is, their orbits must penetrate the inner shells, as shown in Figure 42.12. As it moves closer to the nucleus, the electron is shielded less effectively from the nuclear charge. The result is a decrease in the energy associated with the orbit, in agreement with both the results of the detailed theory and the observations of atomic spectra.

### The Orbital Magnetic Quantum Number $m_l$

Suppose we choose a particular direction $z$ with respect to the atom. (We might single out this direction, for example, by placing the atom in a uniform magnetic field oriented along $z$.) A third quantum rule restricts the magnitude of the component $L_z$ of the angular momentum along this direction:

$$L_z = m_l\hbar, \quad m_l = 0, \pm 1, \pm 2, \ldots, \pm l. \tag{42.27}$$

The quantum number $m_l$ is called the **orbital magnetic quantum number**.

Because the orbiting electron carries electric charge around the nucleus, the atom has an **orbital magnetic moment** $\mu$. Magnetic moment is defined in Section 28.6; see also Problem 28.43. We defer a more detailed consideration of atomic magnetic moments to Section 42.6.

In a magnetic field $\mathscr{B} = \mathscr{B}\hat{\mathbf{z}}$, there is magnetic potential energy

$$U = -\mu \cdot \mathscr{B} = -\mu_z\mathscr{B}, \tag{42.28}$$

which is proportional to $L_z$. This energy must be taken into account in calculating the total energy of the atom.

### Electron Spin and the Spin Magnetic Quantum Number $m_s$

The electron has an **intrinsic angular momentum** independent of the angular momentum that it possesses due to its rotation about the nucleus. This intrinsic angular momentum is often called **spin angular momentum** or, more familiarly, **spin**. The name

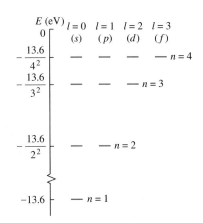

**FIGURE 42.11** Energy-level diagram for hydrogen, showing the energies of states corresponding to various values of $n$ and $l$. For historical reasons, the states corresponding to $l = 0, 1, 2, 3, 4, 5, \ldots$ are given the names $s, p, d, f, g, h, \ldots$. For any $n$, the energy levels for different $l$ are equal. Such levels are called degenerate.

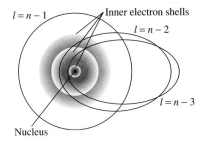

**FIGURE 42.12** Bohr-model sketch of the orbits of outer electrons having the same $n$ but different $l$. The inner electron shells are shown as shaded regions in color.

"spin" arises from analogy with the earth. Because the earth spins on its axis, it possesses an intrinsic angular momentum in addition to the angular momentum due to its revolution about the sun. (The analogy is appealing, but you should not take it too far. The electron is not a tiny earth!)

The existence of electron spin was first proposed in 1925 by Goudsmit and Uhlenbeck. Their proposal was aimed at accounting for the **fine structure** of atomic spectra: Most lines in atomic spectra are split into two or more component lines, separated by wavelength differences of order 0.1 nm.

The spin angular momentum **s** is subject to a quantization rule similar to that for orbital angular momentum, but simpler:

$$s = |\mathbf{s}| = \sqrt{\tfrac{1}{2}(\tfrac{1}{2} + 1)}\,\hbar = \sqrt{\tfrac{3}{4}}\,\hbar. \tag{42.29}$$

Compare this with Equation 42.26. Rather than the variable quantum number $l$, the electron has a fixed intrinsic spin quantum number whose value is $\tfrac{1}{2}$. The electron shares this property with a number of other fundamental particles, including the proton and the neutron. Such particles are called *spin-$\tfrac{1}{2}$ particles*.

Of more immediate interest than the spin is its component along the $z$ direction. The quantization rule is very simple:

$$s_z = m_s\hbar, \quad m_s = \pm\tfrac{1}{2}. \tag{42.30}$$

The quantum number $m_s$ is called the **spin magnetic quantum number**. Its value (only two are possible) affects the total energy of an atom in a magnetic field, for reasons similar to those given in the subsection on $m_l$.

However, a subtler process goes on as well. Taken individually, the magnetic moments associated with $m_l$ and $m_s$ have no effect on energy in the absence of an external magnetic field. However, each of the two magnetic moments "sees" the magnetic field arising from the other. This phenomenon is called **spin-orbit interaction**. Every allowed mutual orientation of **L** and **s** has a particular magnetic potential energy $U$, which must be added to other energy terms in evaluating the total energy of the atom. Although the differences between these spin-orbit energy levels are small (of order $10^{-3}$ eV or less), they account for the fine structure of atomic spectra.

Because protons and neutrons have spin $\tfrac{1}{2}$, the atomic nuclei comprising them possess spin as well. An interaction similar to spin-orbit interaction takes place between the electron magnetic moment and the *nuclear magnetic moment*. Because nuclei are much more massive than electrons, nuclear spin magnetic moments are much smaller than the electron spin magnetic moment (Problems 42.21, 42.44, and 42.46). As a result, the energy differences and the spectral line splittings that these energy differences cause are much smaller than fine-structure splittings. These very small nuclear splittings are called the **hyperfine structure** of the spectrum. Hyperfine structure can be observed directly in atomic spectra, using very high-resolution spectroscopy.

To summarize, the four quantum numbers $n$, $l$, $m_l$, and $m_s$ can be used to account for all of the features of atomic spectra. The rules for carrying out the calculation for any particular spectral line can be very complicated, and we cannot consider them here.

**Goudsmit and Uhlenbeck**

Samuel A. Goudsmit (1902–1978) and George E. Uhlenbeck (1900–1988), Dutch (later American) physicists, played significant roles in the development of the "new" quantum mechanics. During World War II, Goudsmit was assigned to follow close behind the Allied troops as they advanced into Europe and gather information concerning the progress of the German atomic-bomb effort. His book *Alsos* is a fascinating account of this scientific cloak-and-dagger effort and of why the German effort was a flop.

## EXAMPLE 42.7

Show that the maximum number of electrons having the principal quantum number $n$ is $2n^2$.

**SOLUTION:** According to Equation 42.26, $l$ can assume the values $0, 1, 2, \ldots, n - 1$. For each value of $l$, $m_l$ can assume the values given by Equation 42.27, $0, \pm 1, \pm 2, \ldots, \pm l$; there are $2l + 1$ such values.

Consider only those electrons that happen to have $s = +\tfrac{1}{2}$. Summing up the numbers of states for all possible values of $l$,

you have

$$1 + 3 + 5 + \cdots + (2n - 2 + 1).$$

This is an arithmetic progression having $n$ terms. Its sum is

$$\frac{1 + (2n - 1)}{2}\, n = n^2.$$

The number of possibilities for electrons having $s = -\tfrac{1}{2}$ is the same. Thus the total number of possibilities is $2n^2$.

Example 42.7 suggests why successive periods of the periodic table contain increasing numbers of elements. Each shell corresponds to a value of $n$. As $n$ increases, $2n^2$ increases as well. For the first four values of $n$, we have $2n^2 = 2, 8, 18,$ and $32$. For $n \geq 2$, however, this simple argument does not give the correct numbers of elements in the successive periods of the periodic table. This discrepancy is due to the fact that the energy of states corresponding to a "higher" set of quantum numbers is in some cases less than the energy of states corresponding to a "lower" set. Thus, for example, all the states in the $n$th shell may not fill before some of the states in the $(n + 1)$th shell fill. A more detailed argument yields the correct sequence $2, 8, 8, 18, 18, \ldots$ displayed by the periodic table. Nevertheless, when the $n$th shell ultimately fills, it does contain $2n^2$ electrons.

# Quantization of Angular Momentum: The Stern-Gerlach Experiment

In Section 42.5, we sketched the role of orbital and spin angular momenta, as well as the associated magnetic moments, in determining atomic spectra. However, atomic magnetic moments manifest themselves in many other ways, some of them more direct and easier to analyze than atomic spectra.

## *The Bohr Magneton*

We begin by calculating the magnetic moment of a hydrogen atom in its ground state. We use the Bohr model, which yields a surprisingly accurate result.

In the ground state, the electron circles the nucleus in an orbit of radius $r$. If the speed of the electron is $v$, it completes one revolution in a time $\Delta t = 2\pi r/v$. In doing so, it carries charge $-e$ around the orbit. The circulation of charge around the orbit amounts to an electric current

$$i \equiv \frac{dq}{dt} = \frac{-e}{\Delta t} = -\frac{ev}{2\pi r}. \qquad (42.31)$$

According to Equation 28.18, an electric current $i$ encircling an area $A$ gives rise to a magnetic moment of magnitude

$$\mu = iA. \qquad (42.32)$$

For the Bohr orbit of radius $r$, the area is $A = \pi r^2$. The magnetic moment is thus

$$\mu = \pi r^2 i = -\frac{evr}{2}.$$

We can express the current in terms of the angular momentum by using the basic relation $L = mvr$ to eliminate the product $vr$ from this equation:

$$\mu = -\frac{e}{2m} L. \qquad (42.33)$$

According to this equation, the magnetic moment is directly proportional to the angular momentum. The ratio $\mu/L$ is called the **classical gyromagnetic ratio** $\gamma_c$. The observed gyromagnetic ratio has various values for various particles and atoms, usually different from $\gamma_c$.

Up to this point, the argument has been entirely classical. We now introduce the Bohr quantization condition (Equation 41.10b), which we write in the form $L = n\hbar$. Substituting this value of $L$ into Equation 42.33, we obtain

$$\mu = -n\frac{e\hbar}{2m}. \qquad (42.34)$$

The magnetic moment of the hydrogen atom is directly proportional to the principal quantum number. The negative value of $\mu$ is a consequence of the negative charge of the electron. In the important special case $n = 1$, Equation 42.34 yields a constant value. The absolute value $|\mu|$ is represented by the symbol $\mu_B$:

$$\mu_B = \frac{e\hbar}{2m} = 9.274\,015\,4 \times 10^{-24} \text{ J/T}. \qquad (42.35)$$

The quantity $\mu_B$ is called the **Bohr magneton**. All magnetic moments of interest in atomic physics are of this order of magnitude. Consequently, it is often convenient to express atomic magnetic moments as multiples of the Bohr magneton. For any particular atom, we write

$$\mu = g\mu_B. \qquad (42.36)$$

The factor $g$ is called the **Landé g-factor**, after the German (later American) physicist Alfred Landé (1888–1975), who introduced it in 1923.

### The Stern-Gerlach Experiment

A direct way of observing the effects of atomic angular momentum is provided by the **Stern-Gerlach experiment**.* One form of the apparatus is sketched in Figure 42.13. An oven (not shown) vaporizes silver atoms, some of which emerge through a small aperture. Of the emergent atoms, those that pass through a series of slits form a slit-shaped beam. We fix the $x$ axis along the center line of the beam. The beam passes through a magnet whose poles are shaped in such a way that the magnetic field in the pole gap is highly inhomogeneous. In particular, the poles are shaped so as to maximize $\partial \mathscr{B}/\partial z$, the gradient (rate of change per unit distance) of the magnetic field in the $z$ direction. After passing through the magnet, the atoms strike a photographic plate. When the plate is developed, a black image forms where the atoms have struck. With the magnet turned off, the image is a black line parallel to the slits that formed the beam.

**FIGURE 42.13** Geometry of the Stern-Gerlach experiment. The entire apparatus is located in an evacuated chamber. Silver atoms are vaporized in an oven (not shown) and formed into a collimated beam by a series of slits, represented by the slit at the left. The beam of atoms passes between the poles of a magnet; the poles are shaped to produce a very inhomogeneous field. The atoms fall on a photographic plate, where they form an image.

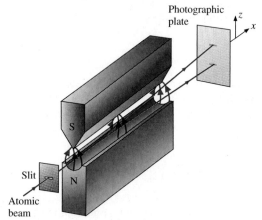

When situated in a magnetic field $\mathscr{B}$, an atom whose magnetic dipole moment is $\boldsymbol{\mu}$ has magnetic potential energy given by Equation 42.28, $U = -\boldsymbol{\mu} \cdot \mathscr{B}$. We use the general relation between potential energy and force to write the $z$ component of the force exerted on the atom:

$$F_z = -\frac{\partial U}{\partial z} = \frac{\partial}{\partial z}(\boldsymbol{\mu} \cdot \mathscr{B}) = \mu_z \frac{\partial \mathscr{B}_z}{\partial z}. \qquad (42.37)$$

The force depends on $\mu_z$, which in turn depends on the orientation of the atom with respect to the $z$ direction. Classically, the atoms can have any orientation, and so the force on various atoms may have all values between the limits $|\boldsymbol{\mu}|\,\partial \mathscr{B}_z/\partial z$ and

---

*First performed in 1922 by the German (later American) physicist Otto Stern (1888–1969) and his colleague Walther Gerlach (1889–1979).

$-|\boldsymbol{\mu}| \, \partial \mathcal{B}_z/\partial z$. Classical physics therefore predicts an image smudged out in the horizontal direction.

The prediction of quantum theory is different. Suppose that the net orbital angular momentum of the atom is zero and its total spin angular momentum is $\frac{1}{2}$. (This is the case for silver.) Then the only possible orientations of the atom with respect to the magnetic field are parallel and antiparallel to the field, corresponding to the magnetic spin quantum numbers $m_s = \pm\frac{1}{2}$. The force is likewise restricted to two possible values. Instead of spreading and making a smudge, the beam splits into two distinct parts, as shown in Figure 42.13. The result is clear evidence of the quantization of angular momentum.

The Stern-Gerlach experiment has been refined and repeated many times, using beams of various atoms and charged particles. If the orbital angular momentum of the atom is not zero, the magnetic spin quantum number may have more than one value, and the image splits into more than two parts. Indeed, the Stern-Gerlach experiment is an important tool for measuring the angular momentum of atoms and other particles.

### EXAMPLE 42.8

In the original Stern-Gerlach experiment, the magnet was designed so that the field gradient along the $z$ axis was a nearly constant $\Delta \equiv \partial \mathcal{B}/\partial z = 1.4 \times 10^3$ T/m over the path along which atoms near the center of the beam moved. The path length through the magnet was $X = 3.5$ cm. The distance from the end of the magnet to the photographic plate was negligibly small. The effective temperature of the oven was $T = 3600$ K. Careful measurements on the original of the photograph shown in Figure 42.14 yield a maximum distance $d = 0.16$ mm between the two parts of the split beam as they strike the photographic plate. **(a)** Express the magnitude $|\mu|$ of the magnetic moment of a silver atom in terms of $\Delta$, $X$, $T$, and $d$. **(b)** Find the magnetic moment of the silver atom.

**SOLUTION:**

**(a)** Find $|\mu|$.

The analysis of the experiment has much in common with the analysis of J. J. Thomson's $e/m$ experiment, shown in Figure 28.20. As the silver atom passes between the magnet poles, it is subjected to a transverse force $F = \mu\Delta$, which accelerates it in the $z$ direction. Call the mass of the atom $M$. Then you can write the acceleration as

$$a = \frac{F}{M} = \frac{\mu\Delta}{M}.$$

Solving this expression for $\mu$, you obtain

$$\mu = \frac{aM}{\Delta}. \tag{42.38}$$

Each half of the split beam is displaced through a maximum distance $d/2$. Because $\Delta$ is constant along the $z$ axis for atoms close to the center of the beam, the acceleration in the $z$ direction is uniform and you can write

$$\frac{d}{2} = \frac{1}{2}\,at^2,$$

where $t$ is the time an atom spends between the magnet poles. This time depends on the speed of the atoms. The speed varies from atom to atom, because the atoms are produced in an oven and therefore conform to the Maxwell-Boltzmann speed distribution (Section 18.5). However, the darkest part of the photographic image is made by the atoms that are most numerous.

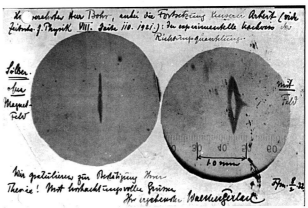

**FIGURE 42.14** Postcard dated 13 February 1922, sent by Walther Gerlach to Niels Bohr. Pasted to the card are enlargements of two images produced by the atomic beam. The photograph on the left was made with the magnet turned off. The one on the right, made with the magnet turned on, clearly shows the splitting of the beam into two discrete parts. (The splitting decreases toward the edges because the magnetic field gradient decreases toward the edges of the magnet.) The scale is established by the 1.0-mm marking handwritten on the right-hand photograph.

These are the atoms whose speed is $v_{mp}$, the most probable speed. According to Equation 18.59, the most probable speed is

$$v_{mp} = \sqrt{\frac{2\kappa T}{M}},$$

where $\kappa$ is Boltzmann's constant. So you can write

$$d = at^2 = \frac{aX^2}{v_{mp}^2} = \frac{aX^2 M}{2\kappa T}.$$

The value of $a$ you need for substitution into Equation 42.38 is

$$a = \frac{2\kappa Td}{X^2 M}.$$

When you use this value, Equation 42.38 becomes

$$\mu = \frac{2\kappa Td}{X^2 \Delta}. \qquad (42.39)$$

Note that the atomic mass $M$ cancels out.

**(b)** Find $\mu$ for silver atoms.

You can now insert the numerical values and obtain

$$\mu = \frac{2 \times 1.38 \times 10^{-23} \text{ J/K} \times 3600 \text{ K} \times 1.6 \times 10^{-4} \text{ m}}{(3.5 \times 10^{-2} \text{ m})^2 \times 1.4 \times 10^3 \text{ T/m}}$$
$$= 9.3 \times 10^{-24} \text{ J/T}.$$

This value is close to 1 Bohr magneton.

---

The result of Example 42.8 is somewhat surprising. The splitting of the beam into a doublet implies that $m_l$ has only the value 0. This in turn implies that $l = 0$ (Equation 42.27) and thus that the orbital angular momentum is $L = 0$ for silver. The magnetic moment must therefore be purely spin magnetic moment. Repeating the experiment with hydrogen atoms in the ground state, for which $L$ is clearly zero according to Equation 42.26, yields the same result.

But the spin angular momentum of the electron is $s = \frac{1}{2}\hbar$. Equation 42.33, which is based entirely on classical considerations, suggests that the Stern-Gerlach experiment should yield $\mu = \frac{1}{2}\mu_B$, rather than $\mu = \mu_B$. To put it another way, the Landé $g$-factor of the electron is 2. A complete understanding of these matters requires a treatment in terms of quantum electrodynamics, and we will not consider the matter further.

## Symbols Used in Chapter 42

| | | | | |
|---|---|---|---|---|
| $A$ | atomic mass, area | | $n$ | quantum number for a particle in a one-dimensional box, principal quantum number |
| $\mathscr{B}$ | magnetic field | | | |
| $c$ | speed of light | | $p$ | momentum |
| $e$ | quantum of charge | | $s$ | spin angular momentum |
| $E, E_0, K, U$ | total energy, rest energy, kinetic energy, potential energy | | $T$ | temperature |
| | | | $Z$ | atomic number |
| $h, \hbar$ | Planck's constant, $h/2\pi$ | | $\Delta$ | magnetic field gradient $\delta\mathscr{B}/\delta z$ |
| $k$ | wave number | | $\epsilon_0$ | permittivity of free space |
| $m$ | electron mass | | $\kappa$ | Boltzmann's constant |
| $l$ | orbital quantum number | | $\lambda$ | wavelength |
| $L$ | orbital angular momentum | | $\mu, \mu_B$ | magnetic moment, Bohr magneton |
| $m_l, m_s$ | orbital magnetic quantum number, spin magnetic quantum number | | $\tau$ | period |
| | | | $\psi, \psi_0$ | wave function, amplitude of wave function |

## Summing Up

Objects that classical physics treats purely as particles have wavelike properties. There is thus a fundamental symmetry between matter and light because light has, in addition to its classical wavelike behavior, particlelike properties that must be described in terms of photons. This symmetry is called the **de Broglie postulate**. The relation between the momentum of any particle (photons included) and its **de Broglie wavelength** is given by Equation 42.3,

$$\lambda = \frac{h}{p}.$$

The de Broglie postulate accounts for the quantization of Bohr orbits. If an electron is to remain in an orbit for more than a very short time, its de Broglie wave must be a standing wave. Consequently, the de Broglie wavelength of the electron must fit into the orbit circumference an integral number of times. Other kinds of constraint on the motion of particles likewise yield quantization. A remarkable property of such quantized systems is the existence in the ground state of a nonzero kinetic energy, called the **zero-point energy**.

The behavior of a particle is linked to its wave function by the **probability density** $\psi^2$. According to the **Born interpretation**, $\psi^2(x)\,dx$ is a measure of the *probability* that an observer will find the particle in the region bounded by $x$ and $x + dx$.

A wave can be restricted to a particular region of space—*localized*—only if its wavelength is not sharply defined. A localized wave, called a **wave packet**, consists of a superposition of waves of various wavelengths. Because, according to the de Broglie postulate, each wavelength corresponds to a different momentum, localization of the wave within a region $\Delta x$ requires *uncertainty* $\Delta p$ in the value of the momentum $p$. The **Heisenberg uncertainty principle** sets a limit on the precision with which $\Delta x$ and $\Delta p$ can be known simultaneously (Equation 42.16a):

$$\Delta x\,\Delta p \gtrsim \frac{h}{2\pi}.$$

There is a similar restriction on the uncertainty $\Delta E$ in the energy of a particle and the uncertainty in the time at which the energy is measured. According to Equation 42.16b, the limit is

$$\Delta E\,\Delta t \gtrsim \frac{h}{2\pi}.$$

Because a knowledge of the wave function of a particle (or a system of particles) makes it possible to determine the probability density $\psi^2$, it is important to determine the wave function and trace its evolution in time. The **Schrödinger equation** is used to accomplish this purpose. Applications of the Schrödinger equation lead to many important results,

among them the following:

1. In principle, the structure of atoms can be fully described in terms of a small number of *quantization rules* and *quantum numbers*.

2. Electrons have an **intrinsic angular momentum**, or **spin**, of magnitude $s$ given by Equation 42.29. Associated with the angular momentum is a magnetic moment. The existence of quantized **orbital** and **spin magnetic moments** and the associated angular momenta is evidenced not only by the *fine structure* and *hyperfine structure* of atomic spectra, but also by the **Stern-Gerlach experiment**.

In addition, we must take into account a law of nature, called the **Pauli exclusion principle**: No two electrons can occupy the same quantum state.

## KEY TERMS

**Section 42.2  de Broglie's Postulate**
de Broglie relation, de Broglie wavelength

**Section 42.3  The Meaning of the de Broglie Wave**
zero-point energy ▪ Born interpretation ▪ probability density ▪ normalization condition

**Section 42.4  The Heisenberg Uncertainty Principle**
wave packet

**Section 42.5  Schrödinger's Equation and Atomic Structure**
Pauli exclusion principle ▪ angular momentum quantum number ▪ degenerate quantum state ▪ orbital magnetic moment, orbital magnetic quantum number ▪ intrinsic (spin) angular momentum, spin magnetic quantum number ▪ fine structure, hyperfine structure ▪ spin-orbit interaction

**Section 42.6  Quantization of Angular Momentum**
classical gyromagnetic ratio ▪ Bohr magneton ▪ Landé g-factor ▪ Stern-Gerlach experiment

## QUERIES

**42.1** *(3) Pinning it down.* The normalization condition for a particle in a one-dimensional box yields Equation 42.14, $\psi_0{}^2 = 2/D$. **(a)** What is the maximum value, $P_{max}\,dx$, of the probability that you will find the particle in a region of length $dx$ inside the box? **(b)** Why is this probability inversely proportional to $D$?

**42.2** *(3) High-level thinking.* Make careful sketches like those of Figure 42.8 for the probability density $\psi^2$ of a particle in a one-dimensional box, for the cases $n = 3$, 4, and 5. Using these sketches, explain qualitatively why $P_{mid} < \frac{1}{2}$ for $n = 3$, $P_{mid} = \frac{1}{2}$ for $n = 4$, and $P_{mid} > \frac{1}{2}$ for $n = 5$.

**42.3** *(4) As sharp as you like.* Why is it possible in principle to express the ionization potential of an atom without uncertainty?

**42.4** *(4) Rethinking big and little.* We usually think of massive objects as large and less massive objects as small. Yet protons and neutrons exist in nuclei, but electrons, with mass about 2000 times smaller, cannot. Why can a massive particle exist in a space too small for a less massive one?

**42.5** *(5) The simplicity of X rays.* The Bohr model accounts fairly well for the optical spectra of atoms having only one electron, such as hydrogen, singly ionized helium, and doubly ionized lithium. But it is not successful in accounting for the optical spectra of even slightly more complex atoms. Nevertheless, it accounts fairly well for X-ray spectra (in particular, for Moseley's law). Explain.

**42.6** *(5) Directional.* An atom for which $L = 0$ interacts isotropically with the outside world; that is, it "looks" the same in all directions from outside. This is not true for atoms having nonzero values for $L$. Explain.

**42.7** *(5) Levels of snobbery.* Imagine that electrons had no spin. What would a helium atom look like if the exclusion

principle **(a)** still held? **(b)** did not hold?

**42.8** *(5) This way and that way.* In the ground state, the two electrons in the helium atom must have opposite spins. Explain.

**42.9** *(G) Hand-waving arguments.* For which physical phenomena do the old and the new quantum theories give the same results? For which does the old theory fail, though the new theory gives a satisfactory account? For which do the two disagree? For which do they agree in part? Make a short list.

# PROBLEMS

## GROUP A

**42.1** *(2) Handy number.* Show that, for nonrelativistic electrons, the relation between wavelength and energy is

$$\lambda \text{ (in pm)} = \frac{1230}{\sqrt{E \text{ (in eV)}}}.$$

**42.2** *(2) On the right wavelength, I.* Find the de Broglie wavelength of an electron that has been accelerated through a potential difference of **(a)** 1 V and **(b)** 1 kV.

**42.3** *(2) On the right wavelength, II.* Find the de Broglie wavelength for a proton that has been accelerated through the potential differences stated in Problem 42.2.

**42.4** *(2) On the right wavelength, III.* Find the de Broglie wavelength for a 1-g ball moving at 0.1 m/s.

**42.5** *(2) Wave as you go by.* The de Broglie wavelength of an electron is 20 pm. Find **(a)** the momentum, **(b)** the energy, and **(c)** the speed of the electron.

**42.6** *(2) Fuzzy TV image?* The electrons that strike the screen of a color TV tube to form an image are accelerated through a potential difference of about 20 kV. Is the wavelength of the electrons likely to be a significant factor in determining the clarity of the image?

**42.7** *(3) Getting excited, I.* **(a)** For the system of Example 42.5, find the probability that the particle lies in the middle half of the box if it is in the excited state $n = 2$. **(b)** Obtain the same result without any calculation at all on the basis of the symmetry of the situation.

**42.8** *(3) Getting excited, II.* For the system of Example 42.5, find the probability that the particle lies in the middle half of the box for the excited states $n = 3$ and $n = 5$.

**42.9** *(3) Little boxes, I.* An electron is confined in a one-dimensional box. What is the length of the box if the ground-state energy is 1 eV?

**42.10** *(3) Little boxes, II.* An electron is confined in a one-dimensional box. The electron makes a transition from the ground state to the next higher state when it absorbs a photon of energy 1 eV. What is the length of the box?

**42.11** *(4) Resolution in the face of uncertainty.* Suppose you know, within 1 μm, the positions of an electron, a proton, and a 1-μg dust particle. If you measure the speeds of all three, what are the minimum uncertainties in each case?

**42.12** *(4) Grafting the uncertainty principle onto the Bohr atom.* A hydrogen atom is in its ground state. **(a)** Consider the atom in terms of the Bohr model. Using the Bohr radius $a_0 = 5.29 \times 10^{-11}$ m, calculate the uncertainty in the speed of the electron. **(b)** Compare this value with the speed of the electron according to the Bohr model. **(c)** The answers to parts **a** and **b** are comparable. What does this tell you about the Bohr model?

**42.13** *(4) And yet it moves.* What is the minimum kinetic energy of an electron confined in a one-dimensional box of length 250 pm?

**42.14** *(4) Between small and big.* An electron is confined in a one-dimensional box of length 1 μm. A measurement shows that its kinetic energy is 4 eV. What is the proportional uncertainty $\Delta v/v$ in its speed? Is your result in accord with the correspondence principle?

**42.15** *(4) Uncertainties of life in the big world.* Using a lightning-fast apparatus that requires only 1 ps ($1 \times 10^{-12}$ s) to take the measurement, you want to measure the energy of a 100-mg ball bearing as it moves past you. **(a)** What is the smallest possible uncertainty in the energy of the ball? **(b)** Suppose the ball is moving at the very low speed of 1 μm/s. What is the uncertainty $\Delta v$ in the speed?

**42.16** *(4) Pinning things down.* The de Broglie wavelength of a particle is $\lambda$. Show that, if the uncertainty in the position of the particle is $\Delta x = \lambda/2\pi$, the uncertainty in its speed is comparable to the speed itself; that is, $\Delta v \simeq v$.

**42.17** *(5) Close but no cigar.* For a hydrogen-atom electron in state $n$, the Bohr theory predicts a value $L_B$ for the angular momentum. The complete theory, based on the Schrödinger equation, predicts a value $L_S$. Find the maximum proportional difference $(L_S - L_B)/L_S$ for **(a)** $n = 1$, **(b)** $n = 4$, **(c)** $n = 100$.

**42.18** *(5) Quantization rules the waves.* An electron is in a state specified by the angular momentum quantum number $l = 3$. What is the magnitude of the orbital angular momentum $L$?

**42.19** *(5) Range of possibilities.* An electron is in a state specified by the principal quantum number $n = 4$. **(a)** What values of $l$ are possible? **(b)** For each possible value of $l$, what values of $m_l$ are possible?

**42.20** *(5) Quantum numbers.* A lithium atom is in the ground state. Give the values of the quantum numbers $n$, $l$, $m_l$, and $m_s$ for each of the three electrons.

**42.21** *(6) Nuclear magneton.* Like the electron, the proton has an intrinsic angular momentum, called the *proton spin I*. Corresponding to the Bohr magneton for electrons is the *nuclear magneton* $\mu_N$. Calculate the value of $\mu_N$. (The magnetic moment of the proton is actually 2.793 times this value.)

**42.22** *(6) Classical electron.* A calculation based on classical electromagnetic theory yields a value $r = 2.8 \times 10^{-15}$ m for the radius of the electron. **(a)** Suppose that the electron were a solid, uniform sphere of this radius. Using the known mass of the electron, calculate what its angular speed $\omega$ would have to be in order to account for the measured spin angular momentum $\hbar$. **(b)** What would the rotational kinetic energy of this electron be? **(c)** What would the mass of the electron be? (Hint: Do you need to make a relativistic calculation?)

## GROUP B

**42.23** *(2) Relativistic wavelengths.* Show that a particle of rest mass $m_0$ and kinetic energy $K$ has a de Broglie wavelength given by

$$\lambda = \frac{h}{\sqrt{2m_0 K\left(1 + \dfrac{K}{2m_0 c^2}\right)}}.$$

**42.24** *(2) Mr. Compton, meet M. de Broglie.* What is the kinetic energy of an electron if its de Broglie wavelength is equal to its Compton wavelength? (See Equation 40.23.)

**42.25** *(2) Mr. Young, meet M. de Broglie.* A parallel beam of 256-eV electrons passes through a pair of narrow slits separated by 50 μm. The electrons then fall on a photographic film located 1.0 m from the slits. What is the distance between adjacent interference maxima on the film?

**42.26** *(3) Balmer in one dimension? I.* Show that an electron confined to a one-dimensional box can absorb and emit photons of wavelength

$$\lambda = \frac{8mcD^2}{h}\frac{1}{n'^2 - n^2},$$

where $n'$, $n = 1, 2, 3, \ldots$ and $n' > n$.

**42.27** *(3) Balmer in one dimension? II.* An electron in a one-dimensional box absorbs and emits light at certain wavelengths, among them 41.5 nm and 22.1 nm. **(a)** Find the longest wavelength of light that can be absorbed by the electron. (Hint: You need consider only the first four energy levels.) **(b)** Which of the following wavelengths can be absorbed: 27.7 nm, 55.4 nm, 23.7 nm? **(c)** What is the length of the box? (Hint: See Problem 42.26.)

**42.28** *(3) Getting excited, III.* For the system of Example 42.5, make a graph of $P_{mid}$ versus $n$.

**42.29** *(3) Getting excited, IV.* For the system of Example 42.5, find the probability that the particle lies in the region $0 \leq x \leq D/5$ for $n = 3$.

**42.30** *(3) It's gotta be in here somewhere! I.* What is the probability that a particle confined in a one-dimensional box will be found in the middle quarter of the box $(3D/8 \leq x \leq 5D/8)$ when **(a)** $n = 1$, **(b)** $n = 2$, **(c)** $n = 3$, and **(d)** $n = 1000$?

**42.31** *(3) It's gotta be in here somewhere! II.* What is the probability that a particle confined in a one-dimensional box will be found in the left one-third of the box $(0 \leq x \leq D/3)$ when **(a)** $n = 1$, **(b)** $n = 2$, **(c)** $n = 3$, and **(d)** $n = 1000$?

**42.32** *(3) Striking a blow for freedom.* A particle of mass $m$ is confined in a one-dimensional box of length $D$. The particle is in the ground state. Show that the magnitude of the force that the particle exerts on either wall is approximately $F = h^2/8mD^3$.

**42.33** *(3) Ideal gas?* A container holds helium gas at 1 atm and 20°C. **(a)** Find the de Broglie wavelength of the molecules having the most probable speed. **(b)** Compare this with the "billiard ball" diameter of the helium atom, about 0.25 nm. **(c)** At atmospheric pressure, helium liquefies at 4.2 K. Are quantum effects likely to be significant in determining the properties of liquid helium? (In fact, helium has remarkable *superfluid* properties, among them zero viscosity and infinite thermal conductivity. Superfluidity is discussed in Section 44.4.)

**42.34** *(4) Energy-time uncertainty.* A particle of mass $m$ moves in the $x$ direction with nonrelativistic speed $v$. **(a)** If the uncertainty in the momentum of the particle is $\Delta p$, show that the uncertainty in its energy (assumed entirely kinetic) can be written $\Delta E = v\,\Delta p$. **(b)** Given the uncertainty $\Delta x$ in the position of the particle, what is the uncertainty $\Delta t$ in the time at which the particle passes an observer? **(c)** Express the uncertainty in terms of $\Delta E$ and $\Delta t$ and thus obtain Equation 42.16b.

**42.35** *(4) Angular momentum-position uncertainty.* A rotating object has angular momentum $L$. Its angular position is specified by the angle $\theta$. Beginning with Equation 42.16a, derive the following form of the uncertainty principle:

$$\Delta L\,\Delta\theta \gtrsim \hbar.$$

**42.36** *(4) Evaluating the quantum number.* Imagine you have a 100-mg ball in a frictionless, V-shaped groove. The ball is confined by two walls located 10 cm apart. You think the ball is at rest. But you have other obligations, and so the best you can do to confirm your opinion is to watch it for an hour. During that time, you are certain that the ball cannot have moved as much as 0.1 mm. **(a)** Set an upper limit on the speed of the ball. **(b)** What is the corresponding momentum? **(c)** What is the upper limit on the quantum number $n$? **(d)** Suppose you want to establish that the value of $n$ is no greater than 100. How long would you have to watch the ball, if you could use a microscope to detect movement over a distance of 1 μm? **(e)** Are you likely ever to see a deviation from constant probability density for the location of the ball?

**42.37** *(4) Uncertain process, I.* The uncertainty principle makes it possible to understand, if not give a direct answer to, two basic questions posed by the Bohr theory: (1) how long does it take the electron in a hydrogen atom to make the transition from one stationary state to another and (2) what happens during this time interval? Consider the transition from the state $n = 2$ to the state $n = 1$. **(a)** What is the energy difference $\Delta E$ between the two states? **(b)** Suppose that the transition takes a finite time $\Delta t$. What is the uncertainty in the energy during this time interval? **(c)** Evaluate $\Delta t$. **(d)** Suppose you tried to observe the transition. To do so, you would have to make a series of measurements, each of which requires a time $\Delta t' \ll \Delta t$. What could you say about the energy of the electron during each interval? **(e)** Explain why question 2 cannot be answered in a meaningful way.

**42.38** *(4) Uncertain process, II.* Repeat the calculation of parts **a** through **c** of Problem 42.37 for a transition from $n = 1001$ to $n = 1000$. Assume that the mean lifetime of the initial state is $10^{-8}$ s. **(a)** Why is it not useful to consider these two levels as discrete? **(b)** How does your result accord with the correspondence principle?

**42.39** *(4) Quantizing the solar system.* **(a)** Estimate the value of the quantum number $n$ for the earth, as it revolves about the sun. **(b)** What is the energy difference $\Delta E$ between the state $n$ in which the earth happens to be and the adjacent state $n + 1$? **(c)** What is the uncertainty in the energy of the earth? Is there any possibility of detecting the quantization of the earth's orbit?

**42.40** *(4) Breaking out.* A particle in its ground state is confined in a one-dimensional box of length $D$. The end walls are suddenly removed, and the wave function of the newly free particle spreads out. **(a)** Using the uncertainty principle, find the time $t$ required for the wave function to spread out over a region of length $qD$, where $q$ is a constant greater than 1. **(b)** Suppose the particle is an electron. Let $D = 100$ pm and $q = 20$. Find $t$.

**42.41** *(4) All roads lead to Rome, more or less, I.* Here is a way to see the intimate connection between the nonzero zero-point energy of quantized systems and the uncertainty principle. A particle is confined in a one-dimensional box of length $D$; its energy $K$ is entirely kinetic. What is the uncertainty **(a)** $\Delta x$? **(b)** $\Delta p$? **(c)** $\Delta K$? **(d)** Compare $\Delta K$ with the ground-state energy $K_1$. Is the ratio $\Delta K/K_1$ very different from 1?

**42.42** *(4) All roads lead to Rome, more or less, II.* You can apply the ideas of Problem 42.41 to an important real system, the hydrogen atom. **(a)** Write the total energy $E = K + U$ for the electron in an arbitrary orbit. **(b)** Using the approximations $\Delta p \approx p$ and $\Delta r \approx r$ for the electron in the ground state, together with the uncertainty principle, write an approximate expression for $E$ as a function of $r$ only. **(c)** For a stable orbit, you must have $dE/dr = 0$. Evaluate the Bohr radius. Compare your result with the value $a_0 = 5.29 \times 10^{-11}$ m. **(d)** Evaluate the ground-state energy $E_1$, and compare your result with the value $E_1 = -13.6$ eV.

**42.43** *(6) The 21-cm cosmic "song."* The field of radio astronomy began in the late 1930s, when communications researchers noted that sensitive microwave antennas aimed at the sky detected "radio noise." This omnidirectional noise turned out to be concentrated at $\lambda = 21$ cm. We know today that the source of this radiation is the atomic hydrogen that is distributed sparsely over interstellar space. In this cold, dark environment, nearly all the atoms are in the ground state $n = 1$, $l = 0$, and they rarely collide with one another. The ground state consists, however, of two very slightly different energy levels. As shown below, the magnetic moments of the electron and the proton can be either parallel or antiparallel.

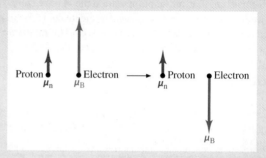

When the electron magnetic moment "flips" from parallel to antiparallel in the magnetic field due to the proton magnetic moment, a 21-cm photon is emitted. The mean lifetime of atoms in the excited state is about $10^7$ years. Nevertheless, there are a lot of atoms in interstellar space because there is so much of it! Consequently, the signal is readily detected with proper equipment. The magnetic moment of the electron is $-\mu_B$; that of the proton is $\mu_p = 2.793\mu_n = 2.793\mu_B/1837$; see Problem 42.21. The magnetic field of a magnetic dipole of moment $\boldsymbol{\mu}$, at distance $r$ in a direction perpendicular to the dipole axis, is

$$\mathcal{B} = -\frac{\mu_0}{4\pi}\frac{\boldsymbol{\mu}}{r^3}.$$

**(a)** What is the magnitude of the magnetic field, due to the proton, that the electron "sees" when making the transition? **(b)** What is the energy $\Delta E$ of the transition? **(c)** What is the average distance $r$ between electron and proton? Compare this distance with the Bohr radius $a_0$. Explain why $r < a_0$. **(d)** Suppose you made the calculation by considering the flipping of the proton in the magnetic field due to the electron. Would it make any difference in the result?

**42.44** *(6) Electron spin resonance.* **(a)** A 1.0-T magnetic field is imposed on a collection of free electrons. Explain why most of the electrons will be oriented in one direction relative to the direction of $\mathcal{B}$. What is that direction? **(b)** What is the energy of the photons required to "flip the spins"—that is, to rotate the electrons to the other orientation? **(c)** What is the wavelength of the photons? In what region of the electromagnetic spectrum does this wavelength lie? The electrons will absorb energy preferentially at this special wavelength, a phenomenon called *electron spin resonance* (ESR) or *electron paramagnetic resonance* (EPR). **(d)** The Landé g-factor is 2 for free electrons. But the value of $g$ can be considerably different from 2 for electrons that are part of chemical compounds. Explain in a qualitative way how this makes possible an investigation into the structure of a compound.

**42.45** *(6) Nuclear magnetic resonance.* (Before working this problem, you should do Problems 42.21 and 42.44.) The easiest way to make an experimental sample containing quite a lot of hydrogen in a small volume is to use water. Suppose you place a quantity of water in a magnetic field of magnitude

$\mathcal{B}$. The magnetic moment of the proton is $2.793\mu_n$, or $1.411 \times 10^{-26}$ J/T. **(a)** Show that the sample will absorb energy strongly from an electromagnetic field whose frequency is $\nu = 4.258 \times 10^7$ Hz/T $\times$ $\mathcal{B}$. This selective absorption, called *nuclear magnetic resonance* or *nmr*, is used routinely to measure magnetic fields precisely; commercial devices called *nmr magnetometers* are available for this purpose. **(b)** In what region of the electromagnetic spectrum will $\nu$ lie for the field of a laboratory magnet? Typical laboratory magnets have fields ranging from 0.1 to 10 T.

The factor that relates $\nu$ to $\mathcal{B}$ varies from nucleus to nucleus. It also varies slightly for any particular nucleus. This slight variation is due to superposition on the external magnetic field of small internal magnetic fields arising from the other components of the molecule in which the nucleus (say, a proton) is located. The proton thus becomes a microscopic probe for measuring the magnetic field inside molecules. This is the basis for the powerful, noninvasive medical diagnostic technique called *nmr imaging*.

**42.46** *(G) Magnetic quantization.* An electron moves with speed $v$ in the plane normal to a uniform magnetic field $\mathcal{B}$. **(a)** Show that the electron moves in a circular orbit. What is the relation between $v$ and the orbit radius $r$? (This will refresh your memory concerning matters discussed in Section 28.4.) **(b)** What is the de Broglie wavelength $\lambda$ of the electron? **(c)** Using the standing-wave condition for a stable orbit, show that the electron is restricted to quantized orbits, and find the radius $r_n$ of the $n$th orbit. **(d)** Show that the kinetic energy of the electron is restricted to the values $E = n\mu_B\mathcal{B}$. **(e)** For $\mathcal{B} = 10$ T, find the minimum allowed orbit radius. **(f)** Find the energy of the electron in the minimum radius.

This effect, called *magnetic quantization*, does not affect any component of velocity the electron may have parallel to $\mathcal{B}$. Magnetic quantization is the basis for the family of phenomena, observed in metals and semiconductors, called the *quantum oscillatory effects*. Most noted among these effects are the *de Haas–van Alphen effect* and the *Shubnikov–de Haas effect*.

## GROUP C

**42.47** *(2) Evening things out.* A particle having nonrelativistic kinetic energy $K$ and mass $m_1$ collides with a stationary particle of mass $m_2$. Imagine an observer in the center-of-mass coordinate system—that is, an observer who moves along with the center of mass of the two particles. Show that, from the point of view of this observer, both particles have the same de Broglie wavelength, equal to

$$\lambda' = \frac{h(1 + m_1/m_2)}{\sqrt{2m_1E}}.$$

**42.48** *(3) Quantized harmonic oscillator without Schrödinger.* The harmonic oscillator—a mass $m$ constrained by a force $F = -k(x - x_0)$—is an excellent model for a wide variety of real systems. Detailed solution of the equation of motion for a harmonic oscillator requires use of Schrödinger's equation. But here is a way to use the uncertainty principle to find an approximate value for the ground-state energy $E_1$. Make the approximation that, in the ground state, momentum $p \simeq \Delta p$ and $\Delta x \simeq x$. (See Problems 42.41 and 42.42.) **(a)** Show that the total ground-state energy of the system is

$$E = K + U \simeq \frac{\hbar^2}{2mx^2} + \frac{kx^2}{2}.$$

**(b)** Find the equilibrium position $x_0$. **(c)** Show that $E_0 \simeq \hbar\sqrt{k/m}$. Detailed calculation using the Schrödinger equation yields $E_0 = \frac{1}{2}\hbar\sqrt{k/m}$.

**42.49** *(4) Uncertain diffraction.* A stream of hydrogen atoms having uniform speed $v = 500$ m/s passes through a slit of width $a$ and produces a diffraction pattern on a screen at a distance $R$ from the slit. In addition to the classical width of the first maximum, $\sin \theta_m = 2\lambda/a$ (see Equation 36.22, which gives the half-width), there is a further widening due to the uncertainty $\Delta p_y$ in the momentum of the atoms in the direction perpendicular to the direction of the stream. (This differs from the classical situation where $\Delta p_y = 0$ and $p_y = 0$ exactly.)

**(a)** Show that the width of the first maximum has a minimum value when the slit width is $a \simeq \sqrt{\hbar R/Mv}$, where $M$ is the mass of the atoms. **(b)** Find the value of $a$ for $R = 1$ m. **(c)** What is the width of the first maximum on the screen?

**42.50** *(3) Particle in a parallelepiped.* A particle is confined in a box of length $D_1$, width $D_2$, and height $D_3$. **(a)** Show that the energy levels of the particle are given by the expression

$$E_{n_1,n_2,n_3} = \frac{h^2}{8m}\left(\frac{n_1^2}{D_1^2} + \frac{n_2^2}{D_2^2} + \frac{n_3^2}{D_3^2}\right),$$

where $n_1$, $n_2$, and $n_3$ can independently take any value 1, 2, 3, . . . . This result has wide application in solid-state physics. **(b)** The levels are degenerate. But find the combinations of $n_1$, $n_2$, and $n_3$ that yield the four lowest possible values of $E$ for a cubic box. What are the values of $E$? **(c)** Find the energy of the sixth level, and show that it is sixfold degenerate (that is, that there are six combinations of $n_1$, $n_2$, and $n_3$ that yield the same energy).

**42.51** *(G) Zeeman effect.* Consider hydrogen atoms located in a magnetic field $\mathcal{B}$. **(a)** Using the Bohr model, calculate the energy difference between ground-state atoms whose orbital angular momenta are parallel and antiparallel to the field. Neglect the intrinsic electron spin. **(b)** What qualitative effect will this energy difference have on the hydrogen spectrum? This effect, predicted on classical grounds by H. A. Lorentz, was observed in the spectrum of a sodium source placed in a strong magnetic field in 1896 by the Dutch physicist Pieter Zeeman (1865–1943). It is called the *normal Zeeman effect*. However, the observations did not agree quantitatively with the predictions. Owing to spin-orbit coupling, the effect observed in weak magnetic fields is much more complicated than the normal Zeeman effect and is called the *anomalous Zeeman effect*. A complete explanation of the Zeeman effect was one of the early triumphs of the new quantum mechanics.

# Quantum Systems I
## Nuclei

In the radioactive decay process, a nucleus emits an energetic particle or a quantum of electromagnetic radiation.

In emitting a particle, the nucleus undergoes a transmutation—it becomes the nucleus of an atom of a different chemical element.

Radioactive decay is a random process; consequently, the rate of decay of a sample of radioactive material is a declining exponential.

Radioactive decay can be understood in terms of the phenomenon of quantum-mechanical tunneling.

The binding energy of a nucleus is related to its mass defect—the difference between its mass and that of all its constituent nucleons.

The binding energy per nucleon is greatest for nuclides near the middle of the periodic table. This fact makes it possible to obtain energy both by nuclear fusion of lighter nucleons and by fission of heavier ones.

*Left:* Loading fuel rods into a nuclear research reactor. The reactor vessel is at bottom; the fuel rods, consisting of enriched uranium pellets clad in a zirconium alloy, fit into the honeycomb structure.

*In small proportions we just beauties see.*

—Ben Jonson

## Introduction

Most of the spectacular scientific and technological advances of the twentieth century have roots in the ideas developed in the preceding three chapters. In this chapter, we consider the understanding of the nucleus made possible by the quantum revolution and some of the applications of that understanding. In Section 43.2, we return to the subject of radioactivity, first considered in Section 39.5. In Section 43.3, we develop a model for radioactive decay; in Section 43.4, we put some detail into the structure of the nucleus.

## Radioactivity

The discovery of X rays in 1895 set off a flurry of investigations aimed at discovering and studying penetrating radiation. In 1896, just a few months after Roentgen's discovery, the French physicist Antoine-Henri Becquerel (1852–1908) laid a lump of a mineral containing uranium on a wrapped, unexposed photographic plate. On developing the plate, he found that it had been blackened, as it would have been if exposed to light. This blackening occurred even when he interposed a sheet of glass to prevent a chemical reaction between the plate and any gas that might be emitted by the mineral. Moreover, the blackening depended only on the amount of uranium present; it was unaffected by the chemical form of the uranium. The penetrating power of this spontaneous radiation indicated that considerable energy was involved. The phenomenon was later given the name **radioactivity**. At first Becquerel's discovery excited relatively little interest. He himself published a series of nine papers over a little more than a year and then went on to other things.

In 1898, Marie Curie (see margin note on p. 1181) revived the study of radioactivity by undertaking *quantitative* investigations. She used a quadrant electrometer, in which the electrostatic repulsion between pairs of plates charged to a large potential difference is measured with a torsion balance. Because radiation ionizes air, she was able to use the rate of discharge of the plates as a measure of the activity of a sample placed near them.

Curie showed that the uranium ore called pitchblende exhibits considerably more radioactivity than can be attributed to the known amount of uranium present. On this basis, she surmised that one or more other, more radioactive elements were present as well. Obtaining a large quantity of pitchblende from which the uranium had already been extracted, she and her husband, Pierre, carried out a long, tedious series of purifications that led to the spectroscopic identification of two new elements, which they called polonium (Po, after her native country) and radium (Ra). Both of these elements were present in tiny amounts, confirming the original supposition that both were much more radioactive than either uranium or thorium (another weakly radioactive element whose presence had been discovered).

By isolating radium and polonium, the Curies not only provided useful sources for further investigation, but made it evident that radioactivity is not an isolated phenomenon. The work of the Curies thus rekindled interest in the study of radioactivity, and beginning in 1898 discoveries came very quickly. By 1915, the following facts were firmly established:

1. Radioactivity is a *nuclear* process.

2. Radioactive emanations—the "stuff" emitted by natural radioactive nuclei—are of three types, all having considerable energy. Typical values are given in parentheses for each type of emanation:
   a. **α particles**, which are helium nuclei (4 MeV);
   b. **β** or **β⁻ particles**, which are electrons (300 keV);
   c. **γ rays**, which are photons of wavelength comparable to or shorter than that of the shortest-wavelength X rays (1 MeV).

Later study of the decay of natural and especially artificial radioactive substances showed that other radioactive processes are possible as well.

3. Emission of α or β particles is a part of a **decay process** in which the atom is **transmuted** from one chemical element to another.

4. Radioactive decay of natural substances is complicated, because many processes are observed at once. There are two reasons for this:
   a. Some kinds of nuclei can decay in more than one way. Usually, but not always, one **decay mode** dominates.
   b. Natural radioactive decay tends to occur in chains, each α or β decay of a **parent** atom resulting in a **daughter** atom that is itself radioactive. A typical chain is the one beginning with the most common isotope of radium:

$$^{226}_{88}\text{Ra} \xrightarrow{\alpha} {}^{222}_{86}\text{Rn} \xrightarrow{\alpha} {}^{218}_{84}\text{Po} \xrightarrow{\alpha} {}^{214}_{82}\text{Pb} \xrightarrow{\beta^-} {}^{214}_{83}\text{Bi} \xrightarrow{\beta^-} {}^{214}_{84}\text{Po} \xrightarrow{\alpha} {}^{210}_{82}\text{Pb} \xrightarrow{\beta^-}$$
$$^{210}_{83}\text{Bi} \xrightarrow{\beta^-} {}^{210}_{84}\text{Po} \xrightarrow{\alpha} {}^{206}_{82}\text{Pb}. \quad \textbf{(43.1)}$$

The decay mode is indicated at each step by the α or β⁻ above the arrow. The superscript accompanying each chemical symbol is the **nucleon number**, or **mass number**, $A$, which is the atomic mass $M$ given to the nearest integer. (This notation is different from that of the preceding two chapters, in which $A$ represents the atomic mass. Up to now, we have not needed to stress the distinction between nucleon number and atomic mass. You will soon see that here $A$ is the number of nucleons—protons and neutrons—in the nucleus.) The subscript below the nucleon number is the atomic number $Z$, which is the number of protons in the nucleus. Because $Z$ identifies the chemical element, the subscript is redundant and is often omitted, thus: $^{226}$Ra is spoken (and sometimes written) radium-226.

Note in Process 43.1 that the emission of an α particle ($A = 4$, $Z = 2$) results in a decrease of $A$ by 4 and of $Z$ by 2. The emission of a β particle ($A = 0$, $Z = -1$) results in no change of $A$ and an *increase* of $Z$ by 1.

In both α and β decay, a number of important quantities are *conserved*:

1. Relativistic mass-energy (see Sections 39.4 and 39.5):

$$c^2 M_{\text{parent}} = c^2 \Sigma M_{\text{daughter}} + \Delta E. \quad \textbf{(43.2)}$$

In this equation, the mass on the left side is the rest mass of the parent atom. The summation on the right side is the sum of the rest masses of the daughter atom and all other particles emitted. The term $\Delta E$ is the total energy released in the decay—most often in the form of kinetic energy of the products.

2. Momentum. When a nucleus emits one or more energetic particles, it recoils. As always, momentum must be conserved in the process (see margin note in Section 39.4).

3. Angular momentum. The vector sum of the spin angular momenta of all the particles produced in the reaction is the same as that for all the particles entering the reaction.

4. Electric charge. Charge conservation is evident in the way $Z$ changes as the nucleus emits an α or β particle.

5. Nucleon number. The sum of the nucleon numbers of all the particles must not change in the decay process.

## Radioactive Half-Life

Decay chains such as that in Process 43.1 can be untangled by means of radiochemical investigations. It then becomes possible to study a single kind of decay in isolation. By good fortune radon (Rn), the first daughter produced in Process 43.1, is a noble gas. It

**Marie Curie**

The Polish-French physicist Marie (or Maria) Sklodowska Curie (1867–1934) is almost certainly the most widely known woman ever to have worked in physics. Working as a governess, she saved enough money to carry on in Paris the advanced studies unavailable to women in Poland. She enrolled at the Sorbonne and graduated at the head of her class. Shortly after her marriage to Pierre Curie (1859–1906), who had already achieved prominence for his ongoing work in magnetism, she began her epoch-making work (in which she was soon joined by her husband when he recognized the extraordinary importance of her work) in a leaky, unheated shed scrounged from the university. The Curies shared a Nobel Prize with Becquerel in 1903. After Pierre's death in a traffic accident, Marie became the first person to win two Nobel Prizes (the second in 1911) and one of only two persons ever to win two Nobel Prizes in science. (The other is John Bardeen; see Chapter 44.) She succeeded to her husband's position, thus becoming the first woman to hold a professorship at the Sorbonne, and she later founded and directed the Curie Institute. In spite of the importance of her work, she was elected to the French Academy only with the greatest difficulty, on account of her sex, and was denied honorary degrees by Harvard and Princeton (though not by Yale) in 1921, in the course of a triumphal American fund-raising tour. She was reinterred at the Invalides, the burial place of the great of France, in 1995, more than sixty years after her death. With this belated recognition, she became the first woman to be so honored on her own merits.

can be collected in pure form simply by sealing a sample of radium in a glass container from which the radon can be drawn. It is particularly easy to work with a sample consisting entirely of a single **nuclear species**, or **nuclide**. For this reason, much early work was done with radon. The quantitative picture we now develop in terms of radon decay is directly applicable to other decay processes.

When $^{226}$Ra decays, the radioactive isotope, or **radionuclide**, produced is $^{222}$Rn, which itself emits $\alpha$ particles. With the passage of time, the **decay rate**, or **activity**, of an initially pure sample of $^{222}$Rn—the number of $\alpha$ particles emitted per second— decreases. When a time interval of 3.8235 days has passed, the activity decreases to one-half its original value. With the passage of each additional interval of 3.8235 days, the decay rate decreases by one-half again. This time interval is called the **half-life** of the radionuclide $^{222}$Rn. The half-life is denoted by the symbol $\tau_{1/2}$.

The existence of a half-life implies that the decay of any particular nucleus is a random event. That is, it is impossible to predict when any particular nucleus will decay. Because the decay of any nucleus can occur at any time, all nuclei in a sample containing many nuclei are equally likely to decay in any given time interval. Thus the decay rate $-dN/dt$, the number of decays per unit time, is proportional to the number $N$ of original nuclei present:

$$-\frac{dN}{dt} \propto N. \tag{43.3}$$

The quantities $N$ and $dN$ are dimensionless. The left side of the proportionality thus has dimension 1/[time]. In order to convert the proportionality into an equation, therefore, we must multiply the right side by a constant factor having dimension 1/[time]. If we call this factor $1/\tau$, we have

$$-\frac{dN}{dt} = \frac{N}{\tau}. \tag{43.4}$$

This differential equation has exactly the same form as the equation that describes the discharge of a capacitor through a resistor (see Equation 32.6 and the equation immediately preceding it). We integrate Equation 43.4 in the same way. First, we rearrange to separate the variables:

$$\frac{dN}{N} = -\frac{dt}{\tau}.$$

We then integrate both sides. If the sample contains $N_0$ nuclei at time $t = 0$ and $N$ nuclei at time $t$, we have

$$\int_{N_0}^{N} \frac{dN}{N} = -\int_{0}^{t} \frac{dt}{\tau}.$$

Carrying out the integration gives us

$$\ln \frac{N}{N_0} = -\frac{t}{\tau}, \tag{43.5a}$$

or

$$N = N_0 e^{-t/\tau}. \tag{43.5b}$$

The quantity $\tau$ is called the **mean life**. The relation between the mean life and the half-life is

$$\tau_{1/2} = \tau \ln 2 = 0.693\tau. \tag{43.6}$$

The derivation of this relation is the subject of Problem 43.26.

Radioactive half-lives of nuclides vary over a vast range. Some half-lives are listed in Table 43.1. A long half-life implies weak radioactivity, because the decay of all the nuclei originally present is distributed over a long time. A short half-life, on the other hand, implies strong radioactivity. By looking at Table 43.1, you can see why Marie Curie was able to detect tiny quantities of polonium and radium in the presence of the much larger quantities of uranium and thorium contained in pitchblende.

## TABLE 43.1 Some Radioactive Half-Lives

| Isotope | Decay Mode | Half-Life | Isotope | Decay Mode | Half-Life |
|---------|-----------|-----------|---------|-----------|-----------|
| $^{8}_{4}\text{Be}$ (beryllium-8) | $\alpha + \alpha$ (fission) | $3 \times 10^{-16}$ s | $^{90}_{38}\text{Sr}$ (strontium-90) | $\beta^{-}, \gamma$ | 28.6 y |
| $^{34}_{11}\text{Na}$ (sodium-34) | $\beta^{-}$ | 20 ms | $^{226}_{88}\text{Ra}$ (radium-226) | $\alpha$ | 1620 y |
| $^{23}_{10}\text{Ne}$ (neon-23) | $\beta^{-}$ | 37.24 s | $^{14}_{6}\text{C}$ (carbon-14) | $\beta^{-}$ | 5730 y |
| $^{1}_{0}\text{n}$ (neutron) | $\beta^{-}$ | 10.25 min | $^{235}_{92}\text{U}$ (uranium-235) | $\alpha$ | $7.038 \times 10^{8}$ y |
| $^{112}_{47}\text{Ag}$ (silver-112) | $\beta^{-}$ | 3.14 h | $^{40}_{19}\text{K}$ (potassium-40) | $\beta^{-}$ | $1.277 \times 10^{9}$ y |
| $^{222}_{86}\text{Rn}$ (radon-222) | $\alpha$ | 3.8235 d | $^{238}_{92}\text{U}$ (uranium-238) | $\alpha$ | $4.468 \times 10^{9}$ y |
| $^{131}_{53}\text{I}$ (iodine-131) | $\beta^{-}, \gamma$ | 8.04 d | $^{232}_{90}\text{Th}$ (thorium-232) | $\alpha$ | $1.39 \times 10^{10}$ y |
| $^{60}_{27}\text{Co}$ (cobalt-60) | $\beta^{-}, \gamma$ | 5.24 y | | | |

EXAMPLE 43.1

The radionuclide $^{41}\text{Ar}$ (argon-41) can be produced by bombarding the common stable nuclide $^{40}\text{Ar}$ with neutrons. You expose a small container of liquid argon to a neutron beam. Removing it from the beam, you use a counter that detects $\beta$ particles emitted by the $^{41}\text{Ar}$. You measure the counting rate $n$ (the number of counts per second) at intervals and tabulate your findings:

| Time (h) | Time (s) | Counting rate $n$ (s$^{-1}$) |
|----------|----------|------------------------------|
| 0 | 0 | 650 |
| 1 | $0.36 \times 10^4$ | 445 |
| 2 | $0.72 \times 10^4$ | 304 |
| 3 | $1.08 \times 10^4$ | 208 |
| 5 | $1.80 \times 10^4$ | 97.5 |
| 7.5 | $2.70 \times 10^4$ | 37.9 |
| 10 | $3.60 \times 10^4$ | 15.2 |

What is the half-life of $^{41}\text{Ar}$?

**SOLUTION:** You do not collect all the radiation emitted because only a fraction of it enters the counter. Nevertheless, for a fixed arrangement of sample and counter, the counting rate at any instant is proportional to the total emission rate $-dN/dt$. According to Equation 43.3, $-dN/dt$ is proportional to $N$. So you have

$$n \propto N;$$

the counting rate is proportional to the number of $^{41}\text{Ar}$ nuclei present. What is true in general is true in particular for $t = 0$:

$$n_0 \propto N_0.$$

Consequently, you can write

$$\frac{n}{n_0} = \frac{N}{N_0}.$$

Making this substitution into Equation 43.5a, you obtain

$$\ln \frac{n}{n_0} = -\frac{1}{\tau}\,t. \tag{43.7}$$

You can write this equation in the form

$$\ln n = \ln n_0 - \frac{1}{\tau}\,t.$$

According to this equation, $\ln n$ depends linearly on $t$, and a plot of $\ln n$ versus $t$ will be a straight line whose slope is $-1/\tau$; this is borne out by the data plot of Figure 43.1. (The slight upward deviation from the line of the last two points is due to the fact that, as the activity of the sample decreases, the contribution of background radiation becomes significant.)

If you use the data in the table to plot a graph like Figure 43.1, you can determine the slope. Using the ordinate intercept and the point marked $P$ at $t = 3.80 \times 10^3$ s, you will find

$$-\frac{1}{\tau} = \frac{2.45 - 6.48}{3.8 \times 10^4 - 0} = -1.06 \times 10^{-4}.$$

This gives you the mean life $\tau = 9.43 \times 10^3$ s. The last step is to use Equation 43.6 to calculate the half-life:

$$\tau_{1/2} = \tau \ln 2 = 6.54 \times 10^3 \text{ s} = 1.82 \text{ h}.$$

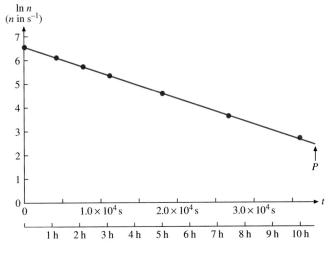

**FIGURE 43.1**

If the half-life of a radionuclide is long, the technique of Example 43.1 cannot be used. This is because the counting rate will not change appreciably over any reasonable time. But it is still possible to determine the half-life if (1) the total number $N$ of atoms of the radionuclide in the sample is known and (2) the counting apparatus is designed to collect all the emission from the sample. The details are the subject of Example 43.2.

## EXAMPLE 43.2

Although most naturally occurring radioactive substances have large atomic numbers $Z$, there are a few for which $Z$ is fairly small. Most notable is potassium-40, which constitutes $1.17 \times 10^{-4}$ of naturally occurring potassium. (The major isotopic components $^{39}K$ and $^{41}K$ are stable.)

You place a 10-g sample of potassium iodide (KI) in an apparatus that counts all the radioactive emanations. The counting rate is $n = 73.0 \text{ s}^{-1}$. What is the half-life of $^{40}K$? Take the atomic mass of potassium to be 39.10 u and that of iodine to be 126.90 u.

**SOLUTION:** You can write Equation 43.4 in the form

$$\tau = -\frac{N}{dN/dt}.$$

In the present case, with all emanations detected, the decay rate is equal to the counting rate: $-dN/dt = n$. So you have

$$\tau = \frac{N}{n}.$$

The task, then, is to calculate $N$, the number of $^{40}K$ atoms present.

First you find the total number $N'$ of potassium atoms present. According to the chemical formula, KI, this number is equal to the number of molecules present. The molecular mass of KI is 39.10 u + 126.90 u = 166.00 u. Thus 166.00 kg of KI contains Avogadro's number of K atoms. In a 10-g sample, you have

$$N' = \frac{6.023 \times 10^{26} \text{ kmol}^{-1} \times 0.01 \text{ kg}}{166.00 \text{ kg/kmol}} = 3.628 \times 10^{22}.$$

Of these potassium atoms, a fraction $1.17 \times 10^{-4}$ are $^{40}K$ atoms. Thus you find

$$N = 3.628 \times 10^{22} \times 1.17 \times 10^{-4} = 4.25 \times 10^{18}.$$

You can now use the relation $\tau = N/n$ to obtain the mean life

$$\tau = \frac{4.25 \times 10^{18}}{73.0} = 5.82 \times 10^{16} \text{ s}.$$

Using Equation 43.6, you obtain

$$\tau_{1/2} = \tau \ln 2 = 4.03 \times 10^{16} \text{ s} = 1.28 \times 10^9 \text{ y}.$$

---

The value of the half-life of a radioactive isotope can have important practical consequences. If the half-life is very long, the activity is low, and the emissions are not likely to be hazardous, at least for short exposures. If the half-life is very short, the activity goes quickly to negligible levels, and the emissions are not likely to be hazardous after a time equal to (say) $10\tau_{1/2}$. It is the intermediate half-lives that present opportunities as well as hazards.

Cobalt-60 emits $\gamma$ rays having energy 1.17 MeV and 1.33 MeV. The half-life of $^{60}Co$, about 5 years, is long enough to make the isotope convenient in industrial X-ray applications where high penetration is needed, and in medical treatments. But the half-life is also short enough that the activity is high, and intense beams of $\gamma$ rays are readily produced. Strontium-90, a component in the waste products of fission reactors, has a half-life of 28.6 years. Again, this value is short enough to imply considerable activity but not short enough to lead to rapid disappearance. Because it behaves chemically very much like calcium, care must be taken to prevent the escape of strontium into the grass-cow food chain that leads to the production of milk, which contains considerable calcium.

### Radioactive Dating

Carbon-14 ($\tau_{1/2} = 5730$ y) is produced continuously in the upper atmosphere. Cosmic rays colliding with air molecules produce a variety of debris, including neutrons. These neutrons react with $^{14}N$ (the common isotope of nitrogen) to form carbon-14 and hydrogen nuclei:

$$^{14}_{7}N + ^{1}_{0}n \rightarrow ^{14}_{6}C + ^{1}_{1}H.$$

The $^{14}C$ combines with oxygen to form $^{14}CO_2$. This carbon dioxide mixes through the atmosphere in a time that is short compared with its half-life. If the production rate of $^{14}C$ is steady (as it is to first approximation), a steady state is reached in which $^{14}C$ is introduced into the atmosphere at the same rate that it decays. As a result, $^{14}CO_2$ constitutes a small but fairly steady fraction—about $1.3 \times 10^{-12}$—of the $CO_2$ in the atmosphere.

As green plants carry out photosynthesis, they absorb carbon dioxide (containing the steady-state fraction of $^{14}CO_2$) from the atmosphere. Because all of the carbon in a plant comes from the atmosphere, the fraction of $^{14}C$ in a living green plant is essentially the same as the fraction in the atmosphere. All other organisms obtain their carbon directly or indirectly from green plants; hence the fraction of $^{14}C$ is essentially the same in all living organisms.

When an organism dies, it ceases to exchange carbon with its environment through metabolic processes. Thus the fraction of $^{14}C$ in the remains of the organism decreases as $^{14}C$ decays, in accordance with Equation 43.5b. By measuring the ratio of $^{14}C$ to the stable isotopes $^{12}C$ and $^{13}C$, the date of death of the organism can be established. Using modern techniques, an object containing organic matter can be dated if its age is less than about six half-lives, or about 35,000 years. The results can be cross-checked by means of independent methods, such as tree-ring dating. These independent methods are used as well to correct for slow variations in the $^{14}C$ production rate.

Carbon-14 dating (sometimes called *radiocarbon dating*) is of great value to archeologists, anthropologists, and historians. Bits of wood, mummified animal and human remains, charcoal, food leavings, and other organic materials have all been used to establish dates for sites or objects of interest.

Many other radioactive decay processes are useful in dating objects whose ages are roughly comparable to the half-life of the radioactive nucleus involved. To give one example, $^{206}Pb$ is the stable isotope that terminates the decay chain that begins with the common isotope of uranium, $^{238}U$. The half-life of $^{238}U$, about $4.5 \times 10^9$ years, is far greater than the half-lives of the other isotopes in the chain. The other half-lives can therefore be neglected. By measuring the ratio of $^{206}Pb$ to $^{238}U$ in a very old sample, its age can be determined. (This technique was first applied in 1906, just ten years after the discovery of radioactivity.) The ratio of $^{208}Pb$ to $^{232}Th$ ($\tau_{1/2} = 1.4 \times 10^{10}$ y) can be used similarly. Geologists find such measurements useful in dating rocks. A detailed absolute geological time scale was first developed in 1914 by the British geologist Arthur Holmes, who related the absolute radioactive time scale derived from igneous rocks to the relative time scale derived from study of fossils in sedimentary rocks with which the igneous rocks were associated. Vastly more detail has been added to Holmes's scale since 1914, but the general features have not needed alteration.

**The Clock That Starts Ticking at Death**

Radiocarbon dating was devised in 1949 at the University of Chicago by the American chemist Willard F. Libby (1908–1980), who won the 1960 Nobel Prize in Chemistry for this work. The original technique required sample quantities of about 10 g. Modern techniques are much more sensitive and can therefore be used to date objects (such as works of art) from which only milligram samples are obtainable.

## The Radioactive Decay Process

In Section 42.3, we developed a quantum-mechanical treatment of the particle in a one-dimensional box. We were aided by the strong resemblance between the particle-in-a-box wave function and the wave function describing the vibration of a stretched string anchored rigidly at its ends. We now expand on some of the ideas of Section 42.3, with a view to explaining radioactive decay.

It is often useful to consider the particle in a box in terms of an energy diagram like that of Figure 43.2. Within the box, the particle is free of all forces. It is convenient to set the potential energy $U$ inside the box equal to zero so that the total energy $E$ of the particle is completely kinetic:

$$E = K + U = K \quad \text{for } 0 \le x \le D.$$

At the walls, the potential energy increases abruptly to infinity. The particle, whose energy is finite, cannot penetrate the *potential energy barrier* imposed by the wall, often called the **potential barrier** for short. Rather, it is trapped in the **potential well** described by the function

$$U = \infty \quad \text{for } x < 0, x > D$$
$$U = 0 \quad \text{for } 0 \le x \le D.$$

**FIGURE 43.2** The simplest possible one-dimensional potential well.

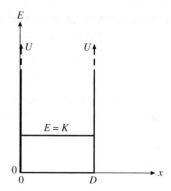

The potential well of Figure 43.2 does not describe real systems. A better approximation to a real confined-particle system is given by the finite potential well shown in Figure 43.3. Here the potential barrier is both of finite height $U$ and of finite thickness $b$.

**FIGURE 43.3** A potential well of finite "height" $U$ and thickness $b$. The classical particle for which $E < U$ is trapped inside the well; the particle for which $E > U$ is free.

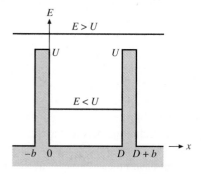

From the classical point of view, a particle confined in the well of Figure 43.3 behaves no differently from a particle confined in the well of Figure 43.2. If its energy is $E < U$, the particle is trapped and can exist only in the region $0 \leq x \leq D$. Only if its energy exceeds the critical value $E = U$ can the particle surmount the barrier and escape from the well. If $E \geq U$, the particle is not confined at all.

One way to describe the fact that a classical particle cannot penetrate into the barrier is in terms of the speed the particle would have if it did penetrate. Because the total energy of the particle is the sum $E = K + U$, its kinetic energy is given by $K = E - U$. We have

$$v = \sqrt{\frac{2K}{m}} = \sqrt{\frac{2}{m}(E - U)}, \tag{43.8}$$

where $m$ is the particle mass. You can see from Figure 43.3 that any particle trapped inside the well has energy $E < U$. Consequently, the quantity under the square-root sign in Equation 43.8 is negative, and the speed $v$ is imaginary. But an imaginary speed has no physical meaning, and so barrier penetration is classically impossible. For a particle whose total energy is $E > U$, the speed is real and the particle is not stopped by the potential barrier. (However, its speed is reduced as it passes over the barrier.)

Now let us take into consideration the wave properties of the particle. A complete treatment requires solution of the Schrödinger equation. But the salient features of the solution emerge when we consider some of the properties of waves.

You learned in Section 42.3 that the particle in the box has associated with it a standing de Broglie wave. A standing wave is the superposition of two traveling waves of equal wavelength and amplitude, moving in opposite directions (Section 21.7). Let us focus on one of the component traveling waves, for which we can write

$$\psi(x, t) = \psi_0 \cos \left( \frac{2\pi x}{\lambda} - \frac{2\pi t}{T} + \phi \right),$$

where $\lambda$ is the wavelength and $T$ the period of the wave. Consider a particular instant when $2\pi t/T = \phi$, so the argument of the cosine is $2\pi x/\lambda$. We then have

$$\psi(x) = \psi_0 \cos \left( \frac{2\pi x}{\lambda} \right).$$

We invoke the de Broglie principle in the form of Equation 42.4, $\lambda = h/mv$, to obtain

$$\psi(x) = \psi_0 \cos \left( \frac{2\pi mv}{h} x \right).$$

We insert the value of $v$ given by Equation 43.8. This gives us

$$\psi(x) = \psi_0 \cos \left[ \frac{2\pi}{h} \sqrt{2m(E - U)} \, x \right]. \tag{43.9}$$

As long as $E > U$, the argument of the cosine is real, and the solution is the familiar sinusoidal one. For the particle of energy $E > U$ in Figure 43.3, this condition holds everywhere. For the particle of energy $E < U$, the argument of the cosine is real inside the well and outside the barrier. Within the barrier, however (that is, for $-b < x < 0$ and $D < x < D + b$), the argument assumes an imaginary value. To make this clear, let us rewrite Equation 43.9 in the form

$$\psi(x) = \psi_0 \cos \left[ i \frac{2\pi}{h} \sqrt{2m(U - E)} \, x \right], \tag{43.10a}$$

where $i \equiv \sqrt{-1}$. We now use an identity that relates the cosine to the hyperbolic cosine:

$$\cos iz = \cosh z, \quad \text{where } \cosh z \equiv \frac{e^z + e^{-z}}{2}.$$

This enables us to rewrite Equation 43.10a in terms of the hyperbolic cosine of a real argument:

$$\psi(x) = \psi_0 \cosh \left[ \frac{2\pi}{h} \sqrt{2m(U - E)} \, x \right]. \tag{43.10b}$$

This equation tells us that *the wave function does not disappear abruptly at the potential barrier*. Rather, it declines in a roughly exponential fashion, as shown in Figure 43.4.* If $\psi$ does not vanish abruptly at the potential barrier, neither does $\psi^2$. But, as you learned in Section 42.3, $\psi^2(x)$ expresses the probability that the particle will be found

*The operation of the smoke detector depends on the radioactive decay of curium, an artificially produced element.*

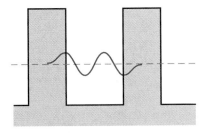

**FIGURE 43.4** A wave function penetrates into a potential barrier.

---

*We have glossed over some important details. The wave functions inside the well and within the barrier must be matched by means of the conditions that $\psi$ and $d\psi/dx$ have continuous values at the barriers.

at $x$. It follows that *the particle may be found inside the potential barrier*. This is very much at variance with the classical prediction.

If the potential barrier is thick enough, the particle will remain trapped in the well. But suppose the barrier is quite thin, as shown in Figure 43.5. Although the amplitude of the wave function diminishes rapidly with distance from the inside edge of the potential well, it does not decline to zero. Therefore, when the wave function is continued outside the barrier, where $U$ is again less than $E$, it must again be represented by a cosine function like that of Equation 43.9, but with a smaller amplitude.

**FIGURE 43.5** Tunneling through a thin potential barrier.

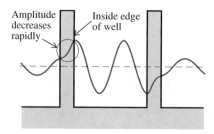

If the wave function is not zero outside the barrier, the probability that the particle will be found outside the well is not zero. That is, the particle has a finite probability of **tunneling** through the barrier. The probability depends in a very sensitive way on the height and thickness of the barrier, but it can be calculated.

## The Nuclear Potential Well

Nucleons are bound inside a nucleus by a potential barrier that has much in common with the barrier of Figure 43.5. However, there are two important differences. First, the nuclear potential well is three-dimensional rather than one-dimensional. Second, the barrier walls are not "square." We know that, for protons or α particles, the outside wall of the barrier has the Coulomb form $U \propto 1/r$ (Equation 25.14b) because the potential barrier outside the nucleus is due to the presence of positive charge $Ze$ within the roughly spherical nucleus. The shape of the inside wall of the barrier is not so easy to infer. This is because the potential energy depends not only on the Coulomb force, but on the short-range *strong nuclear force* that holds the nucleus together. Rather than worry about the details of how $U$ varies with $r$, we simply argue that it varies very rapidly with $r$ and substitute a vertical barrier for the real one. This gives us the potential well shown in Figure 43.6. We do not know where the bottom of this well lies. We therefore choose the zero of potential energy to be the value at $r = \infty$, where the Coulomb force goes to zero. The dashed vertical lines for $U < 0$ indicate our lack of knowledge of the value of $U$ at the bottom of the well.

**FIGURE 43.6** The nuclear potential well. The outside wall of the well conforms to Coulomb's law, $U \propto 1/r$. A nucleon energy level is shown in red inside the well.

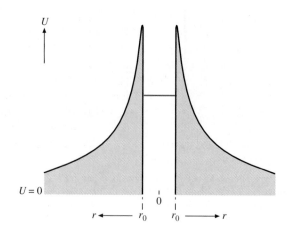

An α particle trapped inside the nucleus is represented by its energy level, shown in red in Figure 43.6. The probability that the α particle will tunnel through the barrier and escape depends very sensitively on its energy, both because an increase in energy means a decrease in $U - E$ and because the thickness of the barrier decreases rapidly with increasing energy. For this reason, the tunneling probability can vary over many orders of magnitude as a consequence of small changes in the structure of the nucleus. The wide range in tunneling probability leads to a corresponding wide range in half-lives, even for closely related nuclei. A crude calculation leads to the following relation between half-life and the kinetic energy $K$ of an emitted α particle:

$$\log \tau_{1/2} \propto \frac{1}{\sqrt{K}}. \tag{43.11}$$

(It does not matter what base is used for the logarithm, because the relation is a proportionality.) In Figure 43.7, the logarithm of the half-life is plotted versus $1/\sqrt{K}$ for the emitted α particle for each isotope of uranium that decays by α emission. The isotope nuclei differ only in $A - Z$, the **neutron number***—a difference that is small compared with the neutron number of any of them, which is about 140. With this in mind, we make the crude assumption that the shapes of the potential wells of the nuclei do not differ greatly from one another and that the various energies of the emitted α particles represent different energy levels within the nuclei from which they were emitted. An increase of α-particle energy from 4.2 MeV (for $^{238}$U) to 6.8 MeV (for $^{227}$U) is associated with a decrease in half-life of 15 orders of magnitude—from $4.51 \times 10^9$ years (one-quarter the age of the universe) for $^{238}$U to 1.3 minutes for $^{227}$U.

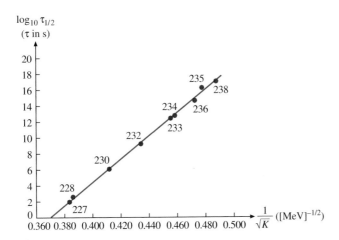

**FIGURE 43.7** Logarithmic plot of the half-life versus the reciprocal of the square root of the energy of the emitted α particle for all uranium isotopes that decay solely by α emission. Each point is labeled with the nucleon number of the isotope it represents.

## Alpha Decay: A Crude Model

With due respect for the dangers of visualizing nuclei in classical terms, imagine an α particle as a little ball trapped inside the potential well of Figure 43.6. Each time it strikes the barrier, it rebounds. The probability that each collision will result in tunneling, so that the particle will emerge, is tiny. But the particle makes many collisions with the walls per unit time. If its energy $E$ is greater than zero, as in Figure 43.6, the particle will eventually tunnel through the barrier. When it does so, it is beyond the range of the short-range nuclear force. The α particle is then subject only to the Coulomb force due to the positive charges on it and the remaining (daughter) nucleus. The α particle is repelled strongly by the Coulomb force and departs the nucleus with considerable energy. As already noted, α particles emitted by radioactive nuclei typically have between 2 and 7 MeV of kinetic energy. The energy released in the decay process is called the **disintegration energy**.

---

*Because $A$ is the total number of nucleons—protons and neutrons—in the nucleus and $Z$ is the number of protons, $A - Z$ is the number of neutrons.

## EXAMPLE 43.3

When $^{238}$U decays by α emission, the disintegration energy is 4.19 MeV. **(a)** Calculate the radius of the uranium nucleus. **(b)** Assume that α particles exist as discrete objects inside the nucleus. Estimate the time between collisions the α particle makes with the potential barrier. **(c)** What is the probability per collision that the α particle will tunnel out of the nucleus?

**SOLUTION:**

**(a)** What is the nuclear radius of $^{238}$U?

According to Equation 41.4,

$$r = r_0 A^{1/3}, \quad \text{where } r_0 = 1.37 \times 10^{-15} \text{ m},$$

you have

$$r = (238)^{1/3} \times 1.37 \times 10^{-15} \text{ m}$$
$$= 8.5 \times 10^{-15} \text{ m} = 8.5 \text{ fm}.$$

**(b)** What is the collision interval?

You don't know the energy at the bottom of the potential well. But make the crude assumption that the kinetic energy of the α particle inside the nucleus is equal to the disintegration energy. You then have, for the "speed" of the α particle inside the nucleus,

$$v = \sqrt{\frac{2K}{m}}.$$

Closely enough for the purposes of this calculation, the mass of an α particle (with its two protons and two neutrons) is four times the proton mass. So you have

$$v = \sqrt{\frac{2 \times 4.19 \text{ MeV} \times 1.6 \times 10^{-13} \text{ J/MeV}}{4 \times 1.67 \times 10^{-27} \text{ kg}}}$$
$$= 1.4 \times 10^7 \text{ m/s}.$$

This is only 5% of the speed of light, so you needn't worry about relativistic corrections.

The time between collisions is

$$t = \frac{2r}{v} = \frac{2 \times 8.5 \times 10^{-15} \text{ m}}{1.4 \times 10^7 \text{ m/s}} = 1.2 \times 10^{-21} \text{ s},$$

a very short time!

**(c)** For each collision, what is the probability that the α particle will tunnel out of the nucleus?

According to Table 43.1, the half-life of $^{238}$U is $4.5 \times 10^9$ y $= 1.4 \times 10^{17}$ s—the time needed for the α particle to have a 50% chance of escaping. During this time, the number of collisions the α particle makes with the walls is

$$\frac{1.4 \times 10^{17} \text{ s}}{1.2 \times 10^{-21} \text{ s}} = 1.2 \times 10^{38}.$$

This is the number of collisions the α particle must make in order to have a 50% chance of escaping. If $1.2 \times 10^{38}$ collisions give a 50% chance, the probability of escape per collision is

$$\frac{0.5}{1.2 \times 10^{38}} \simeq 4 \times 10^{-39}.$$

This probability is very small; nevertheless, it underlies the readily observable radioactivity of $^{238}$U.

---

## Gamma Emission

The emission of γ-ray photons is associated with, and gives further support to, the existence of discrete energy levels within nuclei. As a concrete example, consider the nuclide uranium-232. A sample of $^{232}$U atoms emits α particles having three energies: 5.32, 5.26, and 5.13 MeV. Also emitted are γ-ray photons having energies 0.06 and 0.13 MeV. Note that the γ energies are equal to differences between the α energies:

$$5.32 \text{ MeV} - 5.26 \text{ MeV} = 0.06 \text{ MeV} \quad \text{and} \quad 5.26 \text{ MeV} - 5.13 \text{ MeV} = 0.13 \text{ MeV}.$$

It is reasonable to postulate the processes shown in Figure 43.8. In part *a*, an α particle is emitted directly with energy 5.32 MeV. In part *b*, the α particle falls from

**FIGURE 43.8** Three possible decay modes for $^{232}$U.

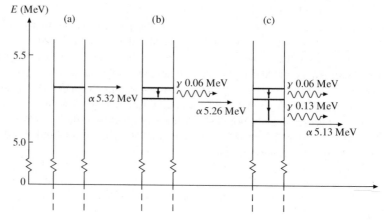

its initial energy level to a lower one within the nucleus, and the γ photon of energy 0.06 MeV is emitted. Subsequently, the α particle is emitted with energy 5.26 MeV. In part c, the α particle makes two successive transitions as the nucleus emits γ photons of energies 0.06 MeV and 0.13 MeV. Then the α particle is emitted with energy 5.13 MeV.

## Beta Decay

Many radionuclides decay by emitting $\beta^-$ particles—that is, electrons. The decay of $^{14}$C is typical:

$$^{14}_{6}\text{C} \rightarrow \,^{14}_{7}\text{N} + \beta^- + \bar{\nu}. \tag{43.12}$$

(The symbol $\bar{\nu}$ denotes an **antineutrino**, a particle we will consider shortly.) The simplest $\beta$ decay is that of the free neutron, which decays with half-life $\tau_{1/2} = 10.25$ min:

$$^{1}_{0}\text{n} \rightarrow \,^{1}_{1}\text{p} + \beta^- + \bar{\nu}. \tag{43.13}$$

A closely related process is **electron capture**, in which a nucleus captures one of its own orbital electrons that has orbital angular momentum $l = 0$. (The Bohr orbits of such electrons pass through the nucleus.) In electron capture, $Z$ decreases by 1 (owing to charge conservation), and the neutron number $A - Z$ increases by 1 (owing to nucleon-number conservation). A typical electron-capture process is

$$^{231}_{92}\text{U} + \beta^- \rightarrow \,^{231}_{91}\text{Pa} + \nu. \tag{43.14}$$

(The symbol $\nu$ denotes a **neutrino**, which we will also discuss shortly. For the purposes of this chapter, we need not distinguish between neutrinos and antineutrinos; we will consider the distinction in Chapter 45.)

When quantitative observations of $\beta^-$ emission were first made in the 1920s, serious anomalies were observed. The energies of the parent and daughter nuclei are fixed; nevertheless, the kinetic energies of the emitted $\beta^-$ particles have all values from zero to some maximum value that depends on the particular emitter. This was the observation that first led Pauli in 1930 to postulate the existence of a particle not yet observed, which was later dubbed the neutrino by Fermi. (See margin notes in Sections 39.4 and 43.4.)

In all $\beta$-emission and electron-capture processes, the presence of the neutrino (or antineutrino) serves to satisfy conservation of energy, linear momentum, and angular momentum. Consider $\beta^-$ emission for concreteness. The neutrino carries away the kinetic energy and momentum not accounted for by the electron. Also, the magnitude of the spin angular momentum of the neutrino is equal to that of the electron. When a nucleus emits an electron and an antineutrino with spins oppositely directed, the angular momentum of the nucleus remains unchanged. The neutrino is very difficult to detect because it possesses no electric charge and has little if any rest mass.

## THINKING LIKE A PHYSICIST

To the casual observer, the act of postulating the existence of a massless, nearly undetectable particle to account for energy, momentum, and angular momentum in $\beta^-$ decay may seem irrational and arbitrary. But the experience of physicists puts very heavy weight on the three conservation principles, and postulating the neutrino is no mere hoping for a miracle. Physicists did indeed feel somewhat uncomfortable during the twenty-two-year period that elapsed before neutrinos were detected experimentally, but nearly all of them had such strong confidence in the conservation principles that they had little doubt that detection would ultimately be achieved. In the meantime, the neutrino took a vital place in the theory of the structure of matter. The confidence of physicists in the conservation principles was strongly vindicated; neutrino observations have been used to study the energy production of the sun, supernova explosions, and many other phenomena.

As you learned in Example 42.6, the uncertainty principle forbids the existence of electrons inside a nucleus. Thus, the electron must come into existence at the instant of β emission. This creation of an electron may seem strange at first glance. But note that very much the same thing happens when an atomic electron makes a transition and a photon is created as the atom emits light. Nevertheless, the fact that free neutrons are not stable but instead decay according to Equation 43.13 suggests that the neutron (and its close relative the proton) are not truly elementary particles, as the electron, the photon, and the neutrino are.

---

## Nuclear Structure

We have already commented on the existence of energy levels in the nucleus (Figure 43.8). These energy levels are measured in MeV, rather than the eV used to measure atomic energy levels. Nevertheless, we are tempted to extend our understanding of the internal structure of nuclei by analogy with the structure of atoms, which is much better understood.

There are considerable difficulties in doing so. Instead of the simple inverse-square Coulomb force that bonds atoms together, we must deal with the nuclear force, whose action is not yet understood in complete detail. Instead of the "planetary" atom, in which the central force exerted by the nucleus dominates, we must deal with a complex many-body problem, with closely packed nucleons exerting forces on one another. In the absence of a detailed analytical theory, we must treat the nucleus in terms of *models*, no one of which describes all aspects of nuclei.

### The Shell Model

We have already considered how orbital electrons dispose themselves around the nucleus in shells, which fill up from the inside out in an orderly way dictated by the Pauli exclusion principle. An outstanding feature of this model is the great stability of atoms having filled shells. These are the Group VIIIA elements—the noble gases $_2$He, $_8$Ne, $_{18}$Ar, $_{36}$Kr, $_{54}$Xe, and $_{86}$Rn. The "magic" atomic numbers 2, 8, 18, 36, 54, 86, . . . that correspond to this stability are thoroughly accounted for by quantum theory. But in fact, the values of these numbers (or most of them) were known long before the reasons for them were fully understood.

A similar observation applies to nuclei. There is a series of **magic nucleon numbers**: 2, 8, 20, 28, 50, 82, 126, . . . . (The word "magic" was first used in this context by Fermi.) Nuclei having these numbers of either protons or neutrons are quite stable against decay either by emission of nucleons or by combinations of nucleons. **Doubly magic** nuclei, having magic numbers of both protons and neutrons, are especially stable. The simplest doubly magic nucleus is the helium nucleus $^4_2$He (the α particle), which has two protons and two neutrons. Other examples are $^{16}_8$O, $^{40}_{20}$Ca, $^{48}_{20}$Ca, and $^{208}_{82}$Pb.

The existence of these magic numbers suggests that nucleons, like atomic orbital electrons, arrange themselves in shells. But this can happen only if the nucleons have defined orbits—or, more strictly speaking, if their wave functions can close on themselves in such a way as to result in standing waves.

How can standing waves be set up in a nucleus, however, where the nucleons are crowded together and we might expect frequent randomizing collisions? The answer lies in the exclusion principle. Two nucleons can collide only if the resulting exchange of energy between them leaves each nucleon with an energy equal to that of an available energy level. But most energy levels in the vicinity of the original nucleon energies are likely already filled by other nucleons. Thus the energy exchange cannot take place, and the nucleons do not collide. (You may be surprised to find, in Section 44.2, that a similar argument made in a very different context explains the inability of dielectric materials to conduct electricity.)

Can quantum theory account for the magic numbers of nuclei, as it accounts for the magic numbers of atoms? Working independently, Mayer and Jensen showed in 1949 that the answer is Yes. Because the nucleus is more complicated and less thoroughly understood than the atom, derivation of the nuclear magic numbers depends on a number of assumptions involving spin-orbit coupling of nucleons under the influence of the nuclear force. However, all of these assumptions are plausible. The theory yields a fairly detailed quantitative picture of the energy levels within the nucleus. It is particularly successful in accounting for the structure of nuclei having large neutron numbers.

## Nuclear Binding Energy

The nucleons constituting a nucleus stay together—that is, the nucleus is stable—only if it is energetically favorable to do so. A measure of nuclear stability is the *binding energy*. The binding energy of a nucleus can be determined by comparing its rest mass with that of its component nucleons. For a nucleus of mass $M$, having nucleon number $A$ and atomic number $Z$, we define the **mass defect** $\Delta$ to be the difference between the mass of the component nucleons and the mass of the nucleus:

$$\Delta \equiv Zm_p + (A - Z)m_n - M. \qquad \textbf{(43.15)}$$

The first term on the right is the total mass of the $Z$ protons, and the second is the total mass of the $A - Z$ neutrons. The **binding energy** $U_b$ is the energy equivalent of the mass defect:

$$U_b = c^2\Delta = c^2[Zm_p + (A - Z)m_n - M]. \qquad \textbf{(43.16)}$$

> ### Sharp Eyes See Inside the Nucleus
>
> Maria Goeppert-Mayer (1906–1972), German-American physicist, and J. Hans D. Jensen (1907–1973), German physicist, shared a Nobel Prize in 1963. In spite of the distinction of her work, Goeppert-Mayer did not hold a mainstream faculty position until 1959, when she was in her fifties; the then-common pretext was that she could not be hired because her husband was a chemistry professor at the same university.

### EXAMPLE 43.4

(a) Calculate the mass defect and binding energy of the $\alpha$ particle ($M_\alpha = 4.001\ 51$ u), given that $m_p = 1.007\ 28$ u and $m_n = 1.008\ 67$ u. (b) Is it possible that an undisturbed $\alpha$ particle will spontaneously split into two deuterons ($M_D = 2.013\ 47$ u)? A deuteron (or "heavy hydrogen" nucleus, ${}_1^2\text{H}$) consists of one proton and one neutron.

**SOLUTION:**
(a) Find $\Delta$ and $U_b$ for the $\alpha$ particle.

The nucleon number of the $\alpha$ particle is 4, and its atomic number is 2. Its constituent nucleons are thus two protons and two neutrons. According to Equation 45.15, the mass defect is

$$\Delta = (2 \times 1.007\ 28\ \text{u}) + (2 \times 1.008\ 67\ \text{u}) - 4.001\ 51\ \text{u}$$
$$= 0.030\ 39\ \text{u}.$$

This much mass—or its energy equivalent—must be supplied to an $\alpha$ particle if it is to break up into four nucleons. The energy equivalent is the binding energy

$$U_b = c^2\Delta = (3.00 \times 10^8\ \text{m/s})^2 \times 0.030\ 39\ \text{u}$$
$$\times 1.661 \times 10^{-27}\ \text{kg/u}$$
$$= 4.54 \times 10^{-12}\ \text{J} = 28.4\ \text{MeV}.$$

In radioactive decay processes, $\alpha$ particles never acquire energies as great as 10 MeV. The value of $U_b$ is much greater, assuring that $\alpha$ particles will not spontaneously disintegrate into their constituent nucleons.

(b) Is the spontaneous decay ${}_2^4\text{He} \rightarrow 2\ {}_1^2\text{He}$ energetically permissible?

Consider the mass difference between two deuterons and an $\alpha$ particle:

$$2m_D - m_\alpha = 2 \times 2.013\ 47\ \text{u} - 4.001\ 51\ \text{u} = 0.025\ 43\ \text{u}.$$

The energy of this much mass must be supplied to break an $\alpha$ particle into two deuterons:

$$U_b = (3.00 \times 10^8\ \text{m/s})^2$$
$$\times 0.025\ 43\ \text{u} \times 1.661 \times 10^{-27}\ \text{kg/u}$$
$$= 3.80 \times 10^{-12}\ \text{J} = 23.8\ \text{MeV}.$$

Although not as much as the 28.4 MeV required to break the $\alpha$ particle into four nucleons, this is still a great deal of energy. We can conclude that the $\alpha$ particle is very stable, consistent with the fact that it is doubly magic.

---

For any nucleus, the binding energy per nucleon is called the **packing fraction** $f$:

$$f \equiv \frac{U_b}{A} = c^2 \frac{Zm_p + (A - Z)m_n - M}{A}. \qquad \textbf{(43.17)}$$

Every nuclide has a characteristic packing fraction. Figure 43.9 is a plot of $f$ versus nucleon number $A$. As you can see, the curve is fairly smooth except at the low end.

**FIGURE 43.9** Plot of packing fraction $f$ versus nucleon number $A$. The red line connects adjacent points.

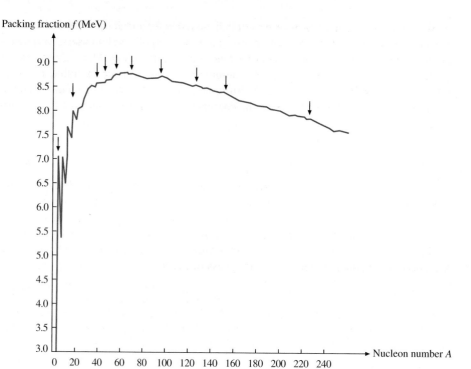

## Nuclear Fusion

Figure 43.9 shows that the packing fraction exhibits sharp peaks (indicated by the arrows) at certain values of $A$. The nuclei for which $f$ peaks in this way are especially stable, because much energy per nucleon must be supplied to break them up. It will not surprise you that the peaks correspond to magic-number nuclei.

The packing-fraction curve has a broad maximum centered about $A = 60$. This maximum has important meaning. A process that combines light (low-$A$) nuclei to synthesize heavier ones is *exothermic*—that is, results in a net release of energy to the environment—up to the point at which the nucleon number of the synthesized nucleus is roughly 60. This type of process is called **nuclear fusion**. A simple fusion process is the reaction

$$\,^2_1\text{H} + \,^2_1\text{H} \rightarrow \,^3_1\text{H} + \,^1_1\text{H} + 3.84 \text{ MeV}; \tag{43.18}$$

two deuterons ($^2$H nuclei) fuse, and the resulting unstable fused nucleus quickly breaks up to yield a triton (a $^3$H nucleus) and a proton. This is one of several processes employed as an energy source for the family of weapons called hydrogen bombs. An earth-based reactor in which fusion energy is released in a controlled manner is the goal of substantial research that has been carried on worldwide for some decades.

Given that the reaction is exothermic, why don't fusion reactions take place spontaneously? The reason lies in the *Coulomb barrier*—the large repulsive electric potential that a positively charged nucleus must overcome in order to get close enough to another nucleus so that the nuclear force can come into play and the two nuclei can interact. At ordinary temperatures, the thermal kinetic energy of the atoms in a gas is far too small to achieve this. But, at very high temperatures (in excess of $10^7$ K), the atoms in the high-energy tail of the Maxwell-Boltzmann distribution (Figure 18.8) begin to have sufficient energy to surmount the Coulomb barrier. Fusion induced by maintaining the reactants at very high temperature is called **thermonuclear fusion**.

Various fusion processes can take place in stars, depending on the mass and age of the star. The process most important in our sun is the chain

$$\,^1_1\text{H} + \,^1_1\text{H} \rightarrow \,^2_1\text{H} + \beta^+ + \nu; \quad \beta^+ + e \rightarrow 2\,\gamma \quad (1.19 \text{ MeV}); \tag{43.19a}$$

$$\,^1_1\text{H} + \,^2_1\text{H} \rightarrow \,^3_2\text{He} + \gamma \quad (5.49 \text{ MeV}); \tag{43.19b}$$

$$\,^3_2\text{He} + \,^3_2\text{He} \rightarrow \,^4_2\text{He} + 2\,^1_1\text{H} \quad (12.85 \text{ MeV}). \tag{43.19c}$$

In the first reaction, two protons fuse to form a deuteron, a positron ($\beta^+$), and a neutrino. The neutrino escapes and plays no further part. The positron very soon collides with an electron in the vicinity; both are annihilated in the collision, and two $\gamma$ photons are created. In the second reaction, the deuteron fuses with another proton, yielding a helium-3 nucleus and another $\gamma$ photon. In the third reaction, two of the helium-3 nuclei produced by means of the first two reactions fuse to yield a helium-4 nucleus and two protons.

If we ignore the neutrinos, the net process is

$$4\,{}^1_1\text{H} \rightarrow {}^4_2\text{He} + 26.21 \text{ MeV}. \qquad (43.20)$$

A great deal of energy is released when four hydrogen atoms combine to form a helium atom.

## The Liquid-Drop Model of the Nucleus

A second nuclear model is useful in many applications, even though it contradicts the shell model in many respects. The model, originally proposed by Niels Bohr and much elaborated by others, envisions the nucleus as a drop of "nuclear liquid" in which the nucleons act like molecules. The drop is held together by the nuclear force acting among the nucleons, analogously to the way a drop of ordinary liquid is held together by the intermolecular forces.

The liquid-drop model is useful in accounting for processes in which the nucleus collides with another particle—say, a neutron—to form a *compound nucleus*. In this compound nucleus, the neutron quickly merges with the rest of the liquid. Any energy the neutron had on entering the nucleus is quickly distributed throughout the nucleus. It is often possible to model the result of the excitation as an oscillation of the liquid drop. This is particularly useful in studying nuclear fission; see Figure 39.10.

## Nuclear Fission

Inspection of Figure 43.9 shows that reactions in which heavy nuclei split into fragments having $A \simeq 60$ also are exothermic. This process, called **nuclear fission**, is considered from an energy point of view in Section 39.5. Here we consider the structural details that make the process possible. Spontaneous fission is rare for most nuclides for the same reason that $\alpha$ emission is rare—the tunneling probability is small. Some nuclides, however, split readily when bombarded with neutrons; this is called **neutron-induced fission**. Because neutrons have zero electric charge, even those having very little kinetic energy can penetrate into a nucleus, unaffected by the Coulomb barrier. Indeed, it is sometimes desirable to use **thermal neutrons**, whose mean kinetic energy is merely the thermal energy $\frac{3}{2}kT$ of the surroundings.

The best-known thermal-neutron-induced fission reaction is that of uranium-235, in which a neutron collides and merges with a ${}^{235}_{92}\text{U}$ nucleus to produce a ${}^{236}\text{U}$ nucleus in an excited state. This excited nucleus breaks into two main **fission fragments**, together with two or three neutrons. These neutrons then produce further fission by colliding with other ${}^{235}\text{U}$ nuclei, in a **chain reaction** (see Figure 39.11). It is impossible to predict how any particular nucleus will break up. However, the most probable processes result in the production of fission fragments having nucleon numbers $A \simeq 95$ and $A \simeq 140$. A typical fission process is

$$ {}^{235}_{92}\text{U} + {}^1_0\text{n} \rightarrow {}^{236}_{92}\text{U}^* \rightarrow {}^{134}_{51}\text{Sb} + {}^{100}_{41}\text{Nb} + 2\,{}^1_0\text{n}. \qquad (43.21)$$

The asterisk indicates that the ${}^{236}\text{U}$ nucleus is in an excited state. As a result, it is unstable and quickly fissions.

Most fission fragments are unstable nuclides. To see why this is so, consider the *neutron-proton ratio* $(A - Z)/Z$. By inspecting the periodic table (Appendix 7), you can see that this ratio is greater for nuclides of large $Z$, such as ${}^{235}\text{U}$, than it is for stable nuclides of medium $Z$. Consequently, the fission fragments usually have too many neutrons to be stable. Typical fission fragments decay by $\beta^-$ emission, converting

**Meitner and Co.**

Nuclear fission was probably seen but not identified for what it was by a number of investigators who, in the mid-1930s, studied nuclei by bombarding them with neutrons. The 1938 discovery that fission was taking place is due to the German chemist Otto Hahn (1879–1968) and his physicist colleague Fritz Strassman (1902–1980). The details of the process were worked out by their former associate, the physicist Lise Meitner (1878–1968), who had fled to Sweden on Hitler's annexation of her native Austria, and by her nephew Otto Frisch (1904–1979), who was a refugee in Copenhagen. See Ruth Lewin Sime, *Lise Meitner: A Pioneer of the Nuclear Age* (University of California Press, 1995).

**The Italian Navigator Reaches the New World**

The details of the chain reaction were worked out by the Italian-American physicist Enrico Fermi (1901–1954) and his collaborators, first in Rome and later in the famous Manhattan Project at Chicago, culminating in the first "atomic pile" in December 1942. Fermi was remarkable among modern physicists in that he excelled both as a theoretician and as an experimentalist.

neutrons into protons and thus transmuting from unstable into stable nuclides. One such decay chain begins with a fission fragment that is an unstable isotope of bromine and ends, five $\beta^-$ decays later, with a stable isotope of zirconium:

$$\,^{90}_{35}\text{Br} \xrightarrow{\beta^-} \,^{90}_{36}\text{Kr} \xrightarrow{\beta^-} \,^{90}_{37}\text{Rb} \xrightarrow{\beta^-} \,^{90}_{38}\text{Sr} \xrightarrow{\beta^-} \,^{90}_{39}\text{Y} \xrightarrow{\beta^-} \,^{90}_{40}\text{Zr}. \qquad \textbf{(43.22)}$$

Although they are secondary to the fission process, such decays yield about 10% of the power output of typical fission reactors.

Because many different unstable fragments are produced in fission processes, these processes involve considerable radioactivity. As already noted, decays having very short or very long half-lives do not present great practical problems. However, intermediate decay times can pose serious handling and disposal problems. In this respect, fusion is a preferable approach to power generation because it does not produce large quantities of lingering radioactive products the way fission does. Fusion has the additional advantage that the raw materials—the various isotopes of hydrogen—are far more common than the heavy metals uranium and plutonium needed for fission. However, the very high temperatures required to overcome the Coulomb barrier in fusion present scientific and technical problems far more difficult than the central problems of fission power generation. The latter have been solved; the former are still the subject of research.

## Symbols Used in Chapter 43

| | | | | |
|---|---|---|---|---|
| $A$ | nucleon number (or mass number), atomic mass | | $U_b$ | binding energy |
| $c$ | speed of light | | $Z$ | atomic number |
| $f$ | packing fraction | | $\Delta$ | mass defect |
| $h, \hbar$ | Planck's constant, $h/2\pi$ | | $\lambda$ | wavelength |
| $i$ | $\sqrt{-1}$ | | $\rho$ | counting rate |
| $k$ | Boltzmann's constant | | $\tau, \tau_{1/2}$ | radioactive mean life, half-life |
| $M$ | atomic weight | | $\phi$ | phase constant |
| $N$ | number of nuclei present | | $\psi$ | wave function |
| $T$ | absolute temperature | | | |

## Summing Up

Natural **radioactive** processes involve emission of energetic charged particles or photons by atomic nuclei. When $\alpha$ **particles** (helium nuclei) or $\beta$ **particles** (electrons) are emitted, the nucleus undergoes **transmutation**, changing its chemical identity. In such transmutations, mass-energy, momentum, angular momentum, electric charge, and nucleon number are conserved.

The radioactive decay of any particular nucleus is an unpredictable event. From this it follows that the **decay rate** of a single nuclear species, or **nuclide**, declines exponentially with time. The number $N$ of remaining **parent nuclei** is given by either of Equations 43.5a and 43.5b,

$$\ln \frac{N}{N_0} = -\frac{t}{\tau} \quad \text{or} \quad N = N_0 e^{-t/\tau}.$$

In this equation, $\tau$ is the **mean life**. It is related to the **half-life** $\tau_{1/2}$ by Equation 43.6, $\tau_{1/2} = \tau \ln 2$.

The radioactive emission of $\alpha$ or $\beta$ particles is a consequence of tunneling. For a description of tunneling, the nucleus can be adequately represented as a potential well whose outer walls have the form $U \propto 1/r$ consistent with Coulomb's law and whose inner walls have the abrupt form consistent with the strength and short range of the strong nuclear force.

The **binding energy** of a nucleus is related to its **mass defect** by Equation 43.16,

$$U_b = c^2 \Delta = c^2[Zm_p + (A - Z)m_n - M].$$

The binding energy per nucleon, $U_b/A$, is called the **packing fraction**, given by Equation 43.17. Generally speaking, the packing fraction is greatest for nuclides near the middle of the periodic table, about $A = 60$. This fact makes it possible to obtain energy both by **nuclear fusion** and by **nuclear fission**.

## KEY TERMS

### Section 43.2  Radioactivity
α particle, β particle, γ ray ▪ decay process, decay mode, transmutation ▪ parent atom, daughter atom ▪ nucleon (mass) number ▪ nuclide, radionuclide ▪ mean life, half-life

### Section 43.3  The Radioactive Decay Process
potential energy barrier, potential well ▪ tunneling ▪ neutron number ▪ disintegration energy ▪ neutrino, antineutrino ▪ electron capture

### Section 43.4  Nuclear Structure
magic nucleon number ▪ doubly magic nucleus ▪ mass defect, binding energy, packing fraction ▪ nuclear fusion ▪ nuclear fission, neutron-induced fission ▪ thermal neutron ▪ fission fragment ▪ chain reaction

## QUERIES

**43.1** *(2) The Rutherford-Royds experiment.* The following experiment dates from 1909. The apparatus is shown below. The thin-walled glass ampoule, containing radon (an α emitter), is placed inside the thick-walled glass apparatus. The apparatus is evacuated, and the valve to the vacuum pump is then closed. After about a week, additional mercury is added to the apparatus until the liquid level reaches the bottom of the capil-

lary tube. A high-voltage discharge is then passed through the capillary tube, and the light emitted is observed with a spectrometer. Explain how the experiment proves that α particles are energetic helium ions.

**43.2** *(2) More or less.* In β⁻ emission, the mass number $A$ of the nucleus does not change. Does the atomic mass $M$ change? If so, does it increase or decrease?

**43.3** *(2) Random is as random does.* In the text, it is argued that the randomness of radioactive decay implies the existence of an exponential decay process with a half-life $\tau_{1/2}$. Make an argument in the opposite direction to show that the existence of an exponential decay process with a half-life $\tau_{1/2}$ implies random decay.

**43.4** *(2) Patience!* Most naturally occurring radioactive isotopes either have very long half-lives or are decay products of isotopes that have very long half-lives. Explain.

**43.5** *(2) No May-September here!* In radioactive dating, the half-life of the nuclide used should be roughly comparable to the age of the event to be dated. Explain.

**43.6** *(2) Ill-dysprosed.* The radioactive isotope dysprosium-156 constitutes about 0.052% of dysprosium. It decays by means of the chain

$$^{156}\text{Dy} \xrightarrow{\alpha} {}^{152}\text{Gd} \xrightarrow{\alpha} {}^{148}\text{Sm} \xrightarrow{\alpha} {}^{144}\text{Nd} \xrightarrow{\alpha} {}^{140}\text{Ce}.$$

The half-lives of the decays are, in order, $>1 \times 10^{18}$ y, $1.08 \times 10^{14}$ y, $7 \times 10^{15}$ y, and $2.1 \times 10^{15}$ y. Why is this decay chain not useful for dating?

**43.7** *(2) Taking the long view.* Nearly all meteorites are of two distinct kinds, called stony and iron. Stony meteorites are

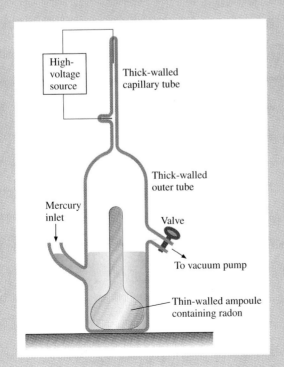

High-voltage source

Thick-walled capillary tube

Thick-walled outer tube

Mercury inlet

Valve

To vacuum pump

Thin-walled ampoule containing radon

rocklike, while iron meteorites consist mainly of iron and nickel. Meteorites of both kinds have been dated by the use of a variety of radioactive decay chains, including $^{238}U \rightarrow {}^{208}Pb$, $^{235}U \rightarrow {}^{207}Pb$, $^{87}Rb \rightarrow {}^{87}Sr$, and $^{40}K \rightarrow {}^{40}Ar$. All give the same age, 4.5 billion years. The same methods yield approximately the same maximum value for the ages of earth rocks. What can you infer about the formation of the solar system?

**43.8** *(3) Positive action.* Some nuclides decay by means of *positron emission.* When a nuclide emits a positron $(\beta^+)$, what happens to the atomic number? the nucleon number?

**43.9** *(3) Getting to the bottom of it.* Suppose that a nuclide is known to be stable and never decays radioactively. What can you say about the energy at the bottom of the nuclear potential well? Take $E = 0$ at $r = \infty$.

**43.10** *(4) Sunbathing.* Why is the sun a copious source of neutrinos?

**43.11** *(4) Alternative energy paths.* Although the chain of reactions given by Equations 43.19a, 43.19b, and 43.19c is the main source of energy for stars like the sun, another significant chain of reactions follows the first two steps (Equations 43.19a

and 43.19b) and then continues

$$^3_2He + {}^4_2He \rightarrow {}^7_4Be;$$

$$^7_4Be + \beta^- \rightarrow {}^7_3Li;$$

$$^7_3Li + {}^1_1H \rightarrow 2\,{}^4_2He.$$

Does this process yield a different amount of energy per unit of mass entering the chain than does the process of Equations 43.19a, 43.19b, and 43.19c? Explain.

**43.12** *(4) Moderation in some things, anyway.* Nuclear reactors depend on a chain reaction in which neutrons produced by fission induce further fission. The neutrons emerge from the fission process with considerable energy. In many reactors (such as those using $^{235}U$), the neutrons are *thermalized* to optimize further fission. This is done by surrounding the fuel elements, in which fission takes place, with *moderators* made of substances that do not absorb neutrons. Carbon (in the form of graphite) is used in some reactors, hydrogen (in the form of water) in others, and deuterium (in the form of heavy water, $D_2O$) in still others. Explain how the moderator works. In particular, why is water a more efficient moderator than graphite?

---

# PROBLEMS

## GROUP A

**43.1** *(2) Mass-energy equivalent.* In calculating the energy of nuclear reactions, it is often desirable to express the equivalence of mass and energy, $E = mc^2$, in terms of MeV and u rather than in J and kg. Show that

$$1\ u \leftrightarrow 931.5\ MeV.$$

You will find this equivalence useful in several of the problems in this chapter.

**43.2** *(2) Counting.* The mean life of a radionuclide is 1.42 h. When you place a sample of the radionuclide in a counting chamber, how long does it take for the counting rate to fall **(a)** by 10%—that is, to 90% of the original rate? **(b)** by 90%?

**43.3** *(2) Still counting.* The counting rate of a freshly made sample of a pure radionuclide decreases by 15% every 27.5 minutes. **(a)** What is the mean life of the nuclide, assuming that the daughter nuclide is stable? **(b)** What is the half-life?

**43.4** *(2) Radon.* You measure the counting rate of a detector placed near a sample of radon. When you come back at the same time the next day, you find that the counting rate has diminished by 18.2%. What are the mean life and half-life of the radon?

**43.5** *(2) Penetrating insight.* You make a cobalt-60 X ray of a steel casting and find that a good photograph requires a 20-s exposure. After exactly three years of service, you recheck the casting with the same setup. How long should your exposure be? The half-life of cobalt-60 is 5.24 years.

**43.6** *(2) Alas, poor Yorick!* Given one billion atoms of $^{209}Po$,

how many will decay in one day? The half-life of $^{209}Po$ is 103 years.

**43.7** *(2) The curie.* The half-life of $^{226}Ra$ is 1620 y. **(a)** How many atoms are there in 1 g of $^{226}Ra$? **(b)** How many atoms decay per second? This decay rate, or activity, is defined to be 1 *curie* (Ci). The curie is not an SI unit, and it is obsolescent.

**43.8** *(2) Middle-aged.* Recent carbon-14 dating of the famous Shroud of Turin indicates that the flax of which the linen cloth is made grew about 1350 C.E. If the proportion of carbon-14 to total carbon obtained from freshly made linen is $k_0$, what is the proportion $k$ of carbon-14 to total carbon obtained from the shroud?

**43.9** *(2) Pretty old.* An archeologist obtains charcoal fragments from a campsite occupied by primitive people in the Ice Age. If the carbon-14 concentration is $6.18 \times 10^{-2}$ of that obtained from modern charcoal, estimate the age of the site.

**43.10** *(2) Alchemist's dream, I.* The platinum isotope $^{190}_{78}Pt$ decays by $\alpha$ emission. **(a)** Find the mass number and atomic number of the daughter nuclide. **(b)** Consult the periodic table to find the chemical name of the daughter.

**43.11** *(2) Chain, I.* The decay chain for the naturally occurring radionuclide $^{232}_{90}Th$ consists of six $\alpha$ decays and four $\beta^-$ decays. What is the final product? (Hint: You do not need to know the order of the decays.)

**43.12** *(2) Chain, II.* The artificial radionuclide $^8_3Li$ decays by $\beta^-$ emission, quickly followed by $\alpha$ emission by the daughter nucleus. What is the final product?

**43.13** (2) *Radiumactivity, I.* The artificial radium isotope $^{227}_{88}$Ra decays by $\beta^-$ emission. **(a)** Find the nucleon number and atomic number of the daughter nuclide. **(b)** Consult the periodic table to find the chemical name of the daughter.

**43.14** (3) *Climb every mountain, I.* Use Coulomb's law to estimate the height of the potential barrier of Figure 43.6 for a uranium nucleus. Express your result in MeV. (Hint: The radius of the uranium nucleus is calculated in Example 43.3.)

**43.15** (4) *Alchemist's dream, II.* The mass of $^{190}_{78}$Pt (see Problem 43.10) is 189.9600 u. The energy of the $\alpha$ particle that it emits in decaying is 3.16 MeV. What is the mass of the daughter nuclide?

**43.16** (4) *Alchemist's nightmare.* In each of the following nuclear processes, what is $x$?

(a) $^{27}_{13}$Al + $^{1}_{0}$n $\rightarrow$ $x$ + $\alpha$.

(b) $^{19}_{9}$F + p $\rightarrow$ $^{16}_{8}$O + $x$.

(c) $^{55}_{25}$Mn + $x$ $\rightarrow$ $^{55}_{26}$Fe + n.

(d) $x$ + p $\rightarrow$ $^{22}_{11}$Na + $\alpha$.

(e) $^{27}_{13}$Al + $\alpha$ $\rightarrow$ $x$ + p.

**43.17** (4) *Radiumactivity, II.* The mass of the daughter nuclide formed in the $\beta^-$ decay of $^{227}_{88}$Ra (see Problem 43.13) is 227.0278 u. The energy of the $\beta$ particle is 1.30 MeV. Find the mass of the parent nuclide.

**43.18** (4) *Stability test.* The most common isotope of potassium is $^{39}_{19}$K, whose mass is 38.963 71 u. The most common isotope of chlorine is $^{35}_{17}$Cl, whose mass is 34.968 85 u. **(a)** Calculate the binding energy and packing fraction for each of these nuclides. **(b)** The mass of the $\alpha$ particle is 4.001 51 u. Is it possible for potassium-39 to decay spontaneously to chlorine-35 by $\alpha$ emission? Why or why not?

**43.19** (4) *Putting a little something in.* When deuterium is bombarded with $\gamma$ rays, the following reaction can take place:

$$^{2}_{1}\text{H} + \gamma \rightarrow {}^{1}_{1}\text{H} + \text{n}.$$

What is the minimum photon energy required to produce the reaction? Take the masses of the deuteron, proton, and neutron to be 2.014 10 u, 1.007 83 u, and 1.008 67 u, respectively.

**43.20** (4) *Fusion efficiency.* In the chain of fusion reactions summarized by Equation 43.20, what fraction of the rest mass entering the reaction emerges as energy?

**43.21** (4) *Fission fragments.* Consider the nuclear fission process

$$^{235}_{92}\text{U} + {}^{1}_{0}\text{n} \rightarrow {}^{127}_{50}\text{Sn} + ? + 3\,{}^{1}_{0}\text{n}.$$

Find **(a)** the atomic number, **(b)** the mass number, and **(c)** the chemical name of the nuclide represented by the question mark.

**43.22** (4) *Fission energy yield, I.* In the fission reaction 43.21, both of the products $^{134}$Sb and $^{100}$Nb are unstable. The former decays by a series of three $\beta^-$ emissions to the stable isotope $^{134}_{54}$Xe, whose mass is 133.9054 u. The latter decays in like manner to the stable isotope $^{100}_{44}$Ru, whose mass is 99.9030 u. Taking into account the mass of the six electrons produced, calculate the total energy yield for the overall process

$$^{235}_{92}\text{U} + {}^{1}_{0}\text{n} \rightarrow {}^{134}_{54}\text{Xe} + {}^{100}_{44}\text{Ru} + 2\,{}^{1}_{0}\text{n} + 6\,\beta^- + 6\,\bar{\nu}.$$

**43.23** (4) *Fission energy yield, II.* In one possible fission process, a 33-keV neutron (rest mass 1.008 67 u) collides with a $^{235}$U nucleus (235.0430 u), and the end products (after a complicated chain of decays) are a lanthanum nucleus (138.906 14 u), a molybdenum nucleus (94.905 84 u), two neutrons, and seven $\beta^-$ particles. **(a)** Calculate the total energy yield. **(b)** Is the kinetic energy of the initiating neutron significant in your calculation?

**43.24** (4) *Plutonium factory.* The artificial nuclide plutonium-239 ($^{239}_{94}$Pu) is produced by bombarding uranium-238 with neutrons, which leads to the following chain of reactions:

$$^{1}_{0}\text{n} + {}^{238}_{92}\text{U} \rightarrow {}^{239}_{92}\text{U};$$

$$^{239}_{92}\text{U} \rightarrow {}^{239}_{93}\text{Np} + \beta^- \quad (1.21 \text{ MeV}), \quad \tau_{1/2} = 23.5 \text{ min};$$

$$^{239}_{93}\text{Np} \rightarrow {}^{239}_{94}\text{Pu} + \beta^- \quad (0.72 \text{ MeV}), \quad \tau_{1/2} = 2.35 \text{ days}.$$

The mass of the uranium-238 nucleus is 238.0508 u; that of the plutonium-239 nucleus is 239.0522 u. **(a)** Is the reaction exothermic or endothermic? If it is exothermic, how much energy appears in the forms of $\gamma$ radiation and the kinetic energy of the daughter nuclei? **(b)** Is there a lower limit on the kinetic energy the neutron must have in order to initiate the reaction?

**43.25** (G) *Ionizing radiation.* $^{214}_{84}$Po emits $\alpha$ particles with energy 7.68 MeV. If such an $\alpha$ particle travels through air, it collides with air molecules, ionizing them. On the average, it takes 34 eV to produce a single ion pair—that is, an ion and the electron removed from it. **(a)** How many ions are produced per $\alpha$ particle emitted? **(b)** If this process is carried out in a chamber in which an electric field exists between two electrodes, the charged particles can be collected. If the electric field is not too large, no further ionization takes place, and the amount of charge collected is a measure of the total energy of the $\alpha$ particles passing through the chamber. What is the charge collected per $\alpha$ particle?

---

## GROUP B

**43.26** (2) *This is your half-life.* Derive Equation 43.5.

**43.27** (2) *Laboratory exercise.* You make the following measurements of the counting rate $n$ of a sample of radionuclide over a period of time:

| $t$ (h) | 0 | 3 | 6 | 9 | 12 | 15 |
|---------|------|------|-----|-----|-----|-----|
| $n$ (s$^{-1}$) | 2160 | 1260 | 760 | 420 | 240 | 180 |

Find the mean life and half-life of the nuclide.

**43.28** (2) *Natural mixture.* In uranium prepared freshly from natural ore, 0.006% is $^{234}$U ($\tau_{1/2} = 21.5 \times 10^5$ y), 0.71% is $^{235}$U ($\tau_{1/2} = 7.1 \times 10^8$ y), and the remainder is $^{238}$U ($\tau_{1/2} = 4.5 \times 10^9$ y). Find the share of the total radioactivity contributed by each isotope. Assume that no daughter products are present.

**43.29** (2) *Potassium-argon dating.* The radioactive isotope potassium-40 has a half-life $\tau_{1/2} = 1.277 \times 10^9$ y, whose order of magnitude is the same as that of the solar system. Most of the potassium in the solar system consists of two other, stable isotopes. Potassium-40 constitutes 118 parts per million of total potassium. Potassium-40 has two decay modes; 89.7% of the nuclei decay by $\beta^-$ emission to $^{40}$Ca. We are interested here in the other 10.3%, which decay by electron capture—equivalent to positron ($\beta^+$) emission—to the stable isotope $^{40}$Ar. Potassium is a very common component of rocks. If $^{40}$Ar is present in molten rock, it can escape readily because it is an inert gas. But once the rock solidifies and cools, the argon is trapped. In the laboratory, you can measure the amount of $^{40}$K currently present by radioactive measurement, and you can measure the amount of $^{40}$Ar by heating the rock and collecting the argon as it escapes. (No other argon is present in the rock.) **(a)** How can you measure the time $T$ that has passed since the rock solidified? Specifically, write an expression for the $^{40}$Ar/$^{40}$K ratio in terms of $\tau_{1/2}$. (Hint: Don't forget that only 0.103 of the potassium decays to argon.) **(b)** If the solidification occurred $2.5 \times 10^9$ years ago, what is the ratio $^{40}$Ar/$^{40}$K? (The potassium-argon dating method was first developed in 1946–47 by the American G. J. Wasserburg, a student of the famous radiochemist H. C. Urey, using improved mass spectrometry techniques developed during World War II in connection with the Manhattan Project.)

**43.30** *(2) Uranium-lead dating.* Consider the $^{238}$U $\rightarrow$ $^{206}$Pb decay chain described in Section 43.2, remembering that the half-life $\tau_{1/2} = 4.5 \times 10^9$ y for $^{238}$U is far greater than that of any other nuclide in the chain. When you make a chemical analysis of a very old rock, you find that $N_U$ atoms of $^{238}$U and $N_{Pb}$ atoms of $^{206}$Pb are present. **(a)** Show that the time that has passed since the formation of the rock is

$$t = \tau \ln\left(\frac{N_{Pb}}{N_U} + 1\right),$$

where $\tau$ is the mean life of $^{238}$U. **(b)** Uranium-containing rocks of many kinds from many parts of the earth have various proportions of $^{206}$Pb to $^{238}$U, but many of the oldest rocks show the same maximum value of $N_{Pb}/N_U = 1$. Estimate the age of the earth.

**43.31** *(2) Geophysical investigation.* You find that the ratio of $^{206}$Pb to $^{238}$U in a rock is 0.32. How old is the rock?

**43.32** *(2) Diminishing returns.* Suppose you can measure the carbon-14 content of a sample of fresh organic matter within ±1%. Within what span of years can you date a sample if its approximate age is **(a)** 300 years? **(b)** 1000 years? **(c)** 3000 years? **(d)** 10 000 years? **(e)** 30 000 years?

**43.33** *(3) Climb every mountain, II.* A uranium-238 nucleus emits a 4.2-MeV $\alpha$ particle. Using the result of Problem 43.14, estimate the thickness of the barrier through which the $\alpha$ particle must tunnel.

**43.34** *(4) Hefty.* **(a)** Calculate the density of the "nuclear fluid" of which the nucleus of $^{235}$U is made. **(b)** If the sun collapsed into nuclear fluid, what would its radius be? (Don't worry; this happens only to stars considerably more massive than the sun.)

**43.35** *(4) Born excited.* The mass of $^{235}$U is is 235.0439 u; that of $^{236}$U is 236.0457 u. **(a)** When a neutron having negligible kinetic energy collides with a $^{235}$U nucleus to produce a $^{236}$U nucleus, how much internal energy must the product nucleus possess? **(b)** This excess energy causes the nucleus to fission and fly apart. The two fragments are then further accelerated by the Coulomb repulsion of their respective positive charges. Assume that the fission has the simplest possible (and hypothetical) form

$$^{236}_{92}U \rightarrow 2 \; ^{118}_{46}Pd.$$

Estimate the radii of the two fragment nuclei, and calculate the additional energy conferred on the two by their mutual repulsion. **(c)** What is the total energy released in the fission? **(d)** Calculate the fraction of the original rest mass released as energy, and compare your result with the value for the fusion reaction of Problem 43.14.

**43.36** *(4) Stability of the $\alpha$ particle, I.* Example 43.4 deals with two possible ways for an $\alpha$ particle to disintegrate and shows that both are energetically unfavorable. There are only two more possibilities consistent with conservation of nucleon number:

$$^4_2He \rightarrow ^1_1H + ^3_1H$$

and

$$^4_2He \rightarrow ^3_2He + ^1_0n.$$

The mass of the hydrogen-3 *(tritium)* nucleus, or *triton*, is 3.015 51 u. The mass of the helium-3 nucleus is 3.016 03 u. Can either of the two possibilities occur spontaneously? If not, how much energy must be supplied in each case? Express your results in MeV.

**43.37** *(4) Stability of the $\alpha$ particle, II.* Another way the $\alpha$ particle might break up is by splitting into two deuterons ($^2$H), each consisting of a proton and a neutron. The mass of the deuteron is $m_d = 2.013\,47$ u. Using values given in Problem 43.36, calculate the energy required to split the $\alpha$ particle into two deuterons.

**43.38** *(4) Running short on fuel?* The sun radiates energy at the rate $3.9 \times 10^{26}$ W. Assume that nearly all of the mass of the sun is initially in the form of hydrogen. What is the lifetime of the sun if it radiates continuously at the current rate and if the fusion reaction is that given by 43.19 and 43.20?

**43.39** *(4) Star stuff.* In the centers of stars, heavier elements are built from hydrogen by a sequence of nuclear fusion reactions. The process provides the energy that the star eventually radiates to space. **(a)** Approximately where in the periodic table does the chain of reactions cease to yield energy? **(b)** How much energy is produced when 40 protons end up as a $^{40}_{20}$Ca nucleus? **(c)** When a large star uses up the "fuel" for nuclear fusion, having converted most of its matter into elements that cannot yield further energy by fusion, it collapses catastrophically and subsequently explodes as a *supernova*. (The entire collapse-explosion process takes about a minute.) Explain how the collapse makes the formation of heavy elements energetically possible.

## GROUP C

**43.40** *(2) Gasworks.* You have 1 g of radium (almost entirely $^{226}$Ra) in an evacuated container. After letting it remain undisturbed for several weeks, you use a vacuum pump to draw

off all the radon. How much radon will you obtain?

**43.41** *(2) Double decay, I.* Radionuclide A decays with mean life $\tau_A$ to yield radioactive daughter nuclide B. Daughter B, in turn, decays with mean life $\tau_B$ to yield nuclide C. Suppose you begin at $t = 0$ with a pure sample containing $N_0$ atoms of A. (a) At any time $t$, call the number of A atoms $N_A(t)$ and the number of B atoms $N_B(t)$. Show that the rate of change of $N_B$ (in other words, the rate of change of the number of B atoms) is given by

$$\frac{dN_B}{dt} = \frac{N_A}{\tau_A} - \frac{N_B}{\tau_B}.$$

(b) Show that this differential equation is satisfied by the solution

$$N_B = N_0 \frac{\tau_B}{\tau_B - \tau_A} (e^{-t/\tau_A} - e^{-t/\tau_B}).$$

(c) Find the time $t_m$ at which the quantity of B is a maximum.
(d) If $\tau_A = 1$ h and $\tau_B = 6$ h, find $t_m$.

**43.42** *(2) Double decay, II.* Consider again the decay process A → B → C described in Problem 43.41. When some time has elapsed, the ratio $N_A/N_B$ approaches a constant value. This condition is called *radioactive equilibrium.* (a) Show that, in radioactive equilibrium, you have

$$\frac{N_B}{N_A} = \frac{\tau_A}{\tau_B}.$$

(b) Generalize this equation to apply to a decay chain 1 → 2 → 3 → ⋯ → $n$ of arbitrary length. (c) The most common isotope of radium, $^{226}$Ra, is the sixth member of the fifteen-member decay chain that begins with $^{238}$U. Freshly mined pitchblende ore contains $3.6 \times 10^{-7}$ parts of radium for each part of uranium. (You can see why the Curies had to work so hard!) The half-life of $^{226}$Ra (which constitutes nearly all of the radium present) is 1620 years. What is the half-life of $^{238}$U?

**43.43** *(2) Double decay, III.* Consider again the decay process A → B → C described in Problem 43.41. If C is a stable nuclide, find an expression for $N_C$, the quantity of C present, as a function of time. Assume that at $t = 0$, $N_A = N_0$, and $N_B = N_C = 0$.

**43.44** *(4) Mirror nuclei.* Here is one way to estimate the size of a nucleus. Mirror nuclei are a pair of nuclei having the same nucleon number $A$ but differing in atomic number $Z$ by 1. Examples are $^3_1$H and $^3_2$He; $^7_3$Li and $^7_4$Be; $^{15}_7$N and $^{15}_8$O. In such a pair, one member has one neutron more and one proton less than the other. Assuming that the strong nuclear force is the same for a neutron as for a proton, the only difference between the binding energies of the two nuclei is that the nucleus with one proton more has greater electrostatic energy and thus a slightly greater mass.

The electrostatic energy of a sphere of radius $r$ having uniformly distributed charge $Ze$ is

$$U = \frac{3}{5} \frac{(Ze)^2}{4\pi\epsilon_0 r};$$

this is the result of Problem 25.40.

Oxygen-15, $^{15}_8$O, decays into nitrogen-15, $^{15}_7$N, by means of a positron-emission process. The total kinetic energy of the emitted particles (a positron and a neutrino) is 1.74 MeV. (a) Find the electrostatic energy difference between the two nuclides. (b) Using this information, calculate the radius of the nitrogen-15 nucleus. (c) Compare your result with that obtained using Equation 41.4, $r = r_0 A^{1/3}$, where $r_0 = 1.37 \times 10^{-15}$ m.

**43.45** *(G) Minimum meaningful mean life.* Although the range of mean lives of radionuclides is tremendous (Table 43.1), there is a limit below which it is no longer meaningful to speak of a nucleus at all. To see why, imagine a nucleus, having a mean life of $10^{-22}$ s, that decays by emitting an α particle. (a) Use the Heisenberg uncertainty principle to calculate the uncertainty in the energy of the α particle. (b) The maximum height of the potential barrier that holds a nucleus together is something like 10 MeV. Does it make sense to speak of a coherent nucleus whose mean life is $10^{-22}$ s or less? Why or why not?

# Quantum Systems II
## Many-Atom Systems

*Left:* A high-temperature-superconducting disk is cooled by liquid nitrogen in a Petri dish. A small permanent magnet floats above it. Because magnetic flux cannot penetrate into a superconductor, a supercurrent induced in the block by the presence of the magnet gives rise to an equal and opposite flux—"expels the flux of the magnet"—and repels the magnet.

*The world is so full of a number of things,*
*I'm sure we should all be as happy as kings.*

—ROBERT LOUIS STEVENSON, *"Happy Thought"*

# Introduction

Although the quantum properties of nature manifest themselves at the microscopic level, these properties have many important consequences at the macroscopic level as well. In this chapter, we study some of these consequences. Section 44.2 treats the properties of electrons in crystalline materials. In Section 44.3, we consider important electronic devices, such as transistors, whose operation is based on the principles developed in Section 44.2. Section 44.4 deals with systems in which the quantized properties of the microscopic world are directly exhibited on a macroscopic scale, as happens in the famous cases of *superfluidity* and *superconductivity*.

# Electrons in Crystalline Solids

In principle, you could build a diamond by bringing carbon atoms together one by one. As you did so, the atoms would arrange themselves in the regular array called a **crystal**. The form of the crystal is dictated by the way in which the atoms bond with one another. Carbon is *tetravalent*—that is, its valence is ±4. One way in which a carbon atom can form chemical bonds is the most symmetrical way possible for a tetravalent atom. All four bonds are equivalent, and they are oriented around the carbon atom at the corners of an imaginary tetrahedron centered on the atom. The same is true of all the atoms in the crystal. As a result, the crystal itself is tetrahedral. This tetrahedral arrangement, called the **diamond structure**, is shown in Figure 44.1. Other elements besides carbon tend to crystallize in the diamond structure; among them are silicon and germanium, which lie in the same column as carbon in the periodic table.

Let us return to our imaginary atom-by-atom construction of a crystal. When the first two atoms come together, there must be a change in the arrangement of their electrons. The two atoms are identical and thus have identical energy levels, the first six of which are occupied by the six electrons of each atom. When the two atoms approach each other, the two sets of electron wave functions overlap; the two atoms no longer have distinct electron clouds but instead merge into a single molecule. The energy levels of the molecule bear some resemblance to the atomic energy levels, but there is an important difference. Because of the exclusion principle, no two electrons having identical wave functions can occupy the same space. Consequently, the energy levels split, as shown in Figure 44.2. In place of the single atomic energy level, there are now two sublevels. (A similar argument underlies the fine-structure splitting that occurs in individual atoms; see Section 42.5.)

When a third atom is brought close to the first two, each atomic energy level must split into three sublevels to accommodate the three sets of electron wave functions that would otherwise be identical. A four-atom molecule requires a fourfold splitting, and so forth.

A crystal contains a very large number of identical atoms. [For instance, a one-carat

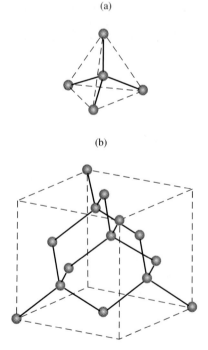

**FIGURE 44.1** The diamond structure. (*a*) Every atom lies at the center of a tetrahedron whose corners are the four neighboring atoms. (*b*) The structure built by repetition of the basic tetrahedron.

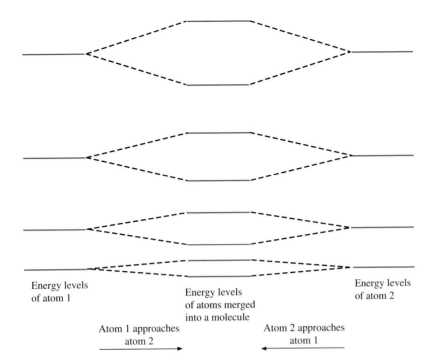

Energy levels of atom 1

Energy levels of atoms merged into a molecule

Energy levels of atom 2

Atom 1 approaches atom 2 →

← Atom 2 approaches atom 1

(200-mg) diamond is not very big but contains about $10^{22}$ atoms.] Each of the original atomic energy levels is now split into so many sublevels, spaced so closely together that it is no longer profitable to think of the levels individually. To a very good approximation, all energies lying in the range between the **band edges**—the uppermost and lowermost sublevels—are accessible to electrons. That is, each sharply defined single-atom energy level has broadened into an **energy band**, as shown in Figure 44.3a. We say that the levels constituting a band are *quasicontinuous*. The ranges of energy between bands, in which no energy levels lie, are called **band gaps** or **forbidden gaps**.

In general, the bands corresponding to higher energy levels broaden more than those corresponding to lower energy levels. This reflects the fact that the outer electrons of neighboring atoms interact more intimately than the inner ones. Indeed, some adjacent bands may *overlap*, as shown in Figure 44.3b.

FIGURE 44.3 Energy-band levels for two hypothetical covalent crystals. Each energy band corresponds to a sharp single-atom energy level. (*a*) The bands do not overlap, but are separated by band gaps. (*b*) Some of the bands overlap.

(a)

(b)

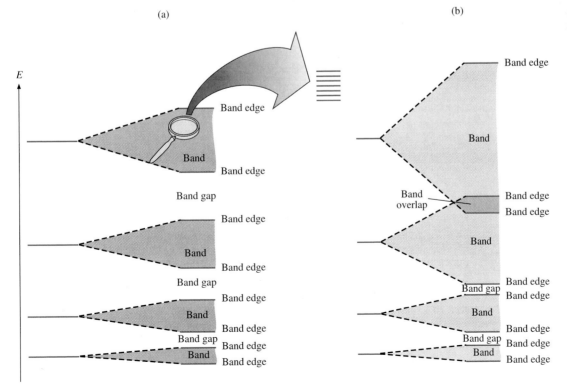

Just as in the hydrogen atom, any energy level in the crystal may be either occupied by an electron or empty. An empty higher level can be filled by exciting an electron from a lower level. But let us focus for the moment on the ground state. In the ground state, all levels up to a certain energy are filled, and all higher levels are empty. Where does the boundary between filled and unfilled levels lie?

## Band Structure of Dielectrics

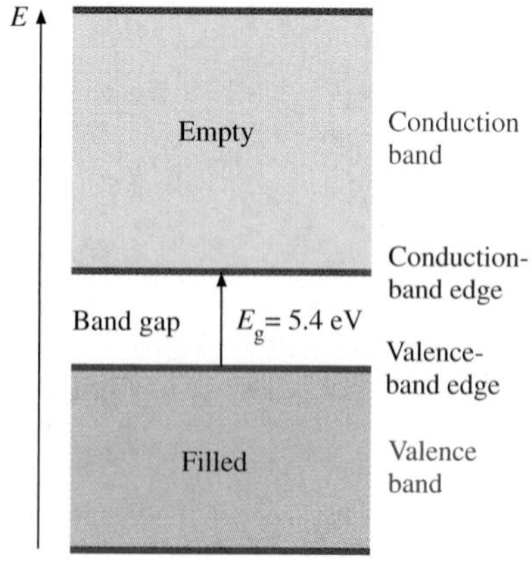

For diamond, the band structure is something like that of Figure 44.3a, in which there is no overlap. If the crystal contains $N$ atoms, each band has $N$ sublevels. There are just enough electrons to fill all the sublevels of all the bands corresponding to the six ground-state energy levels of the carbon atom. Thus the first six bands are full, and the seventh and higher bands, representing excited states, are empty. Figure 44.4 shows the upper-most filled band, called the **valence band**, and the lowermost empty band, called the **conduction band**. For diamond, the band gap between these two bands has ''width'' $E_g = 5.4$ eV.

We can now see why diamond is a poor conductor of electricity—a dielectric. The valence band is completely full of electrons. If charge is to be transferred from one part of the crystal to another, an electron must be set into motion; that is, its energy must be increased slightly. But there are no higher-energy states available in the valence band because every state is filled with an electron and the exclusion principle holds. An electron near the band edge may be moved to a higher-energy state—but only if its energy is increased by 5.4 eV or more. Thermal energy will not suffice; at room temperature, $kT = 0.025$ eV.

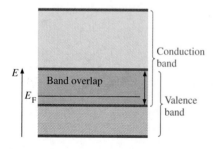

**FIGURE 44.4** Valence and conduction bands of diamond. The band-gap ''width'' is $E_g = 5.4$ eV.

## EXAMPLE 44.1

Suppose an energy range $dE$ just below the edge of the valence band contains $n\, dE$ electrons, where $n$ is the number of electrons per unit (energy·volume). An equal energy range just above the edge of the conduction band contains $n'\, dE$ electrons. Find the ratio $n'/n$ in diamond at room temperature, $T = 300$ K.

**SOLUTION:** Because electrons are subject to the Pauli exclusion principle, their thermal energy is not correctly described by the Maxwell-Boltzmann energy distribution function considered in Chapter 18. However, the Maxwell-Boltzmann distribution is an excellent approximation if $E_g \gg kT$, as is the case here. Specifically, the number of electrons having energy $E$ is proportional to what is called the Boltzmann factor $e^{-E/kT}$ (Problem 18.68). Call the energy at the top of the valence band $E_v$ and the energy at the bottom of the conduction band $E_c$. You have

$$\frac{n'\, dE}{n\, dE} = \frac{e^{-E_c/kT}}{e^{-E_v/kT}} = e^{-(E_c - E_v)/kT},$$

or, because $E_c - E_v = E_g$,

$$\frac{n'}{n} = e^{-E_g/kT}.$$

This gives you

$$\frac{n'}{n} = e^{-(5.4\ \text{eV})/(0.025\ \text{eV})} = 1.6 \times 10^{-94}.$$

Of the electrons present, the fraction excited to the conduction band of diamond at room temperature is truly negligible.

## Band Structure of Metals

Metals are excellent conductors of electricity. Their band structure is typified by Figure 44.3b, in which the valence and conduction bands overlap. Imagine a metal from which all the electrons are first removed and then slowly poured back in. Because of the overlap, the conduction band begins to fill before the valence band is full. When all the electrons are in place, they fill all states up to an energy $E_F$, called the **Fermi level**. In the energy range where the Fermi level lies, states are available immediately above as well as immediately below it (Figure 44.5). It is often useful to think analogously of a *Fermi sea*, of depth $E_F$. Electrons near the surface of the Fermi sea, with energy near $E_F$, have available to them empty states whose energy is only very slightly greater than their present energy. You may think of the small thermal excitation $kT$ of the electrons in terms of a set of ''ripples'' on the surface of the sea. Although the height of the

**FIGURE 44.5** Band overlap in a metal. The Fermi level $E_F$ lies in an energy region with many available states.

ripples is very small compared with the depth of the sea, there is no difficulty in providing electrons with enough energy to carry electric current.

## Band Structure of Semiconductors

A semiconductor is simply a dielectric with a relatively small band gap between the valence and conduction bands. The best-known semiconducting material is silicon, which lies just below carbon in the periodic table and behaves like carbon in many ways. In particular, silicon forms crystals having the diamond structure. However, the band gap is only 1.1 eV, rather than the 5.4-eV gap of diamond.

If we repeat the calculation of Example 44.1 for pure silicon at room temperature, we find

$$\frac{n'}{n} = 7.8 \times 10^{-20}. \tag{44.1}$$

This number is still quite small but is vastly greater than the ratio $1.6 \times 10^{-94}$ obtained for diamond. A more detailed calculation shows that the concentration of current carriers having sufficient thermal energy to be excited across the band gap is about $10^{15}$ m$^{-3}$ and the resistivity is $3 \times 10^4$ $\Omega \cdot$m. For comparison, silver (the metal that is the best room-temperature conductor) has a concentration of current carriers about $10^{12}$ times greater and an electrical resistivity about $10^{12}$ times smaller.

## Electrons and Holes

Every electron that is excited across the energy gap from the valence band to the conduction band leaves behind a vacant energy state in the valence band. This vacant state can be filled by an electron in an adjacent state, but that leaves a new vacant state. To visualize the process, think of the following analogy, which is inexact but adequate for our purposes. You fill a bottle nearly full of water and stopper it. If you turn the bottle upside down, you see a bubble rising to the top. Now, a bubble is really "no water." The bubble doesn't defy gravity in rising to the top. What is happening is that all the water moves down as far as possible. But the process is much easier to visualize in terms of movement of the bubble rather than of the water. The water corresponds to the electrons, the bubble to the vacant state.

With this analogy in mind, let us return to the electrons in the nearly filled valence band. When an electric field is applied, these electrons move in the direction $-\hat{\mathscr{E}}$ because they have negative charge. The vacant state is therefore displaced in the opposite direction, $\hat{\mathscr{E}}$. As far as transport of electric charge is concerned, it is as though a particle having *positive* charge is moving under the influence of the electric field. This imaginary particle is called a **hole**.

In pure silicon (or any other pure semiconductor), every excitation of an electron into the conduction band leaves a hole in the valence band. Thus the concentration $n_i$ of electrons in the conduction band is equal to the concentration $p_i$ of holes in the valence band. Both electrons and holes carry electric current. This electrical conduction mechanism is called **intrinsic conduction**, and a semiconductor that exhibits intrinsic conduction is called an **intrinsic semiconductor**. Figure 44.6 shows the conduction process in a highly schematic way. An electric field $\mathscr{E}$ is imposed on a semiconductor in which one electron is free to carry current in the conduction band and one hole is free to carry current in the valence band. The negative electron moves with velocity $\mathbf{v}_e$ in the direction opposite $\hat{\mathscr{E}}$, while the hole moves with velocity $\mathbf{v}_h$ in the direction $\hat{\mathscr{E}}$. As you learned in Section 27.5, charge carriers of opposite sign moving in opposite directions carry current in the same direction—the direction $\hat{\mathscr{E}}$.

The velocities of the electrons and holes are affected by the periodic variation of electric field that they experience as they move through the regular array of atoms that makes up the crystal. It turns out that this periodic variation can be accounted for by treating the electron as though it has an **effective mass** $m^*$, rather than the free electron mass $m$, and treating the hole similarly. For electrons in most conducting solids, $m^*$ is less than $m$. For silicon, $m^* \simeq 0.2m$. Hole effective masses are typically closer to $m$.

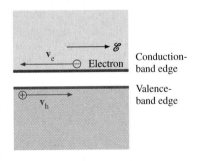

**FIGURE 44.6** Intrinsic conduction by an electron and a hole in a semiconductor crystal subjected to an electric field $\mathscr{E}$.

We can express both $n_i$ and $p_i$ in terms of the Boltzmann factor. By symmetry, the Fermi level must lie in the middle of the band gap, at a "distance" $E_g/2$ from each band edge. In contrast with the situation in metals, no energy states are available to either electrons or holes in the immediate vicinity of the Fermi level. Nevertheless, we can imagine electron-hole pairs coming into being at the Fermi level, the electron going to the conduction band and the hole to the valence band.* Because $E_g/2 \gg kT$, we can write the proportionality

$$n_i = p_i \propto e^{-E_g/2kT}. \tag{44.2}$$

The electrical conductivity is proportional to the number of current carriers present. Equation 44.2 correctly predicts the very rapid increase in the conductivity of an intrinsic semiconductor with increasing temperature.

### Hydrogenlike Impurities

The intrinsic conduction mechanism establishes the minimum conductivity of a semiconductor. However, we are usually interested in semiconductors into which very small, carefully controlled quantities of impurities have been introduced. This process of **doping** can affect the conductivity of a semiconductor very dramatically.

Suppose we dissolve in a crystal of silicon a few parts per million (ppm) of an element from Group V of the periodic table, such as phosphorus, whose valence is $+5$. Four of the five valence electrons are available to participate in the tetrahedral bonding of the diamond structure. Thus the phosphorus atom substitutes for a silicon atom in the crystal. For this reason, it is called a *substitutional impurity*.

The nucleus of the phosphorus atom contains one more proton than do the nuclei of its neighboring silicon atoms. When the phosphorus atom is bound into the crystal, its extra (unbonded) electron still experiences a Coulomb attraction owing to the single extra positive charge on the nucleus. Consequently, the ground state of this electron is a *hydrogenlike* orbit about the nucleus. However, the binding of the electron to its nucleus is much weaker than that of the electron in hydrogen because the dielectric constant $\kappa$ of the solid reduces the electric field that binds the electron to the nucleus. For silicon, $\kappa = 11.7$.

With this point in mind, let us reconsider the equation that expresses the energy of the first Bohr orbit. For hydrogen, Equation 41.16 gives us $E_1 = -me^4/8\epsilon_0^2h^2 = -13.6$ eV. The negative of this quantity, $-E_1$, is the ionization energy of hydrogen. For phosphorus in silicon, the ionization energy must be written in the modified form

$$-E_d = \frac{m^*e^4}{8(\kappa\epsilon_0)^2h^2} = \frac{m^*}{m}\frac{1}{\kappa^2}(-E_1)$$
$$= 1.5 \times 10^{-3} \times 13.6 \text{ eV} \simeq 0.02 \text{ eV}.$$

This quantity is the **donor ionization energy**.

At room temperature, we have $kT = 0.025$ eV. That is, the thermal energy is comparable to the donor ionization energy for phosphorus in silicon. The Boltzmann factor $e^{E_d/kT}$ is therefore comparable to 1 and not a very small number, as it is for intrinsic electrons excited across the band gap (Equations 44.1 and 44.2). Consequently, a substantial proportion of the donor electrons are ionized. They can thus wander freely through the conduction band and can carry current under the influence of an electric field imposed on the semiconductor.

Impurities such as phosphorus that contribute conduction electrons to the semiconductor are called **donor impurities**. A sample containing donor impurities has more conduction electrons than holes—usually many more. Consequently, the electrons in such a sample are called **majority carriers**, and the holes are called **minority carriers**. The donor-doped semiconductor is said to be **n-type**. (The *n* stands for "negative," but the unabbreviated form is never used.)

---

*Just as for a bubble, a hole moving downward implies an increased energy.

The donor atoms act like an array of shallow potential wells from which donor electrons are liberated into the conduction band by their thermal energy. This is shown in Figure 44.7. The Fermi level lies close to the donor energy level; that is, the energy required to ionize an electron from its donor atom to the conduction band is the small donor ionization energy. Consequently, many or most of the donor electrons are ionized and free to move throughout the crystal.

## Acceptors

It is also possible to dope silicon substitutionally with trivalent elements, such as aluminum. An aluminum atom assuming a position in the crystal lattice lacks one of the four electrons required to bind it into a tetrahedron. It must "borrow" this electron from the surroundings, leaving a hole. The positive hole is bound to the aluminum nucleus by a net *negative* charge. This net negative charge is due to the fact that the aluminum nucleus possesses one proton fewer than the silicon nucleus. Thus the hole is bound to the aluminum atom in a hydrogenlike orbit.

As with the electrons in hydrogenlike orbits around donor atoms, the ionization energy $E_a$ of the hole is quite small, and the hole is likely to be ionized at room temperature. It is thus free to move through the crystal. Because impurities such as aluminum remove electrons from the semiconductor, they are called **acceptor impurities**. A sample containing acceptor impurities has more holes than electrons—usually many more. Consequently, the holes in such a sample are the majority carriers, and the electrons are the minority carriers. The acceptor-doped semiconductor is said to be *p*-type (for "positive").

The acceptor atoms act like an array of shallow potential wells from which acceptor holes are liberated into the valence band by their thermal energy. This is shown in Figure 44.8. (Again, remember that as a hole moves downward in the figure, its energy increases.) The Fermi level lies close to the acceptor energy level.

Donor and acceptor impurities have a profound effect on the electrical conductivity of a semiconductor. This is because the donor and acceptor energies are so much smaller than the band-gap energy. At room temperature, only a tiny fraction of the intrinsic electrons are excited across the band gap from the valence band to the conduction band (leaving holes in the valence band). But a large proportion of the donor electrons in an *n*-type semiconductor are excited into the conduction band. Similarly, a large proportion of the acceptor holes in a *p*-type semiconductor are excited into the valence band. We have seen that the room-temperature carrier concentration in intrinsic silicon is only about $10^{15}$ m$^{-3}$. We can increase the electron (or hole) concentration—and thus the conductivity—a millionfold by introducing a donor (or acceptor) impurity at a concentration of $10^6 \times 10^{15}$ m$^{-3}$ = $10^{21}$ m$^{-3}$. Silicon contains about $5 \times 10^{28}$ atoms m$^{-3}$. The necessary impurity concentration is therefore of order 0.1 ppm.

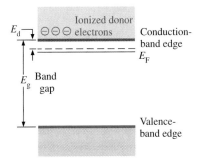

**FIGURE 44.7** The presence of donor impurities introduces a donor level at the small energy $E_d$ below the conduction-band edge. Electrons are readily ionized from the donor level (the hydrogenlike ground state) into the conduction band. The Fermi level lies close to the donor energy.

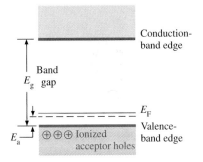

**FIGURE 44.8** The presence of acceptor impurities introduces an acceptor level at the small energy $E_a$ above the valence-band edge. Holes are readily ionized from the acceptor level (the hydrogenlike ground state) into the valence band. The Fermi level lies close to the acceptor energy.

## Semiconductor Devices

**S E C T I O N  44.3**

Suppose you could join a crystal of *p*-type silicon and a crystal of *n*-type silicon together to form a *p-n* **junction**, as shown in Figure 44.9a.* In one form or another, the *p-n* junction underlies the function of almost all semiconductor devices.

On the *n* side of the junction, electrons are the majority carriers. On the *p* side, there are very few electrons. Consequently, the incessant motion of the electrons in the conduction band results in a *diffusion* of more electrons across the junction to the left than to the right.

---

*In practice, this is most commonly done by starting with a sample of (say) *p*-type silicon and using a carefully controlled high-temperature diffusion process to transport a donor dopant into a narrow region near the surface. At sufficiently high concentration, the donor atoms *compensate* and then *overcompensate* for the acceptor atoms, converting the region from *p*-type into *n*-type. The junction thus formed is quite narrow; the transition from *n*- to *p*-type occupies 1 μm or less.

When an electron diffuses into the $p$ region, it soon encounters one of the many holes that are the majority carriers in that region. The encounter results in **recombination** of the electron and the hole—a process in which an electron disappears from the conduction band and a hole disappears from the valence band. The same process takes place in reverse, as holes from the $p$ region diffuse into the $n$ region and recombine with the majority electrons.

The recombination of charge carriers as they move across the $p$-$n$ junction results in a depletion of carriers—both electrons and holes—in the vicinity of the junction. The **depletion region** is shown in Figure 44.9b. On the $n$ side, the depletion results in the appearance of a fixed net positive *space charge* because the relatively positive donor atoms are no longer balanced by the electrons they supply. On the $p$ side, there is a fixed net negative space charge because the relatively negative acceptor atoms are no longer balanced by the holes they supply.

The space charge gives rise to a **contact potential difference** between the two sides of the junction. The presence of the contact potential difference makes possible the smooth transition of the Fermi level from its position near the acceptor level on the $p$ side to its position near the donor level on the $n$ side. You can see from Figure 44.9c why the effect of the contact potential difference is often called "band bending."

## The Junction Diode

Without further embellishments, the $p$-$n$ junction functions as a **rectifier**—a device that permits the passage of electric current in one direction only. The device in Figure 44.10a, called a *junction diode*, is subjected to **reverse bias** by the external source of electromotive force. The negative terminal is connected to the $p$ side and the positive terminal to the $n$ side. The result is an augmentation of the band bending that arises from the contact potential difference, as shown in Figure 44.10b. Even fewer majority carriers than in the unbiased condition can surmount the potential barrier now, and negligible current passes through the junction. Moreover, the depletion region is widened by the application of the reverse bias. Because the depletion region has large electrical resistivity, the current is reduced still further.

In Figure 44.11a, the junction is subjected to **forward bias**. Now the imposed potential difference *reduces* the band bending, as shown in Figure 44.11b. Moreover, the depletion region is narrowed, thus reducing its electrical resistance. The results are an increase in the diffusion of majority carriers across the junction and an increase in the current.

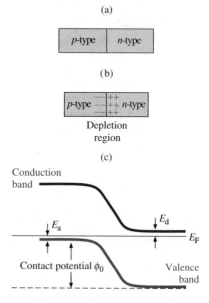

**FIGURE 44.9** (*a*) A $p$-$n$ junction. (*b*) Formation of the depletion region. (*c*) The space charge in the depletion region results in a contact potential difference. Through the depletion region, there is a "band bending" caused by the change in the position of the Fermi level, from $E_a$ above the valence-band edge to $E_d$ below the conduction-band edge.

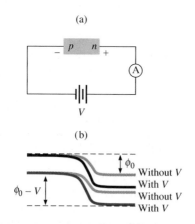

**FIGURE 44.10** (*a*) A $p$-$n$ junction diode on which a reverse bias is imposed. (*b*) Compared with the unbiased state (light-colored curves), the band bending is augmented. There is negligible current across the junction.

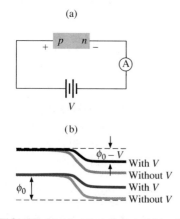

**FIGURE 44.11** (*a*) A forward bias is imposed on the $p$-$n$ junction. (*b*) Compared with the unbiased state (light-colored curves), band bending is reduced. Significant current can pass through the junction.

Because the number of electrons and holes available to carry current is an exponential function of their energy (measured from the Fermi level), the small bending of the bands produced by the applied potential results in a large change in the current. A typical current-versus-voltage curve for a junction diode is shown in Figure 44.12. The ability of the diode to pass current in one direction only has a great variety of applications in electronic circuits.

**Conventional Diode Symbol**

The arrowhead in the symbol

points in the direction in which the diode allows the passage of conventional current. That is, current will pass through the diode when the arrowhead points from + to −.

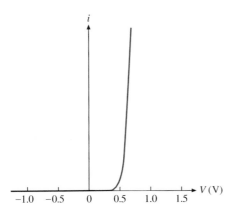

**FIGURE 44.12** A current-versus-voltage plot for a typical *p-n* junction diode.

## The Light-Emitting Diode

The energy released when an electron-hole pair recombines may appear in a variety of forms. Silicon is an example of an *indirect-gap semiconductor*. In such semiconductors, the energy released in recombination increases the vibration of the ions of the crystal lattice. That is, the energy appears as heat. However, in *direct-gap semiconductors*, such as gallium arsenide (GaAs), gallium phosphide (GaP), and indium antimonide (InSb), the energy of recombination is released in the form of photons of energy $E \simeq E_g$. The wavelength of the emitted light thus depends on the band-gap energy:

$$\lambda = \frac{hc}{E_g}.$$

Semiconducting materials are available with band-gap energies corresponding to infrared and visible wavelengths. In particular, GaAs and GaP can be mixed in any proportions. The band gap of the resulting semiconductor can be adjusted to produce any desired wavelength in the near-infrared and much of the visible spectrum by adjusting the arsenic concentration in gallium arsenide phosphide, represented by $x$ in the chemical formula $GaAs_xP_{1-x}$.

By fabricating a *p-n* junction in gallium arsenide phosphide, we can make a **light-emitting diode** (LED for short). By imposing a forward bias on such a diode, we force many carriers across the junction, where they recombine and emit the desired photons. LEDs have wide application in all sorts of displays, such as those on clocks, television sets, and cash registers. They have become so cheap that they have largely supplanted the incandescent light bulbs in automobile high-level stop lights.

If mirrors are placed on the ends of a properly designed LED, it may emit light as a **diode laser**. Diode lasers are used widely in optical-fiber transmission systems and in so-called *optoelectronic* devices.

## The Transistor

The **transistor** was invented in 1947. Though there are many variants, all transistors serve the same basic function. A small applied voltage or current is used to control a much larger voltage or current. Here we describe one of the most common and important types of transistor, the *metal-oxide-semiconductor field-effect transistor*, usually referred to by the acronym **MOSFET** (Figure 44.13). The device is constructed on a base of lightly doped *p*-type silicon crystal, called the **substrate**. (It is also possible to begin with an *n*-type substrate, in which case the device is a *p*-channel MOSFET.) The sub-

**The Transistor: A Direct Application of Science to Technology**

The inventors of the transistor were the American physicists John Bardeen (1908–1991), Walter Brattain (1902–1987), and William Shockley (1910–1989), who worked together at Bell Telephone Laboratories in a project carefully planned toward this goal. They shared a Nobel Prize in 1956. Bardeen subsequently shared in another Nobel Prize (1972) for his contribution to the theory of superconductivity. Transistors have almost completely replaced the older vacuum tubes, which are much bulkier and less reliable, consume much more power, and are unsuitable to many of the applications made possible by the semiconductor revolution.

**FIGURE 44.13** An *n*-channel MOSFET.

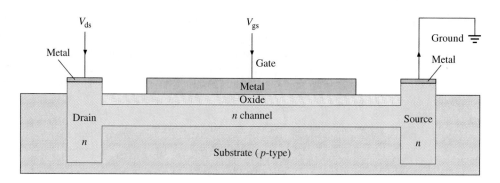

*This bicyclist relieves the tedium of maneuvering through heavy traffic by means of a small, lightweight entertainment system consisting of thousands of the solid-state quantum devices we call transistors.*

strate is coated with an organic material called photoresist, whose solubility in a suitable solvent can be changed by exposure to light. A high-resolution projector is then used to illuminate the photoresist-coated surface in the desired pattern; when the photoresist is dissolved and washed away, the pattern is reproduced as an array of coated and uncoated parts of the semiconductor surface. The entire surface is then coated with an *n*-type dopant, and the crystal is placed in an oven where the dopant diffuses into the uncoated part of the substrate. By this means, two heavily doped *n*-type regions are produced, the **source** and the **drain**. In similar manner, the source and the drain are joined by a much thinner *n*-type layer called the **channel**.

After further treatment with photoresist, the *n* channel is exposed to oxygen at elevated temperature. A thin surface layer of silicon oxidizes to SiO, which is a dielectric. In a final operation, thin layers of metal are deposited on the source, the drain, and the channel. The metal layer above the channel is called the **gate**.

Suppose that the source is grounded and a positive potential $V_{ds}$ is applied to the drain. (Ignore the gate for the moment.) Current will pass from the drain to the source through the channel (Figure 44.14a). The boundary between the *n* channel and the *p*-type substrate is an extended *p-n* junction. Consequently, the lower part of the channel is a depletion region through which little or no current passes. The upper part of the channel is effective in carrying current between drain and source. The entire *n* channel is at a positive potential with respect to ground. Hence, the entire *p-n* junction is reverse biased. But owing to the *iR* drop from the drain to the source, the reverse bias varies along the channel; the bias is large at the drain end and zero at the source end. Consequently, the depletion region is thickest at the drain end and thinnest at the source end. The effective conducting channel is thinnest at the drain end and thickest at the source end.

We can vary the thickness of the depletion region by applying a potential $V_{gs}$ to the gate. Suppose we apply a positive potential, as shown in Figure 44.14b. This positive potential reduces the reverse bias of the substrate-channel *p-n* junction and thus narrows the depletion region. As a result, the effective conducting channel is widened, and its resistance decreases. For fixed $V_{ds}$, the current $i_{ds}$ increases.

Alternatively, we can apply a negative potential to the gate, as shown in Figure 44.14c. This potential increases the reverse bias of the substrate-channel junction and thus widens the depletion region. As a result, the effective conducting channel is narrowed, and channel resistance increases. For fixed $V_{ds}$, the current $i_{ds}$ decreases.

The oxide layer between the gate and the channel is an excellent insulator; a typical resistance is $10^{15}$ $\Omega$, and $10^{18}$ $\Omega$ can be achieved in special applications. Thus a potential applied to the gate results in a very small gate current $i_{gs}$ and very small power dissipation $i_{gs}V_{gs}$. This small input power controls a much greater output power $i_{ds}V_{ds}$. Depending on the desired application, the transistor can be operated either as a proportional controller, in which $i_{ds} \propto V_{gs}$, or as a switch, in which $V_{gs}$ is applied in such a way as to turn $i_{ds}$ on or off.

MOSFETs can be made exceedingly small. By application of modern techniques, it is possible to fabricate large numbers of MOSFETs on a single silicon chip, together with the interconnections needed to make, quite cheaply, a highly complex electronic circuit that dissipates very little power. The technology required to do this has improved rapidly since the mid-1960s and will probably continue to do so. This steady improvement has made the computer revolution possible.

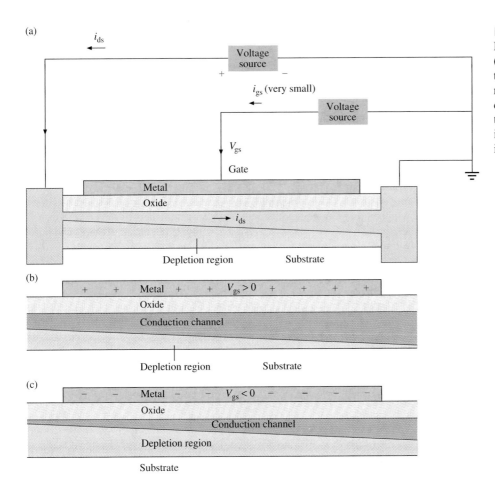

(a)

(b)

(c)

FIGURE 44.14 (a) An n-channel MOSFET used as an amplifier. (b) Application of a positive potential $V_{gs}$ to the gate results in a narrowing of the depletion region in the channel. (c) Application of a negative potential $V_{gs}$ to the gate results in a widening of the depletion region in the channel.

# Superfluidity

Of all the substances that are gaseous at room temperature, helium liquefies at the lowest temperature: 4.22 K ($-268.93°C$) at atmospheric pressure. This is why helium was the last of the once so-called permanent gases to be liquefied.

The structure of the helium atom suggests why helium is so difficult to liquefy. The atom has a closed outer shell, with two electrons. These electrons are not shielded by any inner electrons; they are therefore tightly bound to the nucleus and difficult to ionize. Consequently, the helium atom has very little propensity to combine with other atoms, and the force between it and neighboring helium atoms is very weak. It thus requires only a little thermal energy to disrupt the weakly bound liquid into a gas.

**Kamerlingh Onnes**
Liquefaction of hydrogen (1906) and helium (1908) was first accomplished by the Dutch physicist Heike Kamerlingh Onnes (1853–1926). The renowned low-temperature laboratory at the University of Leiden is named after him. He was awarded the Nobel Prize in 1913.

**EXAMPLE 44.2**

Estimate the energy that binds a helium atom to other atoms in liquid helium.

**SOLUTION:** Although the details may be complex, you know that the energy required to remove a helium atom from the liquid must be roughly comparable to the thermal energy $kT$ at the boiling point. So you have

$$E \simeq kT = 1.38 \times 10^{-23} \text{ J/K} \times 4.22 \text{ K} = 5.8 \times 10^{-23} \text{ J,}$$

or about $3.6 \times 10^{-4}$ eV.

Although liquid helium at its boiling point has very low density for a liquid—about 0.15 that of water—its other properties present no particular surprises. As with all other liquids, its temperature can be reduced by using a vacuum pump to reduce the pressure of the gas above it as it continues to boil. As the temperature of liquid helium is thus reduced, a critical temperature called the **lambda point**, $T_\lambda = 2.18$ K, is reached. At this point, liquid helium begins to display a wide variety of remarkable properties. We list a few here:

1. When flowing through a fine capillary, the liquid exhibits *zero* viscosity.

2. The bubbling observed throughout all ordinary liquids as they boil suddenly disappears in helium as the temperature decreases through the lambda point. This disappearance of bubbling is due to the onset of *infinite* thermal conductivity. Because of the hydrostatic pressure at any point below the surface, the boiling point is slightly lower at the surface than anywhere else. Because heat can be transmitted through the liquid to the surface with infinite ease, only the surface boils.

3. A heat pulse produced within the liquid (say, by passing a small electric current pulse through a resistor immersed in the liquid) does not spread out as heat ordinarily does, the source temperature falling gradually and $dT/dr$ remaining negative everywhere at all times. Rather, the heat pulse propagates *as a pulse* without attenuation, like a sound pulse propagating through a frictionless medium. The propagation speed is about 20 m/s. Because the heat pulse propagates like sound, the phenomenon is called *second sound*.

4. In an open container, the liquid forms a thin film on the walls and migrates up and out to the lowest available point outside the container. If the container is not immersed in another container of liquid helium, the liquid will drip off the bottom. If the container is immersed in another container of liquid helium, the migration will continue until the level is the same in both containers. This is called the *film effect*.

On account of these and other properties, liquid helium below its lambda point is called a **superfluid**. These properties can be accounted for only by quantum theory. That is, liquid helium displays quantum effects, which we usually think of as purely microscopic, on a macroscopic scale.

The key to understanding superfluidity lies in the fact that the spin of a helium atom is zero. (The energy of the atom is minimized when the six spin-$\frac{1}{2}$ particles that constitute the atom—two protons, two neutrons, and two electrons—are aligned so that each particle is aligned with its spin—and thus its spin magnetic moment—antiparallel to that of its corresponding particle.) Particles having spin 0, 1, 2, . . . do not obey the Pauli exclusion principle. Quite to the contrary, any number of such particles may have exactly the same ground-state wave function. Thus, all the particles—here helium atoms—can lie in a single ground state. (Contrast this with the spin-$\frac{1}{2}$ electrons in a semiconductor. Because they conform to the exclusion principle, some of the electrons must have energy comparable to the Fermi energy, which is much greater than the thermal energy $kT$.)

When the temperature of liquid helium is lower than the lambda-point temperature, the liquid begins to "condense" into what is called a *Bose-Einstein liquid* or a *Bose gas*. (Both "liquid" and "gas" are used loosely in this terminology; a distinction is not intended.) At temperatures just below the lambda point, only a small fraction of the liquid is in this state; with decreasing temperature, the fraction increases until all of the liquid is in the Bose-gas state at $T \rightarrow 0$ K. Because the entire liquid is described by a single ground-state wave function, energy can be added only by raising the entire liquid to its first excited state. Although the energy required per atom may be quite small, considerable energy is required to excite the whole liquid.

On this basis, we can understand the zero viscosity, the first superfluid property listed earlier. Viscous effects imply viscous friction. Viscous friction in turn implies that energy is transferred to the liquid in contact with the surface of the capillary tube through which it is flowing. But for a superfluid this is impossible unless a large amount of energy is available. Consequently, the liquid is frictionless.

Second sound has a similar explanation. In normal liquids, a heat pulse will soon result in additional random motion of the molecules. But in superfluid helium, the motion of individual atoms cannot be changed unless large amounts of energy are added. If the energy in the heat pulse is too small to raise the entire liquid to an excited state, the energy must propagate nondissipatively (like sound) rather than dissipatively (like heat).

Other superfluid effects can be explained in terms of the properties of the Bose gas as well.

---

**Bosons**

The Indian physicist Satyendranath Bose (1894–1974) was the first, in 1924, to investigate the properties of large collections of integral-spin particles, called **bosons** for that reason. Einstein encouraged Bose and expanded on his work, deriving the **Bose-Einstein distribution function**, which describes the energy distribution of bosons just as the Maxwell-Boltzmann distribution function describes the energy distribution of classical particles.

---

# Superconductivity

When the temperature of a metal is reduced, its electrical resistivity decreases. At sufficiently low temperature, however, the resistivity of samples of normal metals reaches a limit called the **residual resistivity**.

How can we account for resistivity on a quantum basis? An electron wave of the proper wavelength would pass unimpeded through an ideal crystal, in which each ideal point ion coincided with a node of the wave. But the thermal vibration of the ions and the presence of impurity atoms serve to disturb the perfect periodicity of the ideal crystal and thus cause the electron waves to be *scattered* and lose energy to the crystal lattice.

When an electron wave is scattered and thus contributes momentum and energy to the crystal lattice, the energy transfer must take place in accord with the usual quantum rule that requires transitions between discrete energy levels. Just as isolated atoms absorb and emit light quanta called photons in making transitions from one energy level to another, crystal lattices absorb and emit vibration quanta called **phonons**.* Electrical resistance in solids is a result of the net loss of energy and momentum by the electron wave to the crystal lattice through exchange of phonons. Not surprisingly, the energy shows up as heat; heat is the macroscopic manifestation of random lattice vibration.

Certain metals deviate in a dramatic fashion from the low-temperature behavior just described. As Kamerlingh Onnes first observed in 1911, some metals lose *all* electrical resistivity when the temperature is lowered below some **critical temperature** $T_c$. This phenomenon is called **superconductivity**. Once started, a current circulating through a ring-shaped superconducting sample can continue, without measurable diminution, for an indefinite period of time.

In accordance with Ampère's law, a magnetic field must be associated with this *persistent current*. However, the persistent current will cease if either this magnetic field or an externally imposed one exceeds a value $\mathscr{B}_c$, called the **critical field**.

But there is more than this to superconductivity. Suppose a sample of superconducting material whose temperature is greater than the critical temperature is located in an external magnetic field whose magnitude is less than $\mathscr{B}_c$. Now, suppose the temperature is reduced below the critical temperature. When the transition to superconductivity takes place, the sample will *expel* any magnetic flux inside it, as shown in Figure 44.15a,b. Alternatively, a sample that is already cold enough to exhibit superconductivity will push the flux away from itself when it is moved into a magnetic field; this is shown in Figure 44.15c,d. This active expulsion of magnetic field lines is called the **Meissner effect** [after the German (later American) physicist Karl Wilhelm Meissner (1891–1959) who, with his coworker R. Ochsenfeld, first described it in 1933].

Superconductivity is exhibited in its simplest form in about thirty pure metals—that is, in more than a third of all the metallic elements. In this form, called **type I superconductivity**, both the critical temperature and the critical magnetic field are small. The largest observed values are those for niobium (Nb): $T_c = 9.4$ K, and $\mathscr{B}_c = 0.2$ T. Limited to such values, superconductivity would not have extensive practical applications.

However, a second, more complicated type of superconductivity, called **type II superconductivity**, is observed in several classes of compounds. For many years, the

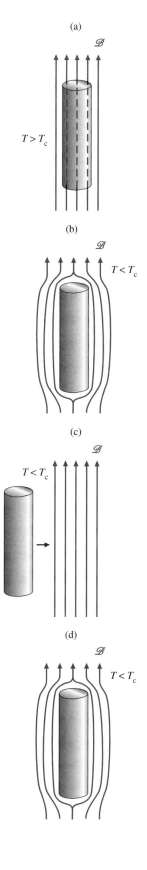

**FIGURE 44.15** The Meissner effect. (*a*) A sample of superconducting material is located in an external magnetic field. Magnetic flux penetrates the sample. (*b*) The temperature of the sample is reduced below $T_c$, and the magnetic flux is expelled. (*c*) A sample of superconducting material whose temperature is lower than $T_c$ is located outside a magnetic field region. (*d*) The sample is moved into the field and distorts the field so as to exclude magnetic flux from its interior.

---

*So called because (1) lattice vibrations travel through the lattice at the speed of sound and (2) they are analogous to photons.

known type II superconductors having the largest values of $T_c$ and $\mathscr{B}_c$ were all Group VB–Group IVA compounds having such chemical formulas as $Nb_3Sn$. The best of these is $Nb_3Ge$, with $T_c \simeq 23$ K. Characterizing the critical field is more complicated in type II superconductors than in type I superconductors. However, these materials can be used to make electromagnets that will produce magnetic fields up to about 15 T. These magnets, which often use a standard storage battery for start-up purposes only, achieve fields about five times as great as those obtainable with conventional iron electromagnets, which require considerable power. However, the low critical temperature still restricts the practical application of type II superconductors to special situations. Nevertheless, type II superconducting magnets have already achieved widespread use in medical magnetic resonance imaging machines.

In 1986, a new class of superconductors was discovered. This class comprises very complex ceramic materials, the first of which was a compound of lanthanum, barium, copper, and oxygen. Although basic understanding of these compounds is limited, they hold great promise both for broadened understanding of the phenomenon of superconductivity and for widespread practical application. The key to their practical application lies in the values of $T_c$. Several compounds already exist that exhibit superconductivity at temperatures well above 77 K, the boiling point of the cheap refrigerant liquid nitrogen.* There is reason to hope for new compounds with $T_c$ above room temperature. In principle, such materials will make possible lossless electric power transmission; compact, powerful magnets; economical power-storage devices; high-speed, low-friction magnetic levitation vehicles; and many other useful devices. Aside from the urgent need for better understanding of the properties of these compounds, however, formidable technical problems remain before widespread practical use can be achieved.

## The BCS Theory of Superconductivity

Although superconductivity was discovered in 1911, it was not until 1957 that a fundamental theoretical understanding was achieved. We can consider only the bare outlines of this beautiful theory here, called the BCS theory [named after the American physicists John Bardeen, Leon N. Cooper (b. 1930), and J. Robert Schrieffer (b. 1931), who collaborated on the theory at the University of Illinois].

Superconductivity bears tantalizing resemblances to superfluidity. In both, the difficulty of adding or removing energy suggests that the system has condensed into a Bose gas. But unlike helium atoms, which are bosons, the electrons that carry electric current in superconducting materials have spin $\frac{1}{2}$ and so are not bosons. The key to the observed behavior—at least in type I superconductors—is in understanding that electrons in a metal interact weakly with one another. This interaction can be described in general terms on the basis of a crude model. (Bear in mind, however, that any entirely satisfactory explanation must be fully quantum mechanical.) As an electron moves through a crystal, it attracts the nearby positive ions that constitute the crystal lattice. The distortion of the lattice propagates as a phonon, which can be absorbed by another electron whose momentum and spin are equal to and opposite those of the first electron. Under proper circumstances, the exchange of momentum between the two electrons, by means of the phonon, results in a weak attraction between the electrons. At high temperature, the attraction is negligible. But, at low temperature, the attraction is sufficient to bind the two electrons together into a **Cooper pair**. The Cooper pair has spin 0 and is a boson. Consequently, the Cooper pairs can condense into a Bose gas.

The formation of a Cooper pair implies that the energy of the pair is less than that of the two electrons separately. The binding energy is quantized, and there is an energy gap between the Cooper-pair ''ground state'' and the excited state in which the two electrons are unbound. What keeps the Cooper pairs together as they carry electric current through the crystal? Any scattering event must raise the energy of the pair at least to the first excited state. If the electric current is not too great, this cannot happen.

---

*$T_c = 125$ K has been confirmed for one compound, with much higher unconfirmed values reported for others.

In the absence of scattering, there is no electrical resistance. Although we will not consider the details here, the Meissner effect arises naturally out of the BCS theory.

**Symbols Used in Chapter 44**

| | | | |
|---|---|---|---|
| $E_c, E_v, E_g$ | conduction-band energy, valence-band energy, band-gap energy | $n$ | electron concentration in valence band or donor states |
| $E_d, E_a$ | donor ionization energy, acceptor ionization energy | $n'$ | electron concentration in conduction band |
| $h$ | Planck's constant | $n_i, p_i$ | intrinsic electron, hole concentration |
| $i$ | electric current | $T$ | absolute temperature |
| $k$ | Boltzmann's constant | $\kappa$ | dielectric constant |
| $m, m^*$ | free electron mass, effective mass | $\lambda$ | wavelength |

**Summing Up**

When identical atoms are assembled to form a **crystal**, the atomic energy levels are spread out into quasicontinuous **energy bands**. In accordance with the exclusion principle, these bands are filled with electrons up to an energy called the **Fermi level**. If the Fermi level lies within a band, the crystal conducts electric current readily and acts as a metal. If the Fermi level lies in a **band gap**, the crystal is a dielectric. A dielectric with a relatively small band gap behaves as a **semiconductor**.

In semiconductors, electric current is carried by electrons excited thermally into the **conduction band** and by **holes** "left behind" in the **valence band**. The conductivity of a semiconductor can be greatly increased by **doping**—the addition of small amounts of **donor** or **acceptor** impurities that introduce **impurity levels** into the band gap near the **band edges**.

Juxtaposition of ***n*-type** and ***p*-type** regions in a semiconductor results in a ***p-n* junction**, which has rectifying properties. Many electronic and optical devices are based on the *p-n* junction, including the **transistor** and the **light-emitting diode**.

At temperatures below the **lambda point**, $T = 2.18$ K, liquid helium exhibits **superfluidity**. Many metals and certain compounds exhibit a related phenomenon, **superconductivity**, at sufficiently low temperatures. Superconductors

have zero electrical resistivity and expel magnetic fields. Both phenomena are based on condensation into a **Bose gas** of particles having spin 0.

### KEY TERMS

**Section 44.2 Electrons in Crystalline Solids**
crystal, diamond structure ▪ energy band, band edge, band gap, valence band, conduction band ▪ Fermi level ▪ hole ▪ intrinsic conduction, intrinsic semiconductor ▪ effective mass ▪ doping ▪ hydrogenlike impurity, donor impurity, acceptor impurity ▪ donor ionization energy, acceptor ionization energy ▪ *n*-type, *p*-type semiconductor ▪ majority carrier, minority carrier

**Section 44.3 Semiconductor Devices**
*p-n* junction ▪ depletion region ▪ contact potential difference ▪ rectifier ▪ reverse bias, forward bias ▪ light-emitting diode, diode laser ▪ transistor ▪ MOSFET, substrate, source, drain, gate

**Sections 44.4 and 44.5 Superfluidity and Superconductivity**
boson, Bose-Einstein statistics ▪ lambda point ▪ residual resistivity ▪ critical temperature, critical magnetic field ▪ Meissner effect ▪ phonon ▪ Cooper pair

**Queries and Problems for Chapter 44**

## QUERIES

**44.1** *(2) Down is up?* When a hole in a semiconductor is excited downward from the Fermi level to a level in the valence band, its energy increases. Using the bubble analogy, explain why the energy of the entire system is increased in the process.

**44.2** *(2)* $3 + \frac{3}{2} \neq 3$. According to Equation 18.42, the molar heat capacity of a dielectric solid is $3R$, where $R$ is the universal gas constant. This result is based on the fact that the molecules of the solid act like harmonic oscillators and have six degrees of freedom. According to the classical free-electron theory of metals, the electrons behave like an ideal gas contained in the metal. According to Equation 18.31, this ideal gas should contribute an additional $\frac{3}{2}R$ to the molar specific heat, for a total value $\frac{9}{2}R$. But, in fact, the molar specific heat of metals is $3R$, just like that of dielectrics. Use the Fermi-sea picture to explain why the electrons do not contribute appreciably to the specific heat.

**44.3** *(2) Indium antimonide.* Indium (In) is trivalent and antimony (Sb) is pentavalent. Explain how the compound InSb can act as a silicon-like semiconductor.

**44.4** *(2) Alchemy.* Which of these elements and compounds—Se, $Al_2S_3$, CdTe, $AgInSe_2$, $CuGaSe_2$, and $CuInSe_2$—are silicon-like semiconductors?

**44.5** *(3) Over the line.* Suppose that you diffuse an increasing concentration of donor atoms into a *p*-type semiconductor. Draw a series of sketches of the band edges and the donor and acceptor states to show how the sample changes from *p*-type through pseudointrinsic to *n*-type.

**44.6** *(3) Diffusion current.* Show that the diffusion of majority carriers in both directions across a *p-n* junction results in a net electric current $i_{diff}$. In which direction is the current? In which direction is the current $i_{drift}$ due to the drift of mi-

nority carriers? When the junction is isolated, what is the relation between $i_{diff}$ and $i_{drift}$?

**44.7** *(3) Diode rectifier.* In the sketch below, the ac source applies a sinusoidal emf across the diode and the resistor. Sketch the emf that appears between terminals *A* and *B*.

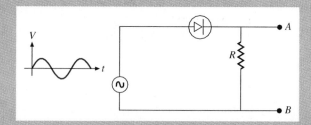

**44.8** *(3) Badly cooked.* To be useful, a rectifier must have a large *backward resistance*. That is, the current driven through it by a reverse bias must be small. When the temperature of a silicon diode is raised above room temperature, the backward resistance decreases quite rapidly (roughly exponentially as a function of *T*) and the diode eventually becomes ineffective as a rectifier. Explain.

# PROBLEMS

## GROUP A

**44.1** *(2) Doping it out.* B, Al, As, In, and Sb can act as substitutional impurities in silicon or germanium. Which are *n*-type and which *p*-type dopants?

**44.2** *(2) A little goes a long way.* You dope intrinsic silicon with 1.0 ppm of arsenic. **(a)** Is the doped silicon *n*- or *p*-type? **(b)** What is the ratio of the room-temperature conduction-electron concentration to that of intrinsic silicon?

**44.3** *(2) Germanium.* The effective mass of conduction electrons in germanium is $m^* = 0.1m$, and the dielectric constant

is $\kappa = 15.8$. What is the ionization energy for phosphorus donor atoms?

**44.4** *(3) Designing an LED.* The band-gap energy of GaAs is 1.4 eV, and that of GaP is 2.3 eV. You wish to design a $GaAs_xP_{1-x}$ diode to emit red light of wavelength 675 nm. To first approximation, what value of *x* should you choose?

**44.5** *(5) Turn-off.* A long, current-carrying niobium wire is maintained at a temperature well below $T_c$. At what current will the wire cease to superconduct?

## GROUP B

**44.6** *(2) Getting warmed up.* Using the method of Example 44.1, find $n'_{600}/n'_{300}$, the ratio of conduction-electron concentration at 600 K to that at 300 K, for pure silicon. By what factor, roughly speaking, is the electrical conductivity of pure silicon increased when its temperature is increased from 300 K to 600 K?

**44.7** *(2) How much excitement can you take?* Using the numerical values given in the last paragraph of Section 44.2, calculate the fraction of all the atoms in intrinsic silicon whose electrons are excited into the conduction band.

**44.8** *(2) Thermal ionization.* A semiconductor is donor doped so that the ratio of donor atoms to host atoms is $n_d/n$. **(a)** Assume that the Fermi level is close to the donor level located $E_d$ below the conduction-band edge, and show that the donor-electron concentration in the conduction band is roughly

$$n'_d = n_d e^{E_d/kT}.$$

**(b)** Let $E_d = 0.020$ eV, and find the proportion of donor atoms that are ionized at room temperature.

**44.9** *(2) Hall coefficient.* If you dope intrinsic silicon with 0.10 ppm of phosphorus, what is the room-temperature Hall coefficient (Equation 28.31)?

**44.10** *(2) Rapid increase.* **(a)** Beginning with Equation 44.2, find the proportional rate of change of electron concentration, $dn_i/n$, as a function of *T* for an intrinsic semiconductor. **(b)** Show that, for intrinsic silicon at room temperature, a 1-C° temperature increase results in approximately a 50% increase in conduction-electron concentration.

**44.11** *(2) Big circles.* In a semiconductor having dielectric constant $\kappa$ and effective mass $m^*$, what is the radius of the ground-state electron orbits for hydrogenlike donor impurities?

**44.12** *(2) Type tester.* The simple device shown (p. 1219)

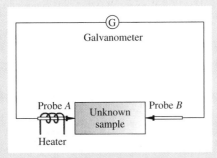

Galvanometer

Probe A    Unknown    Probe B
           sample

Heater

is often used in laboratories to determine quickly whether a semiconductor sample is *n*- or *p*-type. When probe *A* is warmed, the galvanometer indicates the sense of current flow. **(a)** Explain how the device works. **(b)** If the sample is *n*-type, which way will the current flow?

**44.13** *(3) Light-detecting diodes, I.* By exploiting the principle by which light-emitting diodes operate in reverse, you can use a semiconductor diode as a light detector. **(a)** Describe what occurs when a photon strikes the diode—the reverse of the LED process. **(b)** The band-gap energy of GaAs is 1.4 eV. What is the maximum wavelength of photons that can be detected by a GaAs diode? In what region of the spectrum do such photons lie? **(c)** The band-gap energy of InSb is 0.2 eV. What is the maximum wavelength of photons that can be detected by an InSb diode? In what region of the spectrum do such photons lie? (InSb detectors were first used widely during the Korean war of the 1950s in the Sidewinder antiaircraft missile, which guides itself toward the hot tailpipe of a jet engine. The name arises from the fact that the sidewinder rattlesnake observes its prey by means of infrared-sensitive organs located below its eyes.) **(d)** Why must InSb detectors be cooled well below room temperature for best operation?

**44.14** *(3) Light-detecting diodes, II.* What is the minimum wavelength for which diamond ($E_g$ = 5.4 eV) is effective as a radiation detector?

**44.15** *(5) Cooper union?* The critical temperatures of aluminum, mercury, and lead are 1.19 K, 4.15 K, and 7.18 K. Estimate the binding energy of the Cooper pairs in each case.

## GROUP C

**44.16** *(2) Temperature dependence of conductivity for an intrinsic semiconductor.* **(a)** What is the total current carried by the single free electron and hole in Figure 44.6? **(b)** Suppose the concentration of electrons is $n_e$. What is the total current? **(c)** Show that the electrical conductivity of the semiconductor varies with absolute temperature according to the proportionality

$$\sigma(T) \propto e^{-E_g/2kT}.$$

Make the crude assumption that electrons and holes carry current equally well.

**44.17** *(2) Measuring donor ionization energy.* **(a)** You measure the electrical conductivity $\sigma$ of a phosphorus-doped silicon crystal as a function of absolute temperature $T$. Explain how you can obtain the value of $E_d$ from a plot of ln $\sigma$ versus $1/T$. **(b)** In what temperature range, roughly speaking, should you carry out the experiment?

**44.18** *(2) np product.* The location of the Fermi level $E_F$ in a semiconductor band gap can be affected by doping and has a profound effect on the electron concentration $n$ and the hole concentration $p$. Show that, wherever $E_F$ is located, the so-called $np$ product has a constant value

$$np \propto e^{-E_g/kT},$$

which is independent of the location of the Fermi level.

**44.19** *(3) LED design problems.* The earliest pocket calculators (c. 1965) used $GaAs_xP_{1-x}$ LED displays. They were a heavy drain on the batteries, and so it was essential to optimize their efficiency. However, $GaAs_xP_{1-x}$ diodes are most efficient when $x$ is large and the wavelength lies in the infrared. (But infrared diodes are not useful for visible displays.) The efficiency decreases rapidly with decreasing $x$ and therefore with decreasing wavelength $\lambda$ of the emitted light. On the other hand, the sensitivity of the eye as a detector decreases with *increasing* wavelength. In the extreme red, near the limits of the visible region, the eye's sensitivity falls off very rapidly. The optimization of the diode design requires a compromise between the conflicting demands of output efficiency and eye sensitivity.

In the wavelength region of interest, approximate the diode efficiency by the exponential function $\epsilon = ae^{\alpha\lambda}$ and the sensitivity of the eye by the exponential function $\sigma = be^{\beta/\lambda}$. Find the value of $\lambda$ for which the LED should be designed.

Note: The compromise turns out to involve very intense light whose wavelength lies at the extreme limit of the capability of the eye. Although optimal with respect to power drain, the display is very uncomfortable to look at for more than a short time. (It's hard to look at a bright light you can barely see! Besides, the eye has difficulty focusing light of this wavelength.) In later models, $x$ was increased (and $\lambda$ thus decreased) at the expense of additional battery drain. The problem was ultimately solved by abandoning LED displays for liquid crystal displays, which consume much less power.

45

# Ultimate Matter?

——— For every particle there exists a corresponding antiparticle.

——— Ordinary matter consists of particles that can be divided into hadrons and leptons.

——— Protons and neutrons are hadrons. There are many other kinds of hadrons, all of them unstable.

——— Electrons and electron neutrinos are stable leptons; all of the many other leptons are unstable and appear to be excited states of the stable particles.

——— Hadrons are acted on by the strong and weak forces, leptons by the weak force only.

——— Hadrons are further divided into baryons and mesons. Baryons are combinations of three quarks; mesons comprise quark-antiquark pairs; free quarks have not been observed.

——— The four fundamental interactions—strong, weak, electromagnetic, and gravitational—may well be different manifestations of a single unified fundamental force.

*Left:* Faced with the vastness of the universe, the wise person tries not to subdue it but to understand its complex nature in terms of simple principles. The Belgian surrealist René Magritte (1898–1967) has symbolized the growth of human understanding of the universe in which humankind is itself immersed by means of the tiny balloon which, in spite of its frailty, rises ever upward. Beneath are the classical building blocks of the universe—earth, air, fire, and water; in the midground the static, geometric forms that served to quantify the classical universe are mixed with the more dynamical forms required for the modern synthesis. In the foreground, deconstructed and reconstructed, is the guiding principle—the static, classical ideal of beauty made dynamical.

*O we can wait no longer,*
*We too take ship O soul,*
*Joyous we too launch out on trackless seas,*
*Fearless for unknown shores on waves of ecstasy to sail, . . .*
*Chanting our chant of pleasant exploration. . . .*
*O farther, farther, farther sail!*

—WALT WHITMAN, *Passage to India*

# Introduction

There have been times when natural philosophers believed they understood the ultimate structure of matter—that they knew the fundamental building blocks of which everything is made. The followers of Aristotle thought as much when they argued that everything is made up of various proportions of the four elements earth, air, fire, and water. By the late Middle Ages, however, this simple picture had been complicated by the addition of quicksilver (mercury), sulfur, and a number of other substances. But the continual addition of more elements to the list led to its abandonment. If there are too many kinds of building blocks, it is hard to believe they are fundamental.

In the early days of modern chemistry (about 1800), a similar hope arose. But in the course of the nineteenth century, the number of elements grew to almost one hundred—again too many to regard as truly fundamental.

With the discovery of the structure of the atom in the first few decades of the twentieth century, the hope sprang up again. By 1920 or so, it was plausible to argue that all matter was built up of protons and electrons, with photons acting as the carriers of electromagnetic energy. Things became a little more complicated with the discovery or postulation in the early 1930s of the neutron, the neutrino, the positron, and a few other particles. But the list of particles—ten entries more or less—was not so long as to undermine the idea that the particles were fundamental.

The hope of simplicity and completeness at this level was dashed with the discovery of the muon (1937). As I. I. Rabi remarked wryly, "Who ordered this?" Rabi meant that the muon did not seem to have any place in the then-accepted picture of the structure of matter.

Rabi's apprehensions were well founded. In subsequent decades, more and more particles were discovered, ultimately constituting a "particle zoo" with well over a hundred "exhibits." So many different particles cannot be called "fundamental" in any meaningful sense. Keenly aware of this, physicists turned their attention to classifying the particles and then using the classification schemes to predict the existence of other particles still undiscovered. These efforts were analogous to the work inspired in chemistry in the nineteenth century by Mendeleyev's periodic table, which challenged chemists to fill in the gaps. For twentieth-century physicists, as for nineteenth-century chemists, the efforts were highly successful.

But the size of the particle zoo led inevitably to the revival of the old question: What are the fundamental building blocks of the universe? This chapter is concerned with that question and sketches the partial answers that have emerged to date. Section 45.2 concerns *positrons* and other *antiparticles*. In Section 45.3, we consider the *strong* and *weak* nuclear forces, and the particles—*hadrons* and *leptons*—on which they act. Section 45.4 deals with *quarks*, the fundamental particles of which hadrons are constructed.

**Sharp Tongue, Sharp Mind**

Isidor Isaac Rabi (1898–1987), American physicist, won the Nobel Prize in 1944 for his work with molecular beams and especially for his pioneering work in nuclear magnetic resonance. Aside from his extraordinary contributions as a physicist, Rabi was noted for his devastating sarcasm.

In Section 45.5, we examine the connection between each of the four fundamental forces and the *mediating particle* with which that force is associated. Section 45.6 is a brief consideration of ongoing efforts to set our picture of the universe on a unified footing.

# Particles and Antiparticles

**Theoretician's Theoretician**
Paul Adrien Maurice Dirac (1902–1984), British theoretical physicist, won the Nobel Prize in 1933 for his exposition of relativistic quantum mechanics.

In its earliest and still most commonly used form, Schrödinger's equation (Section 42.5) does not take account of the relativistic properties of matter. But, as you have seen, electrons have such small masses that they often exhibit relativistic behavior. In 1928, Dirac recast the quantum theory of the electron in relativistic terms and arrived at two remarkable results. The first pertains to the intrinsic spin angular momentum of the electron (Section 42.5). Dirac showed that this spin arises in a necessary way out of the relativistic treatment. What had been an anomalous property of electrons became a natural property.

The second result, which is more important for our present purposes, is that relativistic quantum theory requires the existence of an **antielectron**, or **positron**, having mass identical with that of the electron but equal and opposite electric charge. Here is a sketch of Dirac's argument. In its relativistic form, Schrödinger's equation predicts, for the free electron, the existence of quantized energy states that extend not from $E = 0$ to $E = \infty$ but from $E = -\infty$ to $E = \infty$. The negative-energy solutions raise several questions. What does negative energy mean? How can we account for the fact that electrons are never observed with negative energies? Indeed, in view of the fact that every electron will tend to the lowest available energy state, $E = -\infty$, how can we observe any electrons at all having finite energy? In response to these questions, Dirac argued that all the energy states up to $E = 0$ are filled with electrons. The exclusion principle then forbids any electron with positive energy from falling to a negative energy state.

Dirac's argument requires the existence of an infinite number of electrons in negative energy states, even in empty space. Although this may strike you as implausible, it presents no inconsistency with what is observed. The negative-energy electrons, like the electrons in the Fermi sea in a filled semiconductor valence band, do not normally interact with their surroundings. To do so, they would have to gain or lose energy and move to another energy state. But all the nearby energy states are filled with other electrons (Figure 45.1a). As long as the electrons are unobservable, they present no problem to consistency or to "common sense" because they do not figure in the results of any experiment. We can imagine any number of negative-energy electrons we like, and it will make no difference when we observe such processes as ionization or the production of light in a gaseous discharge tube. If an electron is to be observed in an ordinary experiment involving modest energies, its energy must be such that there are neighboring vacant energy states into which it can move.

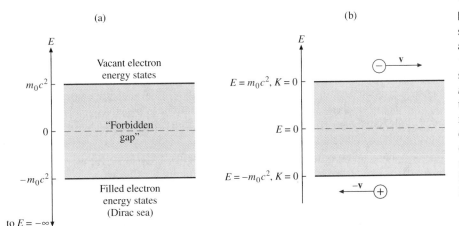

**FIGURE 45.1** (*a*) All of the energy states from $E = -\infty$ to $E = -m_0c^2$ are filled. All of the states from $E = +m_0c^2$ to $E = \infty$ are empty. The states in the range $-m_0c^2 \le E \le m_0c^2$ are forbidden because no electron can have energy less than its rest energy. (*b*) An electron is excited into the region $E > m_0c^2$, analogous to the conduction band in a semiconductor. Left behind is a hole in the negative-energy region; this hole is a positron.

The minimum relativistic energy an observable electron can have is its rest energy, $E = m_0 c^2$. It can have any amount of energy greater than this. The total energy is the sum of the rest energy and the kinetic energy, according to Equations 39.10 and 39.15:

$$E = m_0 c^2 + K. \tag{45.1}$$

For an electron whose energy is not too large, we can use Equation 39.12 to express the kinetic energy in the simple form

$$K = E - m_0 c^2 = \tfrac{1}{2} m_0 v^2. \tag{45.2}$$

As you can see in Figure 45.1b, it is possible to produce a free electron by raising a bound electron from a negative to a positive energy state. To do this requires a minimum energy $\Delta E = 2m_0 c^2$, and the process leaves a "hole" behind in the negative-energy region. The process is analogous to the creation of an electron-hole pair in an intrinsic semiconductor, where the minimum energy required is the band-gap energy $E_g$ (Section 44.2). In the present case, however, the "band gap" energy is much larger than that in the semiconductor; we have

$$\Delta E = 2m_0 c^2 = 2 \times 9.1 \times 10^{-31} \text{ kg} \times (3.0 \times 10^8 \text{ m/s})^2$$
$$= 1.6 \times 10^{-13} \text{ J}, \tag{45.3a}$$
or $\quad\quad \Delta E = 1.0 \text{ MeV}. \tag{45.3b}$

You can see from Figure 45.1b why the hole must have a maximum energy $-m_0 c^2$; the corresponding positron has *minimum* energy $m_0 c^2$ because, as with the semiconductor holes, the energy of a positron is measured downward from $E = 0$.

Now we come to the prediction of the existence of the positron. The hole left behind in the negative-energy region is analogous to a hole in the valence band of a semiconductor. Its existence makes possible the motion of neighboring electrons because it is a vacant state into which a negative-energy electron can make a transition. But, just as in the semiconductor case, it is much easier to consider the motion of the hole rather than that of the many electrons. The hole is a **positron**—a particle with positive charge, mass equal to the electron mass, and spin of magnitude $\tfrac{1}{2}\hbar$. The positron is represented by any of the symbols $e^+$, $\bar{e}$, and $\beta^+$.

In a semiconductor, an electron-hole pair can recombine, resulting in disappearance of both the electron and the hole from the conduction process and the emission of energy $E_g$. For an electron-positron pair, such recombination is called **annihilation**; the energy is released in the form of photons having total energy 1.0 MeV.

An implication of the Dirac picture is that electrons can be created only as members of electron-positron pairs. A photon having energy 1 MeV or more can potentially create such a pair, itself disappearing in the process. However, **pair production** does not occur spontaneously, as Example 45.1 shows.

**Charge Conservation in the Small**

This implication that electrons and positrons are always created and annihilated only in pairs is consistent with our intuitive notion that the total charge of the universe is conserved and equal to zero. Each time a negative electron is created or annihilated, a positive positron is simultaneously created or annihilated, and so there is no net change in the charge of the universe. The argument applies to all other charged particles as well.

## EXAMPLE 45.1

Show that, if it occurred, spontaneous process $h\nu \rightarrow e^- + e^+$ would violate the principle of conservation of momentum.

**SOLUTION:** The photon is a completely relativistic particle, and so its momentum is related to its energy by Equation 39.22, $E = K = pc$. Positrons and electrons, on the other hand, possess rest mass. Their momenta and energies must be represented by the more general Equation 39.20a, $E = \sqrt{p^2 c^2 + m_0^2 c^4}$. In the hypothetical spontaneous process, the squares of the initial momentum $p_i$ and final momentum $p_f$ are found by solving Equa-

tions 39.22 and 39.20a for $p^2$:

$$p_i^2 = \frac{E^2}{c^2} \quad \text{and} \quad p_f^2 = \frac{E^2}{c^2} - 2m_0^2 c^2.$$

Conservation of momentum requires that $p_i^2 = p_f^2$. The result is $m_0 = 0$, which is inconsistent with the nonzero rest masses of the electron and positron. In general, *a photon cannot disintegrate spontaneously to produce a pair of particles having nonzero rest mass.*

Although a photon cannot produce an electron-positron pair *spontaneously*, pair production can occur if a photon of sufficient energy collides with another particle, with which it can exchange momentum. In 1932, Anderson was the first to observe positrons, in the debris produced by the collision of cosmic-ray particles with atoms of air at high altitude. Positrons are readily distinguished from electrons by the sense of the spiral paths they follow in a magnetic field as they are decelerated by repeated collisions with atoms (Figure 45.2).

The paths are made visible by means of the ionization produced by the energetic charged particles as they pass through matter. In photographic film, the ionization "exposes" the film, resulting in a track of black silver particles when the film is developed. In the detectors called cloud chambers, the ionization triggers condensation of water droplets from a supersaturated atmosphere. In the bubble chamber, a similar process leaves a trail of bubbles of vapor in a superheated liquid. In the spark chamber, the ionization promotes sparks as it passes through a series of metal plates, every other one of which is charged to a high potential. Detectors must be made larger and more complex as experiments are undertaken at higher energies; the detectors used with modern accelerators are themselves very large, complicated devices.

The intimate birth-death connection found between electrons and positrons is found for all other particles as well. Every particle has a corresponding **antiparticle**. (The photon is its own antiparticle.) The **antiproton** was first produced in the laboratory in 1955, and many other antiparticles have been produced and observed since.

The greater the rest mass of a particle, the more energy is needed to create it in the laboratory and the larger the detector must be to study its fate. This is one of the major reasons underlying the construction of successive generations of high-energy accelerators (Figure 45.3).

When an electron-positron pair is created, the two particles usually move quickly apart and do not annihilate each other. This is because they usually come into existence with considerable kinetic energy. However, positrons do not live long in our corner of the universe. Electrons are so common in ordinary matter that a positron is certain to encounter one quite soon, and the pair annihilate.

**FIGURE 45.2** The first published photograph of the track left by a positron in a Wilson cloud chamber. The energetic particle enters from the bottom, strikes the 6-mm-thick lead plate in the middle, and leaves at the top. Some of the air molecules struck by the positron are ionized and form nuclei for the condensation of water vapor from the supersaturated air in the chamber. A magnetic field is oriented along the downward normal to the plane of the photograph. The curvature of the positron track is due to the magnetic force exerted on the positron. Taken together with the known direction of the magnetic field (into the page), the sign of the curvature yields the sign of the charge of the positron.

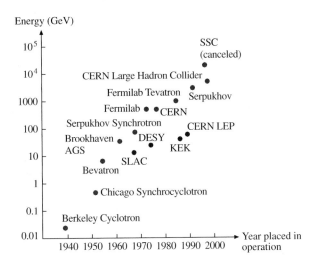

**FIGURE 45.3** Maximum energy output of particle accelerators versus year of initial operation. The red dots represent proton accelerators; the blue dots, electron accelerators. The Bevatron is the Berkeley synchrotron (named before GeV became the standard symbol for $10^9$ eV), SLAC is the Stanford Linear Accelerator, DESY is the German Electron Synchrotron, CERN is the European Center for Nuclear Research (on the Swiss-French border), KEK is the Japanese National Laboratory accelerator, LEP is the CERN Large Electron-Positron Storage Ring, and SSC is the stillborn Superconducting Supercollider. Dates later than 1995 are projected.

## Antimatter

It is possible to conceive of **antimatter**—matter composed entirely of antiparticles. For example, *antihydrogen* can be constructed by combining an antiproton and a positron.

The spectral lines emitted by excited antihydrogen atoms are indistinguishable from the spectral lines emitted by excited hydrogen atoms. Are there regions of the universe where antimatter is the rule and "ordinary" matter the rare exception? Many science-fiction stories have been based on this possibility, but the best current evidence suggests that ordinary matter is predominant in the entire observable universe. The fact that the familiar one of the two possibilities came to pass in the evolution of our universe seems to be a matter of happenstance. But any cosmological theory must account for the evolution of the observed predominance of one kind of matter because this predominance constitutes an asymmetry on the cosmic scale.*

## The Strong and Weak Forces, Hadrons and Leptons, Baryons and Mesons

The existence of an antiparticle for every known particle (except the photon and a few others, which are their own antiparticles) immediately doubles the number of elementary particles. So far, however, we have considered only the particles that constitute ordinary matter, together with their antiparticles. But that is not the end of the story; we must consider other particles as well. Although these other particles are not so obviously present in ordinary matter, a study of their properties, taken together with the properties of the more familiar particles, makes it possible to discern the significant categories into which the particles can be classified.

The force that binds an atomic nucleus, preventing it from decaying by α emission, is called the **strong nuclear force** or simply the **strong force**. This is the short-range force that opposes the mutual Coulomb repulsion of the protons in a nucleus and prevents the nucleus from flying apart. The potential barrier through which α particles tunnel in α decay is a combination of the potentials of the strong force and the Coulomb force (Section 43.3). Particles on which the strong force acts are called **hadrons**, from the Greek root meaning "thick" or "dense." Protons and neutrons are the best-known hadrons. But there are many others as well; hadrons constitute the largest family of particles.

Some particles are not subject to the strong force. They are called **leptons**, from the Greek root meaning "small." Although not subject to the strong force, leptons are subject to the **weak nuclear force**, or **weak force**. In β decay, the emitted particles must overcome the weak force. The weak force operates on hadrons as well as leptons, but for hadrons it is usually overwhelmed by the much stronger strong force.

The strong and weak forces, the familiar Coulomb force (which affects all electrically charged particles), and the very weak gravitational force (which affects all particles having nonzero mass) are the four **fundamental forces** of nature. Unlike the inverse-square Coulomb and gravitational forces, the strong and weak forces are *short-range* forces. They do not have significant effect over distances greater than the diameter of the atomic nucleus—about $10^{-14}$ m. That is why they are not familiar to us on the macroscopic scale.

### The Leptons

Consider first the particles on which the strong force does not act. Of this rather small class of particles, we have made most mention of the electron. We have also had occasion to consider the neutrino, which is always emitted together with the electron in β-decay processes. Here, however, we note that there are several *different* neutrinos. Consequently, we give the name **electron neutrino**, $\nu_e$, to the particle that takes part in β-decay processes. The mass of the electron is small on the atomic scale; see Table 45.1. The rest mass of the electron neutrino may be exactly zero, but this is not yet

*See R. K. Adair, "A Flaw in the Universal Mirror," *Scientific American*, February 1988.

TABLE 45.1 The Leptons

| Particle | Mean Life (s) | Rest Mass | Rest Energy (MeV) | Charge | Spin | Antiparticle |
|---|---|---|---|---|---|---|
| Electron, $e^-$ | Stable | $m_e$ | 0.511 | $-e$ | $\frac{1}{2}$ | $e^+$ |
| Muon, $\mu^-$ | $2.2 \times 10^{-6}$ | $207m_e$ | 106 | $-e$ | $\frac{1}{2}$ | $\mu^+$ |
| Tauon, $\tau^-$ | $3.4 \times 10^{-13}$ | $3490m_e$ | 1784 | $-e$ | $\frac{1}{2}$ | $\tau^+$ |

Note: A neutrino corresponds to each of these charged leptons.

known for certain. However, experimental evidence places a very small upper limit—less than $10^{-35}$ kg, or at most $10^{-5}m_e$—on the rest mass of the electron neutrino. (Because the universe contains a vast number of neutrinos, a nonzero rest mass adds considerable mass to the universe, with important cosmological consequences.)

The **muon** (symbol $\mu$ or $\mu^-$) is much more massive than the electron ($m_\mu = 207m_e$), though its mass is still quite small compared with that of a proton or a neutron. Like the electron, the muon has charge $-e$ and spin $\frac{1}{2}$. Associated with the muon is the **muon neutrino**. The muon and its neutrino are related in the same way as the electron and its neutrino. The muon is unstable, as already noted in Section 35.4. Its mean life is $2.2 \times 10^{-6}$ s, and it decays into an electron, an electron antineutrino, and a muon neutrino, according to the reaction

$$\mu^- \rightarrow e^- + \bar{\nu}_e + \nu_\mu. \qquad (45.4)$$

Owing to the considerable difference in rest mass between the muon and the electron, the kinetic energy of the daughter particles is substantial. Momentum could not be conserved in the decay were it not for the participation of the neutrinos. Moreover, angular momentum could not be conserved if only one neutrino were created in the decay, because each neutrino has spin $\frac{1}{2}$.

## EXAMPLE 45.2

**(a)** Find the total kinetic energy $K_{tot}$ of the daughter particles in Reaction 45.4. **(b)** Find the minimum and maximum possible kinetic energy, $K_e$, of the electron produced in such a reaction.

**SOLUTION:**

**(a)** Find $K_{tot}$ for Reaction 45.4.

You have

$$
\begin{aligned}
K_{tot} &= (m_\mu - m_e)c^2 \\
&= (207 - 1) \times 9.1 \times 10^{-31} \text{ kg} \times (3 \times 10^8 \text{ m/s})^2 \\
&= 1.69 \times 10^{-11} \text{ J} = 105 \text{ MeV}.
\end{aligned}
$$

**(b)** Find the extremal values of $K_e$.

In view of their zero or negligible mass, neutrinos are always relativistic. Momentum is conserved if the two neutrinos move off in opposite directions, with momenta of equal magnitude $p$ and equal energy $K = pc = K_{tot}/2$. In this case, the electron has negligible kinetic energy, as shown in Figure 45.4a. Thus the minimum possible value of $K_e$ is zero.

Momentum is also conserved if the electron and one of the

(a)        (b)

**FIGURE 45.4**

neutrinos move off in opposite directions while the other neutrino carries very little momentum $p$ or energy $pc$ (Figure 45.4b). In this case, the electron is highly relativistic and its rest energy is a negligible proportion of its total energy. (Remember that the rest energy of the electron is about 0.5 MeV.) The situation is then nearly symmetrical in spite of the rest mass of the electron. Hence the electron and the energetic neutrino each have kinetic energy $K = K_{tot}/2 = 53$ MeV, while the other neutrino has negligible energy. Thus $K_e = 53$ MeV is the maximum possible kinetic energy of the electron.

A third pair of leptons is known: the **tauon** $\tau^-$ and its associated **tauon neutrino** $\nu_\tau$. Just as the muon decays into an electron, an electron antineutrino, and a muon neutrino, the tauon can decay into a muon, a muon antineutrino, and a tauon neutrino:

$$\tau^- \rightarrow \mu^- + \bar{\nu}_\mu + \nu_\tau. \qquad (45.5)$$

(Because the mass of the tauon is greater than that of particles which are not leptons, it can decay into such particles as well and does so about half the time.)

The chain of processes by which a tauon can decay into a muon with the release of energy and the muon then decays into a stable electron with the release of energy is reminiscent of the chain of processes by which an electron in an excited state of (say) hydrogen decays into an excited state of lower energy and ultimately into the stable ground state. In this sense, we can regard the tauon and the muon as *excited states* of the electron. When an electron makes a transition between states in a hydrogen atom, the change in its rest mass is negligible because the energy released is quite small, of the order of 1 eV. The large rest-mass differences among the tauon, the muon, and the electron reflect the much larger energy differences between the excited states and the ground state.

So far as is known, the leptons have no internal structure and no measurable dimensions. They appear to constitute a family of truly fundamental particles.

### The Hadrons

The class of hadrons is much larger than the class of leptons. It is divided into two subclasses, the **mesons** (from the Greek root meaning ''medium'') and the **baryons** (from the Greek root meaning ''heavy''). The least massive of the baryons are the now-familiar proton and neutron; all other baryons are more massive. All baryons but the proton are unstable. The mean life of the neutron is about 15 min—a very long time on the scale of microscopic events—but all of the other baryons have mean lives of $10^{-8}$ s or less (in most cases, much less). Some fifty baryons are known. The spins of all baryons are half-integral ($\pm\frac{1}{2}$, $\pm\frac{3}{2}$, and so forth); they are *fermions*.

The mesons were originally so named because the first one discovered, the **pion** ($\pi^-$), had mass $270m_e$, intermediate between those of the electron and the nucleon (the proton and the neutron). Some seventy mesons are known. All are unstable, with mean lives of $10^{-8}$ s or less (in most cases, much less). The spins of all mesons are integral (0, $\pm 1$, $\pm 2$, and so forth); they are *bosons*.

Unlike the leptons, the baryons have finite size and internal structure. The earliest intimation of this structure arose from the *anomalous magnetic moments* of the proton and the neutron. As already noted in Chapter 43, the spin of the proton is 2.79 nuclear magnetons, rather than the 1 nuclear magneton that emerges from the simple picture of a spinning, uniformly charged sphere. On the basis of the same simple picture, we would not expect the uncharged neutron to have any magnetic moment at all. But, in fact, its magnetic moment is $-1.91$ nuclear magnetons. We must infer that, in spite of its overall neutrality, the neutron is built up of components having positive and negative charge, which are not symmetrically distributed through the space occupied by the neutron.

### Conservation Principles

Nearly all information concerning the hadrons is obtained by observing the products of their collisions and their decay modes. For most particles, more than one decay mode is possible, but all have been observed to conform to a set of conservation principles. Among these principles, we have already noted conservation of mass-energy, charge, linear momentum, and angular momentum. These principles are already well founded in the macroscopic realm. But many details of the decay modes that are observed—and of those that are *not* observed—lead to the insight that other conservation principles, whose operation is not observed on the macroscopic scale, also govern the behavior of fundamental particles.

These conservation principles are expressed in terms of quantum numbers. In a process A + B → C + D + E, the sum of each of these quantum numbers for A and B must be equal to its sum over C, D, and E. We now list some of these quantum numbers.

---

**Unstable? Not Very!**

Current theory allows for the possibility that the proton also is unstable, with a mean life in excess of $10^{32}$ y, or $10^{22}$ times the age of the universe. It is no small task to detect the decay of a proton, and no such decay has yet been observed.

**LEPTON NUMBER** The value of the electron-lepton number for electrons and electron neutrinos is 1; for positrons and electron antineutrinos, it is −1. The total value of this number must be conserved in all processes. For instance, its value remains zero in the process by which the neutron decays. With the electron-lepton numbers of the particles shown in parentheses, the process is

$$n \ (0) \rightarrow p \ (0) + e \ (1) + \bar{\nu}_e \ (-1).$$

Similar rules exist for muons and their neutrinos and for tauons and their neutrinos. Note that both electron-lepton and muon-lepton numbers are conserved in Reaction 45.4.

**BARYON NUMBER** The value of the baryon number is 1 for all baryons and −1 for all antibaryons. Should experiment show that the proton is not absolutely stable, this conservation principle would be violated.

**ISOSPIN** Protons and neutrons behave similarly in processes involving the strong interaction; their electric charges are irrelevant in such processes. For this reason, we can regard the two particles as slightly different energy states of the same basic entity, with different orientations of the isospin vector **T** in an abstract space. [The reason for the name "isospin" is that the rules for the quantization of isospin are analogous to those for the quantization of angular momentum in a magnetic field (Section 42.5).] Let us look at three examples of this analogy. The $\Lambda$ particle has isospin 0. This accounts for the fact that the $\Lambda$ is a "singlet" having only a neutral form $\Lambda^0$; there are no positive or negative companion particles, $\Lambda^+$ or $\Lambda^-$. An isospin value $\frac{1}{2}$ leads to a "doublet" like the *nucleon doublet* comprising the proton and the neutron. An isospin value 1 leads to a "triplet" like the one comprising the three $\Sigma$ baryons, $\Sigma^+$, $\Sigma^-$, and $\Sigma^0$.

Conservation principles exist for other quantities as well, among them *hypercharge* and *parity*.

# Quarks

From the foregoing discussion, it is evident that hadrons are not truly fundamental particles. Their variety makes that implausible, while their detailed properties confirm the existence of internal structure. However, all hadrons can be constructed from fundamental particles called **quarks**.

On the basis of observation and of symmetry considerations, the current theory requires the existence of thirty-six kinds of quarks, divided into six major classes whimsically called **flavors** (Table 45.2). Each quark listed in the table comes in three

**Lots of Fun at Finnegan's Wake**

Quarks were so named by the American physicist Murray Gell-Mann (b. 1929). It was (and to some extent still is) the style to give whimsical names to the many new concepts arising from the progress of fundamental-particle physics. Gell-Mann presumably had in mind the fact that all baryons consist of three of the elusive quarks when he recalled a line from *Finnegan's Wake*, the elusive masterpiece of the great Irish writer James Joyce: "Three quarks for Muster Mark."

**TABLE 45.2 The Quarks**

| Name and Symbol | Mass[a] | Electric Charge | Hypercharge | Charm | Bottomness | Topness |
|---|---|---|---|---|---|---|
| up, u | 10 | $+\frac{2}{3}$ | $+\frac{1}{3}$ | 0 | 0 | 0 |
| down, d | 20 | $-\frac{1}{3}$ | $+\frac{1}{3}$ | 0 | 0 | 0 |
| strange, s | 200 | $-\frac{1}{3}$ | $-\frac{2}{3}$ | 0 | 0 | 0 |
| charm, c | 3 000 | $+\frac{2}{3}$ | $+\frac{1}{3}$ | 1 | 0 | 0 |
| bottom,[b] b | 9 000 | $-\frac{1}{3}$ | $-\frac{1}{3}$ | 0 | −1 | 0 |
| top,[b,c] t | $\geq$70 000 | $+\frac{2}{3}$ | $+\frac{2}{3}$ | 0 | 0 | 1 |

[a]In units of electron mass.
[b]The b and t quarks have also been named the "beauty" and "truth" quarks.
[c]No hadron containing the top quark has yet been observed; all such hadrons must be very massive.

"colors," called red, green, and blue after the primary colors of light. To every quark there corresponds an antiquark, making a total of six quarks of each flavor.

The most curious property of quarks is their fractional charge. All quarks have charge $+\frac{2}{3}e$ or $-\frac{1}{3}e$. All hadrons have integral charge, and so every hadron must be constructed of quarks in such a way that their charges add to the proper total.

The other numbers listed in the table—*hypercharge, charm, bottomness,* and *topness*—are quantum numbers representing conservation principles to which quarks (and the particles that are built of them) are subject.

*Every baryon is made up of three quarks.* Only the two least-massive types, u and d, are involved in the structure of the two nucleons. The proton is made up of the combination uud. Adding the charges listed in Table 45.2, you can see that the proton charge has the correct value $\frac{2}{3} + \frac{2}{3} - \frac{1}{3} = 1$, in units of the fundamental charge $e$. Similarly, the hypercharge adds to 1, the correct value for a nucleon. (For nucleons and other "nonstrange" baryons, the hypercharge is equal to the baryon number.) The charm, bottomness, and topness are all equal to zero.

The neutron is made up of the combination udd. This combination yields electric charge $\frac{2}{3} - \frac{1}{3} - \frac{1}{3} = 0$, as required; the hypercharge is again 1 and the other quantum numbers 0. For the $\Sigma^-$ particle ($m = 2335m_e$, or about 1.27 times the proton mass), the combination is dds, yielding electric charge $-1$ and hypercharge 0. More massive particles often contain still more massive quarks.

*Every meson is made up of one quark and one antiquark.* The $\pi^+$ meson, one of the least massive and longest lived of the mesons, consists of an up quark, u, and an antidown, $\bar{\text{d}}$. The charge is $\frac{2}{3} + \frac{1}{3} = 1$.

Both baryons and mesons fall into groups, in accord with the particular quarks they contain. In a group consisting of the same quarks, the differences among the members of the group are accounted for on the basis of the relative orientation of the quarks, which affects their mass-energy. There is a close analogy to the way in which atomic energy levels having the same principal quantum number $n$ are distinguished by the differences in their angular momentum quantum numbers. The quark picture thus reduces the numerous hadrons to systems of energy levels that can accommodate excited states of a few basic types of particle.

Free quarks have not yet been observed, in spite of substantial efforts to do so. The large force between quarks is due to their "color charge," a property associated with their color. Unlike the strong nuclear force, which is very strong but falls off rapidly with distance, the color force between quarks does not diminish, and may even increase, with separation. Thus quarks within a particle are bound together very tightly. Associated with this force are hypothetical particles, called *gluons* because they are the "glue" that holds matter together.

Many nuclear processes can be reanalyzed in terms of what happens to the constituent quarks during the process. Consider, for instance, the simplest β-decay process, in which a neutron decays spontaneously into a proton, an electron, and an antineutrino:

$$\text{n} \rightarrow \text{p} + \text{e}^- + \bar{\nu}. \tag{45.6a}$$

If we substitute for the proton and the neutron their constituent quarks, this process becomes

$$\text{udd} \rightarrow \text{uud} + \text{e}^- + \bar{\nu}. \tag{45.6b}$$

This new view of the process has the advantage that it expresses the process entirely in terms of fundamental particles—quarks and leptons.

Next, notice that one u and one d quark appear unaltered on both sides of the arrow. We therefore remove them from the process to obtain the more fundamental view

$$\text{d} \rightarrow \text{u} + \text{e}^- + \bar{\nu}. \tag{45.6c}$$

That is, the core of the β-decay process is the transformation of a down quark into an up quark. According to Table 45.2, the d quark is more massive than the u quark; this explains why the neutron is more massive than the proton and why the decay products have considerable kinetic energy. (However, do not try to calculate this energy by simple subtraction of the mass of the u from the mass of the d. On account of the binding energy involved when quarks are bound into hadrons, the masses of bound quarks are not those given in Table 45.2.)

Although the quark picture has been remarkably successful in accounting for the existence and the properties of the hadrons, the multiplicity of quarks (as mentioned earlier, thirty-six at present count—six quarks and their antiquarks, each in three different colors) makes one wonder whether even the quarks are fundamental. (Indeed, the existence of still other quarks has been postulated.) Arguments have been proposed in which quarks themselves are built up of still more fundamental particles, variously called "preons" or "rishons." Although intriguing, these arguments are highly speculative at present. Nevertheless, one wonders today—as physicists must have wondered after the discovery of the muon in the late 1930s, and chemists before them in the mid-nineteenth century—whether the search for truly fundamental particles will ever end.

# Fundamental Forces and Their Mediating Particles

An electron exerts a Coulomb force on a positron in spite of the fact that the space between them is empty. How can this be? We have usually dealt with this question in terms of the electric field that exists in the space surrounding any charge. The field approach makes it possible to carry out all kinds of calculations and theoretical developments, but it does not close the question.

The problem was evident to René Descartes in the seventeenth century. It was largely for this reason that Descartes postulated a space filled with matter, through which forces could be transmitted. Newton preferred to take the view that space is largely empty and proposed that some forces act over a distance.

Only in the present century has it been possible to resolve the difficulty in a definitive way. For the electromagnetic force, the solution is complete and exceedingly precise; the theory that carries the solution is called **quantum electrodynamics** or QED for short. In the spirit of quantum theory, QED takes a dual view. The conceptual and mathematical difficulties of the theory place it beyond the scope of this book. Nevertheless, we can sketch some of the basic ideas.

Associated with every force is a **mediating particle** that carries the force. We can make a classical analogy as follows. Two children stand on frictionless ice and play catch. Every time one of them throws the ball and the other catches it, the children are accelerated away from one another. Thus the game of catch can be interpreted in terms of a repulsive force that exists between the children, *mediated* by the ball that moves back and forth between them.

**QED**

The crucial contributions to the development of quantum electrodynamics were made by the Japanese physicist Sin-itiro Tomonaga (1906–1979) and the American physicists Julian Schwinger (1918–1994) and Richard Feynman (1918–1988). They shared a Nobel Prize in 1965 for this work.

For the electromagnetic force, the photon plays the role of the ball. An additional complication, for which there is no classical analogue, is that the momentum exchange resulting from the transfer of the photon can be either repulsive or attractive, depending on the circumstances.

Why do we not see the mediating photon? The reason lies in the uncertainty principle. The photon spends sufficiently little time en route from one particle to the other that the uncertainty in its energy is comparable with its energy—that is, $\Delta E \geq h/\Delta t \geq E$. We call such a photon *virtual*, and it is not directly observable. Virtual mediating particles are considered in more detail in Problem 45.20.

QED is well established in part because it is easy to do experiments involving the electromagnetic force. The theories for the three other fundamental forces are less complete. However, we can safely presume that each force possesses its own mediating particle. The particle associated with the gravitational force is the **graviton**. Like the photon, it has zero rest mass and zero charge. However, the gravitational force is very weak, and the graviton has not yet been observed experimentally.

The particle that mediates the strong nuclear force is the gluon, mentioned earlier in connection with quarks. The gluon has a range comparable to the nuclear diameter—of order $10^{-15}$ m—and it does not exist as a free particle. The particle that mediates the weak force is the **intermediate boson**. Various types of this particle, called the $W^+$, the $W^-$, and the $Z^0$, were detected in 1984. The range of the weak force is even less than that of the strong force—about $10^{-18}$ m.

The theory developed in analogy to QED to account for the strong force is called **quantum chromodynamics** or QCD for short. QCD is complicated by the fact that there are three "colors" to be considered, in contrast with only two kinds of electric charge in QED. For this and other theoretical reasons, and because of the much greater difficulty of experiments involving the strong force, QCD is not nearly as well developed as QED.

## SECTION 45.6    Toward Fundamental Unity

A major long-term goal of modern physics is to find a way to describe all four fundamental forces as special cases of a single fundamental force. The difficulties of this project are formidable, in view of the very different properties of the forces, summarized in Table 45.3.

### TABLE 45.3  Comparison of the Fundamental Forces

| Force | Relative Strength | Range | Mediating Particle |
| --- | --- | --- | --- |
| Strong | 1 | $10^{-15}$ m | gluon |
| Electromagnetic | $10^{-2}$ | $\infty$ | photon |
| Weak | $10^{-6}$ | $10^{-18}$ m | $W^+$, $W^-$, $Z^0$ bosons |
| Gravitational | $10^{-38}$ | $\infty$ | graviton |

Nevertheless, progress has been made toward this goal. In 1968, Glashow, Salam, and Weinberg* proposed an **electroweak theory** that describes the electromagnetic and weak forces as different aspects of a single **electroweak force**. This unification may be seen as taking one step further the work of Maxwell and Einstein, who unified the electric and magnetic forces. The great triumph of the electroweak theory was its prediction of the (very large) mass of the intermediate bosons.

The next step is to unify the electroweak and strong forces. A theory that does this is called a *grand unified theory* or GUT for short. At present, GUTs are at best incomplete, but progress continues at a rapid rate. The final step constitutes unification of all four forces. A theory that does this is called, only half jokingly, a "theory of everything." The development of such theories remains in the speculative stage.

Ultimate theories of matter are intimately connected with theories of the origin and structure of the universe. This is true in part because the complications introduced by the fourfold multiplicity of fundamental interactions melt away at exceedingly high temperatures. It is just such conditions that existed in the very early universe. The interactions that took place during the first brief moments of the universe have left their signatures in a variety of ways that can be seen by astronomers when they observe very distant astronomical objects. Conversely, the properties of fundamental particles and their interactions govern the evolution of the universe and have done so from the earliest moments. There is esthetic delight in contemplating the way in which the study of the smallest things in the universe has merged with the study of the largest things.

> *. . . Then Janus looks both ways anew, his sight*
> *Amended by the Glass of Knowledge. Past*
> *From Future sundered by his gate no more,*
> *His portal opened wide, now Great and Small*
> *Are one world whole.*

---

*Sheldon Glashow (b. 1932) and Steven Weinberg (b. 1933), American physicists, and Abdus Salam (b. 1926), Pakistani physicist.

A relativistic treatment of quantum theory leads to the conclusion that there exists an **antiparticle** for every elementary particle. The **positron** is the antiparticle of the electron. The positron has the same spin and mass as the electron, but opposite charge $+e$. Electron-positron pairs can be created from photons and can be annihilated to produce photons.

Ordinary matter consists of particles that fall into two general categories called **hadrons** and **leptons**. Both hadrons and leptons are acted on (very weakly) by the gravitational force, and electrically charged particles of both types are acted on by the electromagnetic force. In addition, hadrons are acted on by both the **strong force** and the **weak force**, but leptons are acted on by the weak force only. The strong force is mainly responsible for holding nuclei together against the Coulomb repulsion arising from their positive charge. The weak force comes into play in β-decay processes.

In addition to the **nucleons**—the proton and the neutron—there exist many other hadrons. All of these hadrons are unstable and quickly decay into less massive particles.

The electron and its associated **electron neutrino** $\nu_e$ are the only stable leptons. The **muon** and the **tauon** are unstable and may be regarded as excited states of the electron. There are neutrinos associated with the muon and the tauon. Leptons appear to be truly fundamental particles.

Hadrons are not truly fundamental particles. All hadrons consist of combinations of **quarks**. There are two subclasses of hadrons: **baryons** and **mesons**. Baryons are combinations of three quarks (or antiquarks in the case of antibaryons); mesons consist of quark-antiquark pairs. The construction of hadrons from quarks is subject to a set of conservation principles. The same conservation principles underlie the processes by which hadrons interact and decay.

Free quarks, with their remarkable fractional electric charges, have not been observed. The strong force, mediated by **gluons**, confines them within the bounds of the hadrons of which they are part.

Each of the four fundamental forces is mediated by a particle. The electromagnetic force is mediated by photons. The photons are not detected on account of their very short lifetimes as independent particles; consequently, they are called **virtual photons**.

The gravitational force is mediated by **gravitons**. Owing to the weakness of the gravitational interaction, gravitons have not yet been detected experimentally. The mediating particle for the weak interaction is called the **intermediate boson**.

Efforts to discern the unity underlying the four fundamental interactions have been partially successful. The electromagnetic and weak interactions may be regarded as two aspects of a single **electroweak interaction**. Current effort is directed toward a grand unified theory that will unify the strong and electroweak interactions. A successful "Theory of Everything" that unifies all four interactions lies further in the future.

### KEY TERMS

**Section 45.2 Particles and Antiparticles**
antielectron (positron) ▪ annihilation, electron-positron pair ▪ antiproton ▪ antimatter

**Section 45.3 The Strong and Weak Forces, Hadrons and Leptons, Baryons and Mesons**
strong nuclear force, weak nuclear force ▪ electron neutrino ▪ muon, muon neutrino ▪ tauon, tauon neutrino ▪ pion

**Section 45.4 Quarks**
flavors, colors

**Sections 45.5 and 45.6 Fundamental Forces and Their Mediating Particles; Toward Fundamental Unity**
quantum electrodynamics ▪ quantum chromodynamics ▪ graviton ▪ electroweak theory, electroweak force

---

## QUERIES

**45.1** *(2) Curling up.* In Figure 45.2, the radius of curvature of the positron track above the lead plate is less than the curvature below the lead plate. Explain.

**45.2** *(3) You can't have everything!* The negative tauon, whose mass is greater than that of the proton, can decay in a process of which an electron is one of the end products. The proton does *not* decay into a positron according to the process

$$p \rightarrow e^+ + \text{energy}.$$

Why does this process not take place?

**45.3** *(3) Cosmic Tinkertoy.* Taking leptons and quarks together, how many kinds of fundamental particles are required to construct the world of ordinary matter?

**45.4** *(3) End of the line.* When a baryon decays, the chain of decays may be quite long and complex. However, there is always a proton among the stable end products. When a meson decays, there is never a proton among the end products unless there is also an antiproton or antineutron, even though the mass-energy of many mesons is greater than that of a proton. Explain.

## PROBLEMS

### GROUP A

**45.1** *(2) Annihilation, I.* A positron and an electron annihilate to form two $\gamma$-ray photons. **(a)** What is the minimum energy of each photon? **(b)** What is the corresponding photon wavelength?

**45.2** *(2) Annihilation, II.* A positron and an electron annihilate to form two photons. Each photon has energy 1.35 MeV. What is the total kinetic energy of the positron and the electron just before they collide?

**45.3** *(2) Creation, I.* What is the minimum energy required to create a proton-antiproton pair?

**45.4** *(2) Creation, II.* What is the minimum energy required to create a neutron-antineutron pair?

**45.5** *(3) Neutron decay.* Free neutrons decay (with mean life 898 s) according to the process

$$n \rightarrow p + e^- + \bar{\nu}_e.$$

**(a)** What is the total kinetic energy released in the decay? **(b)** What conservation rules require that the electron antineutrino be one of the products?

**45.6** *(3) Death and transfiguration.* The neutral pion, $\pi^0$, usually decays into two photons. If the rest mass of the $\pi^0$ is $264.2m_e$, what is the energy of each photon? Neglect the kinetic energy of the $\pi^0$.

**45.7** *(3) Kaon kollapse.* The neutral $K^0$ meson, or kaon, can decay in either of the following two modes:

$$K^0 \rightarrow \pi^+ + \pi^- \quad \text{or} \quad K^0 \rightarrow \pi^0 + \pi^0.$$

Assume that the kaon has negligible kinetic energy. For each decay mode, find the kinetic energy of the daughter pions. The mass of the kaon is $974m_e$; the mass of the $\pi^0$ is $264m_e$; the mass of the $\pi^+$ and the $\pi^-$ is $274m_e$.

**45.8** *(3) Yes or no.* Which of the following processes are forbidden by the principle of conservation of lepton number? **(a)** $n \rightarrow p + e^- + \nu_e$; **(b)** $\pi^+ \rightarrow \mu^+ + e^- + e^+$; **(c)** $\pi^- \rightarrow \mu^- + \nu_\mu$; **(d)** $p + e \rightarrow n + \nu_e$; **(e)** $\mu^+ \rightarrow e^+ + \nu_e + \bar{\nu}_\mu$; **(f)** $K^- \rightarrow \mu^- + \bar{\nu}_\mu$. (The kaon is a meson; see Problem 45.7.)

**45.9** *(3) No-no.* What conservation principles do the following forbidden processes violate? **(a)** $\pi^- + p \rightarrow \Omega^- + K^+ + K^-$ (the $\Omega^-$ is a baryon; the K's are mesons); **(b)** $K^- + n \rightarrow \Omega^- + K^+ + K^0$; **(c)** $\pi^- \rightarrow \mu^- + e^+ + e^-$; **(d)** $\mu^- \rightarrow e^- + \nu_e + \bar{\nu}_\mu$. (Hint: Compare with Reaction 45.4.)

**45.10** *(4) Building blocks.* What component quarks are needed to make up an antiproton, $\bar{p}$? an antineutron, $\bar{n}$?

**45.11** *(5) Weak, weaker, weakest.* An electron is located an arbitrary distance from a positron. What is the ratio of the electric force to the gravitational force exerted on the electron by the positron?

### GROUP B

**45.12** *(3) Art is long, life short, I.* You observe the tracks of 100-MeV negative pions in a large detector. You find that, on the average, the pions travel 11 m before decaying. **(a)** What is the directly observed mean life $\tau'$ of the pions? **(b)** What is their proper mean life $\tau$—the mean life as observed by an observer moving along with them?

**45.13** *(3) Art is long, life short, II.* Negative pions decay according to the process

$$\pi^- \rightarrow \mu^- + \bar{\nu}_\mu.$$

Assuming that the initial kinetic energy of the pion is negligible, use the principles of conservation of mass-energy and momentum to find the kinetic energy of the daughter $\mu^-$ and the energy of the neutrino.

**45.14** *(3) Making $\pi$'s without an oven.* One way to produce $\pi^+$ mesons in the laboratory is to use an accelerator to make an energetic proton collide with a proton at rest in a hydrogen target. One of the reactions that takes place is

$$p + p \rightarrow p + n + \pi^+.$$

The mass of the $\pi^+$ is $273m_e$. What minimum kinetic energy must the accelerator give the proton if this reaction is to proceed? (Mass-energy and momentum must be conserved.)

**45.15** *Extra precision.* Rework Example 45.2, improving the accuracy of the result by taking the rest mass of the daughter electron explicitly into account.

**45.16** *Fuzzy, I.* The $\Lambda^0$ particle has a mean life $\tau = 2.63 \times 10^{-10}$ s. **(a)** What is the maximum precision with which the rest energy of the particle can be measured? (This is called the *energy width.*) **(b)** The mass of the $\Lambda^0$ is $2180m_e$. Calculate the mass uncertainty $\Delta m/m$.

**45.17** *Fuzzy, II.* The $\phi^0$ meson has energy width (see Problem 45.16) 4.2 MeV. Estimate the mean lifetime of the $\phi^0$.

**45.18** (3) *Little Bang.* The Large Magellanic Cloud is located at a distance $R = 4.9 \times 10^{17}$ m away from us. In 1987, a supernova explosion was observed in the cloud. Among the many types of matter and radiation emitted copiously were visible-light photons and neutrinos. **(a)** Assume that the neutrinos have energy $E$ and rest mass $m_0$, and write an expression for the time interval between the arrival at the earth of the photons and that of the neutrinos. **(b)** This interval was measured and found to be no more than 1.5 s. Set an upper limit on the mass of the neutrino. (Hint: See Problem 38.9.)

## GROUP C

**45.19** (2) *Why colliders?* In the Berkeley Bevatron experiment that first produced an antiproton in the laboratory, a beam of energetic protons came to rest in a solid sample containing protons to produce the reaction

$$p + p \rightarrow p + p + p + \bar{p}.$$

**(a)** Impose the requirement that momentum as well as energy be conserved, and calculate the minimum energy of the incident proton that can produce the reaction. **(b)** Suppose that, instead of producing a collision between a moving proton and a stationary one, we produce a collision between two protons of equal energy moving in opposite directions. What is the minimum energy required to produce the reaction of part **a**? Your conclusion should make clear the justification for building modern colliding-beam accelerators, in which two particle beams of equal energy, moving in opposite directions, are the sources of the particles that collide.

**45.20** (3) *The Yukawa particle.* In 1935, the Japanese physicist Hideki Yukawa (1907–1981) argued that the strong nuclear force was mediated by the exchange of a then-unknown virtual particle, later identified as the pion. He argued by analogy with the photon, which mediates the Coulomb force. Yukawa was able to predict the mass of the pion, $m_\pi$, with fair accuracy, using the following approach. **(a)** A mediating particle must be undetectable because its spontaneous emission and absorption violate the principle of conservation of energy. What is the uncertainty in the energy of the system if a particle of rest mass $m_\pi$ is continually created and annihilated? **(b)** Use the uncertainty principle to express the maximum time $\Delta t$ that such a particle can exist if it is to be undetectable. **(c)** What is the maximum distance the particle can travel? **(d)** The range of the strong force is something like $2 \times 10^{-15}$ m. What is the mass of Yukawa's particle? Compare your result with the value later observed experimentally, about $270m_e$. **(e)** Explain why the photon must have zero rest mass. **(f)** The rest mass of the intermediate boson is about $90m_p$. What is the range of the weak force, which it mediates?

(Historical note: Yukawa received a Nobel Prize in 1949 for his prediction. For some years, the Yukawa particle was believed to be the muon, discovered in cosmic radiation by Carl Anderson and Seth Neddermeyer (1907–1988) in 1937. But experiments just after World War II showed that the muon interacted only weakly with nucleons and could thus not be the mediating particle for the strong force. Finally, in 1947, the British physicist C. F. Powell (1903–1969) and the Italian physicist G. P. S. Occhialini (1907–1993) discovered the pion in cosmic-ray debris. Powell was awarded a Nobel Prize in 1950 for this and related work.)

**45.21** (3) *Threshold energy.* An accelerator is used to produce an energetic beam of particles $A$, which collide with a stationary target consisting of particles $B$. This results in the production of particles $1, 2, 3, \ldots, n$, according to the process

$$A + B \rightarrow 1 + 2 + 3 + \cdots + n.$$

The rest masses of the particles are $M_A, M_B, m_1, m_2, m_3, \ldots, m_n$. Show that the minimum kinetic energy of the $A$'s must be

$$K = \frac{(m_1 + m_2 + m_3 + \cdots + m_n)^2 - (M_A + M_B)^2}{2M_B} c^2.$$

(Hint: Mass-energy and momentum must be conserved; see Problem 45.19.)

**45.22** *Scattering.* A neutral pion decays into two $\gamma$ photons. Let the rest mass of the $\pi^0$ be $m_\pi$, and suppose that the initial kinetic energy of the $\pi^0$ is $K$. **(a)** Show that the angle $\theta$ between the daughter $\gamma$'s is given by the relation

$$K = m_\pi c^2 \left( \csc \frac{\theta}{2} - 1 \right).$$

**(b)** Show that the energy $E_\gamma$ of each of the $\gamma$'s is

$$E_\gamma = \frac{1}{2} m_\pi c^2 \sin \frac{\theta}{2}.$$

**(c)** Find the values of $K$ and $E_\gamma$ if $\theta = 60°$.

# APPENDIX 1

# Measuring Angles in Radians

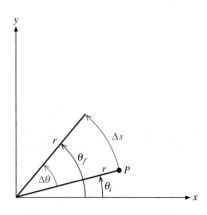

**FIGURE A1.1**

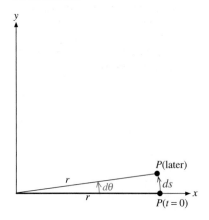

**FIGURE A1.2**

Angles can be expressed in a number of ways, as in degrees (°) or revolutions (rev). But it is most often convenient in mathematical and physical analysis to express them in terms of **radians** (rad), the SI unit of angle measure.

As shown in Figure A1.1, a circle of arbitrary radius $r$ is drawn centered on the vertex of an arbitrary angle $\Delta\theta$. The angle subtends an arc whose length (measured along the arc) is $\Delta s$. The angle $\Delta\theta$, expressed in radians, is defined to be the ratio of the length $\Delta s$ of the arc to the radius $r$:

$$\Delta\theta \equiv \frac{\Delta s}{r}. \qquad \textbf{(A1.1)}$$

This definition is just as valid when the arc length and the angle are infinitesimal. Hence, in the case shown in Figure A1.2, the infinitesimal angle $d\theta$ is given by the ratio

$$d\theta = \frac{ds}{r}. \qquad \textbf{(A1.2)}$$

Because the angle $\Delta\theta$ or $d\theta$ is thus defined as the ratio of two lengths, it is itself a dimensionless quantity. (The same is true of angles measured in other units, but that is not so easy to see.) As noted in Chapter 6, it follows that the unit of angular velocity, rad/s, can equally well be expressed as reciprocal seconds ($\text{s}^{-1}$). Similarly, as noted in Chapter 11, it follows that the unit of angular acceleration, rad/s², can equally well be expressed as reciprocal seconds squared ($\text{s}^{-2}$).

## EXAMPLE

**(a)** Express in radians the value of the angular displacement equal to one complete revolution, or 360°. **(b)** The shaft of an electric motor rotates at 3600 rev/min. Express this angular velocity in SI units. **(c)** Find the factor for converting angular velocities from rev/min into rad/s.

**SOLUTION:**

**(a)** Express in radians the value of the angular displacement equal to one complete revolution.

When the system of Figure A1.1 experiences one complete revolution, the typical point $P$ describes a circle whose circumference is $\Delta s = 2\pi r$. Inserting this value of $\Delta s$ into Equation A1.1 gives $\Delta\theta = \Delta s/r = 2\pi r/r = 2\pi$. So one revolution (or 360°) is equal to $2\pi$ radians.

**(b)** Express the angular velocity 3600 rev/min in SI units.

Using the result of part **a**, you have

$$\omega = 3600\,\frac{\text{rev}}{\text{min}} \times \frac{2\pi\,\text{rad}}{1\,\text{rev}} \times \frac{1\,\text{min}}{60\,\text{s}}$$
$$= 377.0\,\text{rad/s}.$$

**(c)** Find the factor for converting angular velocities from rev/min into rad/s.

The principles underlying the calculation in part **b** can be expressed more generally by writing, for any angular velocity $\theta = N$ rev/min,

$$N\,\text{rev/min} = N\,\frac{\text{rev}}{\text{min}} \times \frac{2\pi\,\text{rad}}{1\,\text{rev}} \times \frac{1\,\text{min}}{60\,\text{s}}$$
$$= \frac{2\pi}{60}\,N\,\text{rad/s} = 0.1047N\,\text{rad/s}.$$

# Derivation of the Work Integral

In Figure A2.1, a body moves along an arbitrary path $s$ that begins at $s_i$ and ends at $s_f$. As it moves, the body is acted on by a force whose magnitude and direction both vary in an arbitrary way. Because of this variation, it is not possible to define the work by means of Equation 7.5c, $W = \mathbf{F} \cdot \mathbf{s}$. Nevertheless, the force is doing work on the body, and we must find a way of expressing that work.

We begin by approximating the path by a large number of small, straight segments $\Delta \mathbf{s}$. As noted in Section 7.3, each segment has vectorial properties even though the path as a whole does not. This is because each segment has a definite direction $\Delta \hat{\mathbf{s}}$ as well as a length $\Delta s$:

$$\Delta \mathbf{s} = \Delta \hat{\mathbf{s}} \, \Delta s. \tag{A2.1}$$

As the body moves along the path, the force $\mathbf{F}$ varies. However, we have chosen the segments $\Delta \mathbf{s}$ short enough so that $\mathbf{F}$ does not vary much over any particular segment. Thus, over each particular segment $\Delta \mathbf{s}_j$, the force does not vary much from the value $\mathbf{F}_j$ measured at, say, the middle of the segment. The small amount of work $\Delta W_j$ done on the body by the force as the body moves through the segment $\Delta \mathbf{s}_j$ is therefore given by

$$\Delta W_j \simeq \mathbf{F}_j \cdot \Delta \mathbf{s}_j. \tag{A2.2a}$$

This is just Equation 7.5c, rewritten so that it applies specifically to the present case. Because $\Delta \mathbf{s}_j = \Delta \hat{\mathbf{s}}_j \, \Delta s_j$, we have

$$\Delta W_j \simeq (\mathbf{F}_j \cdot \Delta \hat{\mathbf{s}}_j) \, \Delta s_j.$$

The quantity in parentheses in this approximation has the value $F_j \cos \theta_j$, where $\theta_j$ is the angle between the force and the direction of motion at segment $j$. Approximation (A2.2a) thus becomes

$$\Delta W_j \simeq F_j \cos \theta_j \, \Delta s_j. \tag{A2.2b}$$

Compare this with Equation 7.5a, $W = |\mathbf{F}| \cos \theta \, |\mathbf{s}|$.

Approximation A2.2b applies equally well to each of the many segments that make up the path. Thus the work done by the force over the entire path is closely approximated by the *sum* of the $\Delta W_j$:

$$W \simeq \Delta W_1 + \Delta W_2 + \cdots + \Delta W_j + \cdots + \Delta W_{n-1} + \Delta W_n$$

or $\quad W \simeq F_1 \cos \theta_1 \, \Delta s_1 + F_2 \cos \theta_2 \, \Delta s_2 + \cdots + F_j \cos \theta_j \, \Delta s_j + \cdots$

$$+ \, F_{n-1} \cos \theta_{n-1} \, \Delta s_{n-1} + F_n \cos \theta_n \, \Delta s_n. \tag{A2.3a}$$

This approximation can be written in the condensed form

$$W \simeq \sum_{j=1}^{n} F_j \cos \theta_j \, \Delta s_j. \tag{A2.3b}$$

Approximation A2.3 in either form suggests a direct experimental way of evaluating $W$, at least in principle. If we measure $\mathbf{F}_j$ at each of the $n$ midpoints of the segments $\Delta s_j$, we can evaluate each of the terms in the equation and then evaluate the sum.

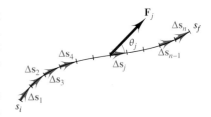

FIGURE A2.1 A body moves along the arbitrary path shown. The path may be approximated by a series of short, straight segments, of which $\Delta \mathbf{s}_j$ is a typical one. When the body lies within the segment $\Delta \mathbf{s}_j$, it is acted on by a force $\mathbf{F}_j$ which makes an angle $\theta_j$ with $\Delta \mathbf{s}_j$.

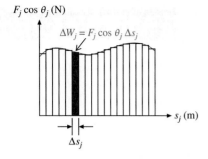

$F_j \cos \theta_j$ (N)

$\Delta W_j = F_j \cos \theta_j \, \Delta s_j$

$s_j$ (m)

$\Delta s_j$

**FIGURE A2.2** The work $\Delta W$ done on the body as it passes over the segment $\Delta s_j$ is approximated by the area of the shaded rectangle. This area is expressed algebraically by Equation A2.2b. Over the entire path, the work $W$ done is approximated by the sum of the areas of the rectangles, expressed algebraically by Equation A2.3b. The work $W$ is given exactly by the area under the red curve. Evaluating Equation A2.4b is equivalent to the geometric process of measuring the area under the curve.

Figure A2.2 is a graphic representation of Approximation A2.3. According to the approximation, $\Delta W_j$ is the product of the height $F_j \cos \theta_j$ of the $j$th narrow rectangle in the figure with its width $\Delta s_j$. Thus $\Delta W_j$ is the *area* of the $j$th rectangle. The total work $W$, which is the sum of the $\Delta W_j$, is approximated by the sum of the areas of these rectangles. The approximation can be improved by using a larger number of smaller segments $\Delta s_j$, but at the cost of more labor.

This method of directly measuring the areas of the small rectangles and adding them up is useful on occasion. But another method is useful much more frequently and is exact. We can often express $F \cos \theta$ as a mathematical function of the position $s$ measured along the curve from the starting point. Suppose that the red curve in Figure A2.2 represents the graph of such a function. Because the function is continuous, we know its value for *every* value of $s$, at least in principle. We can therefore evaluate and sum the products $F_j \cos \theta_j \, \Delta s_j$, no matter how narrow we choose the segments $\Delta s_j$.

Accordingly, we divide the path into an infinite number of infinitesimal segments $d\mathbf{s}$. That is, we evaluate the sum of Approximation A2.3b in the limit where $\Delta s$ approaches zero:

$$W = \lim_{\Delta s \to 0} \sum_j \Delta W_j = \lim_{\Delta s \to 0} \sum_j F_j \cos \theta_j \, \Delta s_j. \qquad \textbf{(A2.4a)}$$

This is an equation rather than an approximation because the error that comes from using a fixed value of $F_j \cos \theta_j$ over each interval $\Delta s_j$ vanishes as the intervals become infinitesimal. The values of $j$ over which the sum is evaluated still range from 1 to $n$, just as in Approximation A2.3. But now $n$ approaches infinity as $\Delta s$ approaches zero.

When the limiting process is carried out, the right side of Equation A2.4a is written

$$W = \int_{s_i}^{s_f} F(s) \cos \theta(s) \, ds. \qquad \textbf{(A2.4b)}$$

The quantity on the right side of the equation is called the **integral** of the function $F \cos \theta$ over the specified path $s$ beginning at $s_i$ and ending at $s_f$; the quantities $s_i$ and $s_f$ are called the **limits of integration**. The notations $F(s)$ and $\theta(s)$ remind us that both $F$ and $\theta$ are variables whose values are known as functions of $s$. The value of the integral is exactly the area under the red curve of Figure A2.2. Equation A2.4b is identical with Equation 7.13b.

Over each infinitesimal segment $ds$, an amount of work $dW$ is done. The value of $dW$ is implicit in Equation A2.4a; it is

$$dW = \lim_{\Delta s \to 0} \Delta W = F \cos \theta \, ds, \qquad \textbf{(A2.5)}$$

where $dW$, $F$, and $\theta$ are all functions of $s$. This equation is analogous to the finite approximation A2.2b. It follows immediately that Equation A2.4b can be written in the compact form

$$W = \int_{s_i}^{s_f} dW, \qquad \textbf{(A2.6)}$$

which is Equation 7.12. We have thus shown that the work done on a body by a varying force as the body moves over an arbitrary path can be represented by a definite integral.

# Selected Mathematical Relations

## Exponentials and Logarithms

$$x^0 = 1; \quad x^1 = x$$

$$x^{-a} = \frac{1}{x^a}; \quad x^a x^b = x^{a+b}; \quad \frac{x^a}{x^b} = x^{a-b}$$

$$\log xy = \log x + \log y; \quad \log \frac{x}{y} = \log x - \log y;$$

$$\log x^a = a \log x$$

### Inverse relation between exponentiation and taking the logarithm:

$$\log_a a^x = x; \quad a^{\log_a x} = x$$

### Change of base:

In general, $\log_b a \log_a x = \log_b x$

In particular, for $\log x \equiv \log_{10} x$ and $\ln x \equiv \log_e x$,

$$\log x = \log e \ln x = 0.4343 \ln x;$$
$$\ln x = \ln 10 \log x = 2.303 \log x$$

## Quadratic Formula

If $ax^2 + bx + c = 0$, then $x = \dfrac{-b \pm \sqrt{b^2 - 4ac}}{2a}$

## Pythagoras's Theorem and Trigonometry

### For a right triangle:

$$x^2 + y^2 = r^2 \text{ or, equivalently, } \left(\frac{x}{r}\right)^2 + \left(\frac{y}{r}\right)^2 = 1$$

### Definitions of the trigonometric functions:

$$\cos \theta = \frac{x}{r} \qquad \sec \theta = \frac{1}{\cos \theta} = \frac{r}{x}$$

$$\sin \theta = \frac{y}{r} \qquad \csc \theta = \frac{1}{\sin \theta} = \frac{r}{y}$$

$$\tan \theta = \frac{y}{x} = \frac{\sin \theta}{\cos \theta} \qquad \cot \theta = \frac{1}{\tan \theta} = \frac{x}{y}$$

### From the second form of Pythagoras's theorem and the definitions of cos θ and sin θ,

$$\cos^2 \theta + \sin^2 \theta = 1$$

### Other trigonometric identities:

$$\cos x = \sin (90° - x) = -\cos (180° - x)$$

$$\sin x = \cos (90° - x) = \sin (180° - x)$$

$$\cos (-x) = \cos x; \quad \sin (-x) = -\sin x;$$
$$\tan (-x) = -\tan x$$

$$\cos (x \pm y) = \cos x \cos y \mp \sin x \sin y$$

$$\sin (x \pm y) = \sin x \cos y \pm \cos x \sin y$$

$$\sin x \pm \sin y = 2 \sin \frac{x \pm y}{2} \cos \frac{x \mp y}{2}$$

$$\cos x + \cos y = 2 \cos \frac{x + y}{2} \cos \frac{x - y}{2}$$

$$\cos x - \cos y = -2 \sin \frac{x + y}{2} \sin \frac{x - y}{2}$$

$$\cos \frac{x}{2} = \sqrt{\frac{1 + \cos x}{2}}; \quad \sin \frac{x}{2} = \sqrt{\frac{1 - \cos x}{2}}$$

### Signs of sin x, cos x, and tan x in the four quadrants:

The quadrants in which the functions are positive are conveniently remembered by means of the mnemonic *Act Stupid, Take Chances.* All three functions are positive in the first quadrant, the *s*ine in the second quadrant, and so forth.

## General Triangles

Law of sines: $\dfrac{A}{\sin \alpha} = \dfrac{B}{\sin \beta} = \dfrac{C}{\sin \gamma}$

Law of cosines:

scalar form, $C^2 = A^2 + B^2 - 2AB \cos \gamma$

vector form, $C^2 = A^2 + B^2 + 2\mathbf{A} \cdot \mathbf{B}$

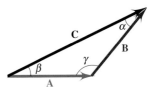

## Areas

Area of a triangle of base $b$ and altitude $h$:   $A = \frac{1}{2}bh$

Area of a rectangle or parallelogram of base $b$ and altitude $h$:   $A = bh$

Area of a regular polygon having $n$ sides of length $l$:
$A = \frac{1}{4}nl^2 \cot(180°/n)$

Area of a circle of radius $r$:   $A = \pi r^2$

Area of an ellipse with major axis $a$ and minor axis $b$:
$A = \pi ab$

Surface area of a sphere of radius $r$:   $A = 4\pi r^2$

## Volumes

Volume of a parallelepiped of base area $A$ and altitude $h$:
$V = Ah$

Volume of a sphere of radius $r$:   $V = \frac{4}{3}\pi r^3$

Volume of a circular cylinder of radius $r$ and altitude $h$:
$V = \pi r^2 h$

Volume of a cone of base radius $r$ and altitude $h$:
$V = \frac{1}{3}\pi r^2 h$

## Selected Series Expansions

### Binomial expansion:

$$(1 \pm x)^n = 1 \pm nx + \frac{n(n-1)x^2}{2!}$$
$$\pm \frac{[n(n-1)(n-2)x^3]}{3!} + \cdots$$

Any series $(x \pm y)^n$, with $y < x$, can be written

$$x^n\left(1 \pm \frac{y}{x}\right)^n,$$

and the preceding formula applied.

### Taylor series:

$$f(x + h) = f(x) + hf'(x) + \frac{h^2}{2!}f''(x) + \frac{h^3}{3!}f'''(x) + \cdots$$

$$e^x = 1 + x + \frac{x^2}{2!} + \frac{x^3}{3!} + \cdots$$

$$\ln x = (x-1) - \frac{1}{2}(x-1)^2 + \frac{1}{3}(x-1)^3 - \cdots$$

$$\sin x = x - \frac{x^3}{3!} + \frac{x^5}{5!} - \frac{x^7}{7!} + \cdots$$

$$\cos x = 1 - \frac{x^2}{2!} + \frac{x^4}{4!} - \frac{x^6}{6!} + \cdots$$

Equations 2.18, 2.19, 2.20, 6.4a, and 6.4b give some basic rules for differentiation; for a short table of integrals, see Table 7.1.

# Selected Fundamental Physical Constants

| Quantity | Symbol | Value | Units | Uncertainty (in parts per million) |
|---|---|---|---|---|
| Speed of light in vacuum | $c$ | 299 792 458 | $\text{m·s}^{-1}$ | defined |
| Permeability of free space | $\mu_0$ | $4\pi \times 10^{-7}$ | $\text{N·A}^{-2}$ or $\text{T·m·A}^{-1}$ | defined |
| Permittivity of free space | $\epsilon_0 \equiv \mu_0^{-1}c^{-2}$ | $8.854\ 187\ 817\ldots \times 10^{-12}$ | $\text{C}^2\text{·N}^{-1}\text{·m}^{-2}$ | defined |
| Universal gravitational constant | $G$ | $6.672\ 59 \times 10^{-11}$ | $\text{N·m}^2\text{·kg}^{-2}$ | 128 |
| Planck's constant | $h$ | $6.626\ 075\ 5 \times 10^{-34}$ | $\text{J·s}$ | 0.60 |
| | $\hbar \equiv h/2\pi$ | $1.054\ 572\ 66 \times 10^{-34}$ | $\text{J·s}$ | 0.60 |
| Atomic mass unit (unified) | $u \equiv \frac{1}{12}m(^{12}\text{C})$ | $1.660\ 540\ 2 \times 10^{-27}$ | kg | 0.59 |
| Avogadro's number | $A$ | $6.022\ 136\ 7 \times 10^{26}$ | $\text{kmol}^{-1}$ | 0.59 |
| Boltzmann's constant | $k$ | $1.380\ 658 \times 10^{-23}$ | $\text{J·K}^{-1}$ | 8.5 |
| Gas constant | $R \equiv Ak$ | 8314.510 | $\text{J·kmol}^{-1}\text{·K}^{-1}$ | 8.4 |
| Stefan-Boltzmann constant | $\sigma$ | $5.670\ 51 \times 10^{-8}$ | $\text{W·m}^{-2}\text{·K}^{-4}$ | 34 |
| Wien's constant | $b \equiv \lambda_{\max}T$ | $2.897\ 756 \times 10^{-3}$ | $\text{m·K}$ | 8.4 |
| Elementary charge | $e$ | $1.602\ 177\ 33 \times 10^{-19}$ | C | 0.30 |
| Faraday's constant | $F$ | $96\ 485.309 \times 10^3$ | $\text{C·kmol}^{-1}$ | 0.30 |
| Electron mass | $m_e$ | $9.109\ 389\ 7 \times 10^{-31}$ | kg | 0.59 |
| | | $5.485\ 799\ 03 \times 10^{-4}$ | u | 0.023 |
| Electron charge/mass ratio | $e/m_e$ | $1.758\ 819\ 62 \times 10^{11}$ | $\text{C·kg}^{-1}$ | 0.30 |
| Proton mass | $m_p$ | $1.672\ 623\ 1 \times 10^{-27}$ | kg | 0.59 |
| | | 1.007 276 470 | u | 0.012 |
| Neutron mass | $m_n$ | $1.674\ 928\ 6 \times 10^{-27}$ | kg | 0.59 |
| | | 1.008 664 904 | u | 0.014 |
| Rydberg constant | $R_\infty \equiv \mu_0^2 c^3 m_e e^4/8h^3$ | 10 973 731.534 | $\text{m}^{-1}$ | 0.0012 |
| Bohr radius | $a_0 \equiv \mu_0 c e^2/8\pi h R_\infty$ | $0.529\ 177\ 249 \times 10^{-10}$ | m | 0.045 |
| Bohr magneton | $\mu_B \equiv e\hbar/2m_e$ | $9.274\ 015\ 4 \times 10^{-24}$ | $\text{J·T}^{-1}$ | 0.34 |
| Nuclear magneton | $\mu_N \equiv e\hbar/2m_p$ | $5.050\ 786\ 6 \times 10^{-27}$ | $\text{J·T}^{-1}$ | 0.34 |
| Electron magnetic moment | $\mu_e/\mu_B$ | 1.001 159 652 193 | dimensionless | $1 \times 10^{-5}$ |
| Proton magnetic moment | $\mu_p/\mu_N$ | 2.792 847 386 | dimensionless | 0.023 |
| Neutron magnetic moment | $\mu_n/\mu_N$ | 1.913 042 75 | dimensionless | 0.24 |

Abridged and slightly modified from E. R. Cohen and B. N. Taylor, *Rev. Mod. Phys.* **57**, 1121 (1987).

# SI Base, Supplementary, and Derived Units

## SI Base and Supplementary Units

| Quantity | Unit | Symbol |
|---|---|---|
| *Base Units* | | |
| length | meter | m |
| time | second | s |
| mass | kilogram | kg |
| electric current | ampere | A |
| temperature | kelvin | K |
| amount of substance | mole | mol |
| luminous intensity | candela | cd |
| *Supplementary Units* | | |
| plane angle | radian | rad |
| solid angle | steradian | sr |

## Selected SI Derived Units

| Quantity | Unit Name | Symbol | Equivalent |
|---|---|---|---|
| *Mechanical Units* | | | |
| frequency | hertz | Hz | $s^{-1}$ |
| speed, velocity | meter per second | m/s | |
| acceleration | meter per second squared | $m/s^2$ | |
| angular velocity | radian per second | rad/s | |
| angular acceleration | radian per second squared | $rad/s^2$ | |
| force | newton | N | $kg \cdot m/s^2$ |
| stress, pressure | pascal | Pa | $N/m^2$ |
| work, energy, heat | joule | J | N·m |
| power | watt | W | J/s |
| momentum, impulse | newton-second | N·s | kg·m/s |
| angular momentum, angular impulse | newton-meter-second | N·m·s | $kg \cdot m^2/s$ |
| entropy | joule per kelvin | J/K | $kg \cdot m^2/s$ |
| *Electromagnetic Units* | | | |
| electric charge | coulomb | C | A·s |
| electric flux | newton-meter squared per coulomb, volt-meter | $N \cdot m^2/C$, V·m | |
| electric field | newton per coulomb, volt per meter | N/C, V/m | |
| electric potential, emf | volt | V | J/C |
| electric circulation | newton-meter per coulomb, volt | N·m/C, V | |
| resistance | ohm | Ω | V/A |
| conductance | siemens | S | A/V, $\Omega^{-1}$ |
| capacitance | farad | F | C/V |
| inductance | henry | H | Wb/A, V·s/A |
| magnetic flux | weber | Wb | V·s |
| magnetic field | tesla | T | $Wb/m^2$, A/m |
| magnetic circulation | tesla-meter, ampere | T·m, A | |

# Non-SI Units

| Quantity | Name | Symbol | SI Equivalency |
|---|---|---|---|
| *Non-SI Units in Use with SI* | | | |
| time | minute | min | $1\ \text{min} = 60\ \text{s}$ |
| | hour | h | $1\ \text{h} = 60\ \text{min} = 3600\ \text{s}$ |
| | day | d | $1\ \text{d} = 24\ \text{h} = 86\ 400\ \text{s}$ |
| | year | y | $1\ \text{y} = 365.24\ \text{d} = 31\ 556\ 736\ \text{s}$ |
| plane angle | degree | ° | $1° = \dfrac{\pi}{180}\ \text{rad}$ |
| | (arc) minute | ′ | $1' = \left(\dfrac{1}{60}\right)° = \dfrac{\pi}{10\ 800}\ \text{rad}$ |
| | (arc) second | ″ | $1'' = \left(\dfrac{1}{60}\right)' = \dfrac{\pi}{648\ 000}\ \text{rad}$ |
| area | hectare | ha | $1\ \text{ha} = 10^4\ \text{m}^2$ |
| volume | liter | L | $1\ \text{L} = 10^{-3}\ \text{m}^3$ |
| mass | tonne | t | $1\ \text{t} = 10^3\ \text{kg}$ |
| energy | electron-volt | eV | $1\ \text{eV} = e \times 1\ \text{V} = |e|\ \text{J} \simeq 1.60 \times 10^{-19}\ \text{J}$ |
| | kilocalorie | kcal, Cal | $1\ \text{kcal} = 4186\ \text{J}$ |
| *Selected Metric Non-SI Units* | | | |
| length | Ångstrom unit | Å | $1\ \text{Å} = 10^{-10}\ \text{m} = 0.1\ \text{nm}$ |
| | centimeter | cm | $1\ \text{cm} = 10^{-2}\ \text{m}$ |
| | astronomical unit | AU | $1\ \text{AU} = 149.5 \times 10^9\ \text{m}$ |
| | light year | ly | $1\ \text{ly} = 9.460 \times 10^{15}\ \text{m}$ |
| | parsec | pc | $1\ \text{pc} = 3.084 \times 10^{16}\ \text{m} = 3.26\ \text{ly}$ |
| mass | gram | g | $1\ \text{g} = 10^{-3}\ \text{kg}$ |
| force | dyne | dyn | $1\ \text{dyn} \equiv 1\ \text{g·cm/s}^2 = 10^{-5}\ \text{N}$ |
| pressure | dyne per centimeter squared | dyn/cm$^2$ | $1\ \text{dyn/cm}^2 = 0.1\ \text{Pa}$ |
| energy | erg | erg | $1\ \text{erg} \equiv 1\ \text{dyn·cm} = 10^{-7}\ \text{J}$ |
| magnetic field | gauss | G | $1\ \text{G} = 10^{-4}\ \text{T}$ |
| *Selected Nonstandard Units* | | | |
| length | inch | in. | $1\ \text{in.} \equiv 2.54\ \text{cm}$ |
| | foot | ft | $1\ \text{ft} = 12\ \text{in.} \equiv 0.3048\ \text{m}$ |
| | statute mile | mi | $1\ \text{mi} = 5280\ \text{ft} \equiv 1.609\ \text{km}$ |
| | nautical mile | nmi | $1\ \text{nmi} = \dfrac{2\pi}{360 \times 60}\ \text{mean earth radius} = 1.852\ \text{km}$ |
| mass | slug | slug | $1\ \text{slug} \equiv 14.59\ \text{kg}$ |
| | pound mass | lb, lbm | $1\ \text{lbm} = 0.4536\ \text{kg}$ |
| force | kilogram force | kgf | $1\ \text{kgf} \equiv g_{\text{std}}\ \text{N} = 9.807\ \text{N}$ |
| | poundal | pdl | $1\ \text{pdl} = 0.138\ \text{N}$ |
| | pound force | lb, lbf | $1\ \text{lbf} \equiv 4.448\ \text{N}$ |
| pressure | atmosphere | atm | $1\ \text{atm} \equiv 1.013 \times 10^5\ \text{Pa}$ |
| | mm Hg (Torr) | Torr | $1\ \text{Torr} = \tfrac{1}{760}\ \text{atm} = 133.3\ \text{Pa}$ |
| | "kilogram-force" per square centimeter | kg/cm$^2$, kgf/cm$^2$ | $9.807 \times 10^4\ \text{Pa}$ |
| | pound per square inch | psi, lb/in.$^2$ | $1\ \text{psi} = 6895\ \text{Pa}$ |
| energy | foot-pound | ft·lb | $1\ \text{ft·lb} = 1.356\ \text{J}$ |
| | British thermal unit | Btu | $1\ \text{Btu} = 1055\ \text{J}$ |
| | kilowatt-hour | kWh | $1\ \text{kWh} = 3.6 \times 10^6\ \text{J}$ |
| power | horsepower | hp | $1\ \text{hp} \equiv 550\ \text{ft·lb/s} = 745.7\ \text{W}$ |

# Periodic Table of the Elements

**Key**

| 1 | — Atomic number |
|---|---|
| H | — Symbol |
| Hydrogen | — Name |
| 1.0079 | — Atomic mass |

Note: The atomic mass value given is for naturally occurring proportions of isotopes. Values in parentheses are mass numbers for the most stable isotope.
‡Reported but not confirmed; no name proposed.

→ Stable region?

| Group IA | IIA | | IIIB | IVB | VB | VIB | VIIB | | VII | | IB | IIB | IIIA | IVA | VA | VIA | VIIA | Noble Gases VIIIA |
|---|---|---|---|---|---|---|---|---|---|---|---|---|---|---|---|---|---|---|
| 1 H Hydrogen 1.0079 | | | | | | | | | | | | | | | | | | 2 He Helium 4.0026 |
| 3 Li Lithium 6.941 | 4 Be Beryllium 9.01218 | | | | | | | | | | | | 5 B Boron 10.81 | 6 C Carbon 12.011 | 7 N Nitrogen 14.0067 | 8 O Oxygen 15.9994 | 9 F Fluoride 18.9984 | 10 Ne Neon 20.179 |
| 11 Na Sodium 22.98977 | 12 Mg Magnesium 24.305 | | | | | | | | | | | | 13 Al Aluminum 26.9815 | 14 Si Silicon 28.0855 | 15 P Phosphorus 30.97376 | 16 S Sulfur 32.06 | 17 Cl Chloride 35.453 | 18 Ar Argon 39.948 |
| 19 K Potassium 39.0983 | 20 Ca Calcium 40.08 | 21 Sc Scandium 44.9559 | 22 Ti Titanium 47.88 | 23 V Vanadium 50.9415 | 24 Cr Chromium 51.996 | 25 Mn Manganese 54.9380 | 26 Fe Iron 55.847 | 27 Co Cobalt 58.9332 | 28 Ni Nickel 58.70 | 29 Cu Copper 63.546 | 30 Zn Zinc 65.38 | 31 Ga Gallium 69.72 | 32 Ge Germanium 72.59 | 33 As Arsenic 74.9216 | 34 Se Selenium 78.96 | 35 Br Bromine 79.904 | 36 Kr Krypton 83.80 |
| 37 Rb Rubidium 85.4678 | 38 Sr Strontium 87.62 | 39 Y Yttrium 88.9059 | 40 Zr Zirconium 91.22 | 41 Nb Niobium 92.9064 | 42 Mo Molybdenum 95.94 | 43 Tc Technetium 98.906 | 44 Ru Ruthenium 101.07 | 45 Rh Rhodium 102.9055 | 46 Pd Palladium 106.4 | 47 Ag Silver 107.868 | 48 Cd Cadmium 112.41 | 49 In Indium 114.82 | 50 Sn Tin 118.69 | 51 Sb Antimony 121.75 | 52 Te Tellurium 127.60 | 53 I Iodine 126.9045 | 54 Xe Xenon 131.30 |
| 55 Cs Cesium 132.9054 | 56 Ba Barium 137.33 | 57–71 *Rare earths | 72 Hf Hafnium 178.49 | 73 Ta Tantalum 180.9479 | 74 W Tungsten 183.85 | 75 Re Rhenium 186.207 | 76 Os Osmium 190.2 | 77 Ir Iridium 192.22 | 78 Pt Platinum 195.09 | 79 Au Gold 196.9665 | 80 Hg Mercury 200.59 | 81 Tl Thallium 204.37 | 82 Pb Lead 207.2 | 83 Bi Bismuth 208.9804 | 84 Po Polonium (209) | 85 At Astatine (210) | 86 Rn Radon (222) |
| 87 Fr Francium (223) | 88 Ra Radium 226.0254 | 89–103 †Actinides | 104 Rf Rutherfordium (261) | 105 Ha Hahnium (262) | 106 Sg Seaborgium (263) | 107 Ns Neilsbohrium (262) | 108 Hs Hassium (265) | 109 Mt Meitnerium (266) | 110 ‡ (269) | 111 ‡ | | 114 | | | | | |

*Lanthanides 6

| 57 La Lanthanum 138.9055 | 58 Ce Cerium 140.12 | 59 Pr Praseodymium 140.9077 | 60 Nd Neodymium 144.24 | 61 Pm Promethium 145 | 62 Sm Samarium 150.4 | 63 Eu Europium 151.96 | 64 Gd Gadolinium 157.25 | 65 Tb Terbium 158.9254 | 66 Dy Dysprosium 162.50 | 67 Ho Holmium 164.9304 | 68 Er Erbium 167.26 | 69 Tm Thulium 168.9342 | 70 Yb Ytterbium 173.04 | 71 Lu Lutetium 174.967 |
|---|---|---|---|---|---|---|---|---|---|---|---|---|---|---|

†Actinides 7

| 89 Ac Actinium 227.0278 | 90 Th Thorium 232.0381 | 91 Pa Protactinium 231.0359 | 92 U Uranium 238.029 | 93 Np Neptunium 237.0482 | 94 Pu Plutonium (244) | 95 Am Americium (243) | 96 Cm Curium (247) | 97 Bk Berkelium (247) | 98 Cf Californium (251) | 99 Es Einsteinium (254) | 100 Fm Fermium (257) | 101 Md Mendelevium (258) | 102 No Nobelium 259 | 103 Lr Lawrencium 262 |
|---|---|---|---|---|---|---|---|---|---|---|---|---|---|---|

Periods → 1 2 3 4 5 6 7

# Answers to Odd-Numbered Problems

Answers are not given for problems requiring a lengthy explanation.

## Chapter 38

**38.1** $V = \sqrt{\frac{3}{4}}c$.  **38.3** $V = 0.991c = 2.97 \times 10^8$ m/s.
**38.5** $\Delta t_0 = 7.22$ μs.  **38.7** (a) $\Delta t = 1.8 \times 10^{-8}$ s.
(b) The "stationary" clock.  **38.9** (b) $t = 1.075 \times 10^{-5}$ s.
(c) $t = 3.40 \times 10^{-5}$ s.  (d) $t = 1.075 \times 10^{-4}$ s.
**38.15** (a) $L_0 = 1.66$ m.  (b) $L = 1.66$ m.
**38.17** It appears $8.16 \times 10^{-14}$ m shorter, or $2 \times 10^{-7}$ times the wavelength of blue light.
**38.19** (a) $0.357c$.  (b) 7.2 ly.  (c) 4.32 ly; 5.4 y.
(d) 3.14 ly; 3.49 y.  **38.21** (b) For small $V$, $V/c \rightarrow 0$.
**38.25** $v'_x = v_x - V_x$, $v'_y = v_y - V_y$, $v'_z = v_z - V_z$.

**38.27** (a) $t = \sqrt{\dfrac{1 + \dfrac{V}{c}}{1 - \dfrac{V}{c}}}$ years.  (b) $V = \dfrac{c}{\sqrt{2}}$.

**38.33** (a) $\dfrac{\Delta L}{L} = 1 - \sqrt{1 - \dfrac{V^2}{c^2}}$.

(b) Fitzgerald did not explain why such a contraction should take place. It could not be due simply to ether pressure, because all materials contracted equally.

**38.35** (a) $\nu_e = \dfrac{\sqrt{1 - \dfrac{V^2}{c^2}}}{\Delta t}$.  (b) $\Delta x = \dfrac{v \, \Delta t}{\sqrt{1 - \dfrac{V^2}{c^2}}}$.

(e) $v' = v \sqrt{\dfrac{1 - \dfrac{V}{c}}{1 + \dfrac{V}{c}}}$.  (f) $\dfrac{\Delta \nu}{\nu} = \sqrt{\dfrac{0.25}{1.75}} - 1 = -0.622$.

**38.37** (a) $\Delta \nu / \nu = 2.22 \times 10^{-7}$.  (b) $\Delta \nu = 18.5$ Hz.

## Chapter 39

**39.1** $v = 4.21 \times 10^7$ m/s.  **39.3** $v/c = 0.548$.
**39.5** (a) $v = c - 1.005 \times 10^{-18}$ m/s;
(b) $m = 1.11 \times 10^{-17}$ kg;  (c) $p = 3.33 \times 10^{-9}$ kg·m/s.
**39.7** $K = 1.503 \times 10^{-10}$ J $= 9.383 \times 10^8$ eV.
**39.9** $E = 1.165 \times 10^8$ J; the energy released is equivalent to more than 27 kg of TNT.
**39.11** (a) $E = 2.5 \times 10^{-10}$ J $= 1.6 \times 10^9$ eV $= 0.6 m_0 c^2$.
(b) $K = 1.0 \times 10^{-10}$ J $= 6.3 \times 10^8$ eV.

**39.13** (a) $E = 1.02 \times 10^6$ eV;  (b) $E = 1.88 \times 10^9$ eV.
(c) Neutrinos and antineutrinos are essentially massless and cannot collide at $v \ll c$ because they always travel at $c$.
**39.15** $\Delta m / \Delta t = 6.3 \times 10^7$ kg/y.
**39.17** $\Delta M / M = 1.498 \times 10^{-10}$.
**39.19** (a) $\Delta M = 2.14 \times 10^{-1}$ kg.  (b) $\Delta M / M = 9.1 \times 10^{-4}$.
**39.21** $p = c \sqrt{m^2 - m_0^2}$.  **39.23** $V = 1.21 \times 10^9$ V.
**39.25** (a) $\omega_r / \omega_c = 0.808$.
(b) $\omega_r = 6.154 \times 10^7$ rad/s; $\omega_c = 7.62 \times 10^7$ rad/s.
**39.27** $E^2 - p^2 c^2 = m_0^2 c^4$.  **39.29** (b) $R = 1.687 \times 10^{-15}$ m.
**39.31** 3 $^4$He $\rightarrow$ $^{12}$C: $E = 1.119 \times 10^{-12}$ J. $^{12}$C + $^4$He $\rightarrow$ $^{16}$O: $E = 1.133 \times 10^{-12}$ J. 2 $^{12}$C $\rightarrow$ $^4$He + $^{20}$Ne: $E = 7.551 \times 10^{-13}$. 2 $^{12}$C $\rightarrow$ $^{24}$Mg: $E = 2.232 \times 10^{-12}$ J. 2 $^{16}$O $\rightarrow$ $^4$He + $^{28}$Si: $E = 1.550 \times 10^{-12}$. 2 $^{16}$O $\rightarrow$ $^{32}$S: $E = 2.649 \times 10^{-12}$ J. 2 $^{28}$Si + 2 e $\rightarrow$ $^{56}$Fe: $E = 2.829 \times 10^{-12}$ J.

**39.33** (c) $M'_0 = m_0 \sqrt{2 \left( \dfrac{K}{m_0 c^2} + 2 \right)}$.

(d) $M_0 > 2 m_0$ because the initial kinetic energy that is classically converted to heat energy in the inelastic collision of macroscopic bodies is converted into mass in the inelastic collision of two microscopic bodies. That is why the rest mass is greater than $2 m_0$.  **39.35** $\mathscr{B} = 4.945$ T.
**39.37** (a) $E = (m_i - m_f) c^2$.
(b) $|p| = E/C = (m_i - m_f)c = (m_i - m_f)v + m_i \, dv$.
(d) $v/c = 24/26 = 0.923$.
**39.41** (a) $K_0 = 6 m_0 c^2 = 5.63 \times 10^9$ eV.  (b) $\eta = \frac{1}{3}$.
(c) **100%.**

## Chapter 40

**40.1** $I = 5.95 \times 10^6$ W/m$^2$.  **40.3** $A = 3.36 \times 10^{-5}$ m$^2$.
**40.5** $\lambda_{max} = 483$ nm; bluish.
**40.7** (a) $\lambda_{max} = 2.9 \times 10^{-11}$ m; X rays or γ rays.
(b) $E = 6.86 \times 10^{-15}$ J $= 42$ keV.
**40.9** (a) $\nu_0 = 4.589 \times 10^{14}$ Hz; $\lambda_0 = 654$ nm; red.
(b) $V_0 = 1.21$ V.  **40.11** (a) $h = 6.60 \times 10^{-34}$ J·s.
(b) $\phi = 2.48$ eV.  **40.13** $E = 4.74 \times 10^{-19}$ J $= 2.95$ eV.
**40.15** $E = 6.63 \times 10^{-34}$ J.  **40.17** (a) $K_{max} = 1.28$ eV.
(b) $v = 6.71 \times 10^5$ m/s.
**40.19** $n/At = 3.54 \times 10^{21}$ photons/m$^2$·s.
**40.21** (a) Li: $\nu_0 = 5.75 \times 10^{14}$ Hz; Na: $\nu_0 = 5.57 \times 10^{14}$ Hz; K: $\nu_0 = 5.31 \times 10^{14}$ Hz; Rb: $\nu_0 = 5.07 \times 10^{14}$ Hz; Cs: $\nu_0 = 4.59 \times 10^{14}$ Hz.
(b) Li: $\lambda_0 = 520$ nm; Na: $\lambda_0 = 520$ nm; K: $\lambda_0 = 565$ nm; Rb: $\lambda_0 = 590$ nm; Cs: 654 nm.
**40.23** (a) $K_{max} = 0.50$ eV $= 8 \times 10^{-20}$ J.  (b) $\phi = 2.26$ eV.
(c) $\nu_0 = 5.46 \times 10^{14}$ Hz; $\lambda_0 = 550$ nm.
**40.25** (b) $\lambda_{min} = 1.38 \times 10^{-11}$ m.

**40.27** (a) $E = 1.65 \times 10^{-15}$ J. (b) $p = 5.52 \times 10^{-24}$ kg·m/s.
**40.29** $\theta = 80°$. **40.31** $\Delta E/E_{max} = -0.0672$.
**40.35** $I_{675}/I_{520} = 0.864$.
**40.37** (a) $I_{1000}/I_{520} = 0.440$. (b) $I_{200}/I_{520} = 0.0430$.
(c) $I_{1.2\,cm}/I_{520\,nm} = 1 \times 10^{-16}$.
**40.39** (a) 278 K or 5°C. This is very reasonable for an average
temperature over the earth. (b) $\lambda_{max} = 10.4$ m, infrared.

**40.41** (a) $\lambda_{max} = \epsilon\left(\dfrac{\sigma}{I}\right)^{1/4}$. (b) $\lambda_{max} = 2.14$ μm.

(c) $T = 1360$ K. **40.43** (a) $P_T = 2.66 \times 10^{28}$ W.
(b) $T = 5180$ K. (c) $A = 6.52 \times 10^{20}$ m$^2$.
(d) $R = 7.20 \times 10^9$ m.
(e) The radius of Capella is ten times that of the sun.
**40.45** (b) $a = ch/k; b = 2\pi c^2 h$. **40.47** (b) $\lambda = 517$ nm.

**40.49** $n = \dfrac{4\pi\epsilon_0 r(h\nu - \phi)}{e^2}$. **40.51** $\phi = 1.9$ eV.

**40.53** $n/V = 1.18 \times 10^{13}$ photons/m$^3$.
**40.55** (a) $n = 2.3 \times 10^{16}$ s$^{-1}$. (b) $T = 2.55 \times 10^{-13}$ N·m.
(c) $\theta = 0.636°$. **40.57** (a) $r = 2.3 \times 10^{-6}$ m.
**40.59** $\epsilon = 4.10 \times 10^{-14}$ J $= 250$ keV.

**40.61** (a) $\dfrac{1}{T^3} = \dfrac{1}{T^3} + \dfrac{3\sigma}{\chi}\left(\dfrac{36\pi}{\rho^2 M}\right)^{1/3} t.$

(b) $t = \dfrac{7\chi}{30 T_L{}^3}\left(\dfrac{\rho^2 M}{36\pi}\right)^{1/3}$. (c) $t = 1.92 \times 10^6$ s $= 533$ h.
**40.65** (a) $p = \frac{1}{3}n\langle\epsilon\rangle A$; $A$ is the area of the wall. (b) $\Pi = \frac{1}{3}n\langle\epsilon\rangle$.

(c) $u = n\langle\epsilon\rangle$. **40.67** (a) $h\nu + m_0 c^2 = \dfrac{m_0 c^2}{\sqrt{1 - v^2/c^2}}$.

(b) $\dfrac{h\nu}{c} = \dfrac{m_0 v}{\sqrt{1 - v^2/c^2}} + \dfrac{h\nu'}{c}$.

**40.67** (a) $h(\nu - \nu') = (m - m_0)c^2$. (b) $\dfrac{h}{c}(\nu - \nu') = p$.

(c) If $\nu' = 0$, $v_{electron} = c$, which is impossible.

## Chapter 41
**41.3** 122 nm, 656 nm, 1.88 μm,
4.05 μm, 7.46 μm, 12.4 μm, 19.1 μm.

**41.5** No. Note that $\dfrac{1}{4^2} - \dfrac{1}{7^2} > \dfrac{1}{5^2}$. **41.7** (a) $1.81 \times 10^{-15}$ m.

(b) Lithium nucleus radius $= 2.61 \times 10^{-15}$ m.
**41.9** (a) $2.19 \times 10^6$ m/s. (b) Not necessary; $v/c = 0.73\%$.
(c) Less; no. **41.11** $7.19 \times 10^{-9}$ m. **41.13** 21 pm.
**41.15** (a) 62 pm.
(b) Any such X-rays generated will be absorbed by the air if the
child sits sufficiently far from the screen. **41.17** (b) 17.2 kV.
**41.19** (a) 145 pm. (b) 8.55 kV. **41.21** 1.79 eV.
**41.23** 1.15 μm. **41.25** 5.35 MeV. **41.27** 92 nm.
**41.33** α line: 164 nm, ultraviolet; series limit: 91.1 nm,
ultraviolet. **41.35** $R_{He^+} = 4R_H$.
**41.39** (a) $r_n = 2.56 \times 10^{-12}$ m.
(b) $E_n = -207 \times 13.6$ eV $\times (1/n^2)$. (c) 0.587 nm; X-ray.
**41.41** (a) $-1.84 \times 10^{-14}$ J. (b) $r = 5.77 \times 10^{-13}$ m, $v = 2.0 \times 10^8$ m/s $\simeq \frac{2}{3}c$; $\sqrt{1 - (v^2/c^2)} \approx 0.7$, relativistic correction
is clearly needed. **41.43** 60.5 keV. **41.45** 62.8 pm.
**41.49** (a) 10.2 eV. (b) $7.9 \times 10^4$ K. (c) $T_\beta > T_\alpha$.
**41.51** (b) 12.5 T. **41.53** $4.388 \times 10^7$ m$^{-1}$.
**41.55** (a) The difference in reduced masses results in a slight
difference in the energy levels and thus in the wavelength of the
emitted radiation. (b) 0.179 nm. **41.57** (a) 44.9. (b) 487 pm.

## Chapter 42
**42.3** (a) 28.7 pm. (b) 0.905 pm. **42.5** (a) 3.31 kg·m/s.
(b) $6.02 \times 10^{-16}$ J $= 3.77$ keV. (c) $3.64 \times 10^7$ m/s.
**42.7** (a) $\frac{1}{2}$. (b) See Figure 42.8, $n = 2$ case.
**42.9** $6.13 \times 10^{-10}$ m.
**42.11** Electron: 116 m/s; proton: $6.32 \times 10^{-2}$ m/s;
dust: $1.06 \times 10^{-19}$ m/s. **42.13** $9.65 \times 10^{-19}$ J $= 6.03$ eV.
**42.15** (a) $1.06 \times 10^{-22}$ J. (b) $1.06 \times 10^{-12}$ m/s.
**42.17** (a) $-\infty$, (b) $-0.15$, (c) $-0.0050$.
**42.19** (a) $l = 0, 1, 2, 3$.
(b) $l = 0, m_l = 0; l = 1, m_l = 0, \pm 1; l = 2, m_l = 0, \pm 1, \pm 2;$
$l = 3, m_l = 0, \pm 1, \pm 2, \pm 3$. **42.21** $5.05 \times 10^{-27}$ J/T.
**42.25** 1.54 μm. **42.27** (a) 11.1 nm. (b) 27.7 nm.

(c) 31.8 nm. **42.29** $P = \dfrac{1}{5} - \dfrac{1}{6\pi}\sin\dfrac{6\pi}{5} = 0.231$.

**42.31** $P = \dfrac{1}{3} + \dfrac{1}{2n\pi}\sin\dfrac{2n\pi}{3}$; hence $P = $ (a) 0.196, (b) 0.402,

(c) $\frac{1}{3}$, (d) 0.3335. **42.33** (a) $\lambda = 0.0902$ nm.
(b) $\lambda \simeq 35\%$ of atomic diameter $d$.
(c) Yes. Here $\lambda = 0.753$ nm, or about $3d$.
**42.37** (a) $\Delta E = 10.2$ eV. (b) $\Delta E = \hbar/\Delta t$.
(c) $\Delta t = 1.30 \times 10^{-16}$ s.
(d, e) For any shorter time interval $\Delta t'$, the uncertainty in energy
$\Delta E' = \hbar/\Delta t'$ will be greater than the energy difference $\Delta E$
between the states, and it makes no sense to try to specify the
state in which the electron lies. **42.39** (a) $n = 2.5 \times 10^{75}$.
(b) $\Delta E = 2.1 \times 10^{-40}$ J.
(c) For any two adjacent states, $\Delta E/E \simeq 8 \times 10^{-76}$, a difference
for which detection is not a plausible possibility.
**42.41** (a) $\Delta x = D/2$; that is, $x = D/2 \pm D/2$.
(b) $\Delta p = h/\pi D$. (c) $\Delta K = h^2/2\pi m D^2$.
(d) No, $\Delta K/K_1 = 4/\pi = 1.3$; the uncertainty in the ground-state
energy is about the same as the energy.
**42.43** (a) $\mathscr{B} = 9.53 \times 10^{-3}$ T.
(b) $\Delta E = 9.5 \times 10^{-25}$ J $= 5.92$ μeV.
(c) $r = 3.0 \times 10^{-11}$ m. The Bohr orbit corresponding to $l = 0$ is
a straight line of length $2a_0$ passing through the nucleus; hence
$r < a_0$. (d) No.
**42.45** (b) 42.6 MHz $\leq \nu \leq$ 426 MHz; this is the radio-
frequency range. **42.49** (b) $a = 11$ μm.
**42.51** (a) $\Delta E = 2en\hbar\mathscr{B}/\pi m$. (b) The spectral lines will split.

## Chapter 43
**43.3** (a) 169 min. (b) 117 min. **43.5** 30 s.
**43.7** (a) $2.67 \times 10^{21}$. (b) $3.62 \times 10^{10}$. **43.9** 23,000 y.
**43.11** $^{208}_{82}$Pb. **43.13** (a) $A = 227, Z = 89$. (b) Actinium (Ac).
**43.15** 185.9551 u. **43.17** 227.0297 u. **43.19** 2.24 MeV.
**43.21** (a) $Z = 42$; (b) $A = 106$; (c) Molybdenum (Mo).
**43.23** (a) 203.9 MeV. (b) Not significant.
**43.25** (a) $2.26 \times 10^5$. (b) $3.62 \times 10^{-14}$ C.
**43.27** $\tau = 5.45$ h, $t_{1/2} = 3.78$ h.

**43.29** (a) $\dfrac{^{40}\text{Ar}}{^{40}\text{K}} = \dfrac{0.103 \times (1 - e^{-T/\tau})}{e^{-T/\tau}}$. (b) 0.30.

**43.31** $1.8 \times 10^9$ y. **43.33** $5.5 \times 10^{-14}$ m.
**43.35** (a) 6.39 MeV. (b) $6.72 \times 10^{-15}$ m, 227 MeV.
(c) 233 MeV. (d) 0.1%; much less than in the fusion reaction.
**43.37** 23.8 MeV. **43.39** (a) Co–Ni. (b) 197 MeV.
(c) The collapse results in extremely high temperatures, so
endothermic reactions can take place.

**43.41** (a) $N_A/\tau_A$ = rate of A → B; $N_B/\tau_B$ = rate of B → C.
(c) $t_m = \dfrac{\ln(\tau_B/\tau_A)}{\dfrac{1}{\tau_A} - \dfrac{1}{\tau_B}}$. (d) 2.15 h.

**43.43** $N_C = N_0\left[1 - \dfrac{\tau_A e^{-t/\tau_A} - \tau_B e^{-t/\tau_B}}{\tau_A - \tau_B}\right]$.

**43.45** (a) 6.6 MeV.
(b) No, because the uncertainty in the energy of the α particle is greater than the binding energy of the ''nucleus.''

## Chapter 44
**44.1** B, Al, and In are *p*-type; As and Sb are *n*-type.
**44.3** $5.4 \times 10^{-3}$ eV.
**44.5** $1 \times 10^6\, R$ A, where $R$ is the radius of the wire.
**44.7** $2 \times 10^{-14}$. **44.9** $1.84 \times 10^{-3}\,\mathrm{m}^3/\mathrm{C}$.

**44.11** $r_D = \kappa^2\, \dfrac{m_e}{m^*}\, r_B$.

**44.13** (a) A photon is annihilated in the depletion region near the junction, creating an electron-hole pair.
(b) 889 nm, near infrared. (c) 6200 nm, far infrared.
(d) $kT_{\text{room temp}} \simeq 0.025$ eV. It is important that $E_g \gg kT$.

**44.15** Aluminum: $1.03 \times 10^{-4}$ eV; mercury: $3.58 \times 10^{-4}$ eV; lead: $6.20 \times 10^{-4}$ eV.
**44.17** (a) The slope of the line is $-(|E_d|/k)$.
(b) Choose $T$ so that $kT \simeq |E_d|$. **44.19** $\lambda = \sqrt{\beta/\alpha}$.

## Chapter 45
**45.1** (a) 0.511 MeV. (b) $\lambda = 2.43$ pm. **45.3** 1.876 GeV.
**45.5** (a) 0.78 MeV.
(b) Conservation of electron-lepton number, conservation of momentum, conservation of angular momentum.
**45.7** $K^0 \to \pi^+ + \pi^-$: 218 MeV; $K^0 \to \pi^0 + \pi^0$: 228 MeV.
**45.9** (a) Charge; (b) charge; (c) muon-lepton number;
(d) electron-lepton number, muon-lepton number.
**45.11** $2.3 \times 10^{39}$. **45.13** $K_{\mu^-} = 4.12$ MeV, $E_{\bar{\nu}_\mu} = 29.8$ MeV.
**45.15** 52.3 MeV. **45.17** $1.6 \times 10^{-22}$ s.
**45.19** (a) 5630 MeV. (b) 938 MeV.

# Photo and Illustration Credits

# Index

This index covers the contents of *Physics for Scientists and Engineers* and *Modern Physics for Scientists and Engineers*.

magnetic orbits, 770
magnetic pole strength, 765, 791
magnetic poles, 762–765
magnetic quantization, 1177
magnetic remanence, 763
magnetism, 812
   Gauss's law for, 787
magnetization, induced, 763
magneto, 835, 836
magnetohydrodynamic (MHD) generator, 791
magnetometer
   nmr, 1177
   rotating-coil, 846
magneton, nuclear, 1175
magnetoresistance, longitudinal, 785
magnets, 762–765
magnification
   angular, 1028
   lateral, 1010
   longitudinal, 1011, 1039
magnifier, simple, 1028–1029
magnitude, 22
Maiman, Theodore H., 1140
major axis, 366
major radius, of toroid, 809
majority carrier, 1208, 1209, 1212
Malus, Etienne Louis, 957
Malus's law, 958–960
mantle, 419
many-atom systems, 1204–1217
Marconi, Guglielmo, 929
Mariotte, Abbé Edme, 441
Mariotte bottle, 441
Marsden, Ernest, 680, 1126
maser, 1140
Maskelyne, Nevil, 391
mass, 57–58, 237, 239
   atomic, 1134, 1136
   effective, 758, 790, 1207
   gravitational role of, 68
   inertial manifestation of, 382, 1070–1073
   inertial role of, 68
   reduced, 247
   relativistic, 1073
   rest, 1073
   zero rest, 1082, 1107
mass defect, 1193
mass-energy conservation, 1108
mass flux, 426
mass number, 1181
mass spectrometer, 771
mass-spring system, 150–155, 192–194, 202–203, 877
master grating, 993
matching, impedance, 756
matter, wavelike properties of, 1152–1173
Maxwell, James Clerk, 496, 557, 913, 1115, 1232
Maxwell-Boltzmann distribution, 495–497, 1194
Maxwell-Boltzmann heat distribution, 504
Maxwell's demon, 557
Maxwell's equations, 917–919, 924
   in vacuum, 919
Maxwell's generalization of Ampère's law, 915, 918

mean, 25
mean acceleration, 32, 115
mean free path, 503, 740
mean free time, 503, 739
mean life, 1182
mean power, 182
mean radius, 366
mean solar day, 4
mean speed, 497
mean value, 187
mean value theorem of calculus, 187
mean vector velocity, 115
mean velocity, 25–27, 114
measurement, 2–6
   MKS system and, 58
mechanical advantage, 193
mechanical energy, 191, 200
   conservation of, 199–203
   linear momentum and, 250–260
mechanical equivalent of heat, 461–467
mechanical waves, 568
   classical, 17
   relativistic, 1070–1087
mediating particle, 1231
Meissner, Karl Wilhelm, 1215
Meissner effect, 1215
Meitner, Lise, 1195
Mendeleyev, Dmitri I., 1134
merit, figure of, 752
meson, 1228, 1230
metals, band structure of, 1206–1207
meteoroids, 387
meter, 4–5
metric system, 3
Michell, John, 375, 765
Michelson, Albert Abraham, 974, 1005, 1065
Michelson interferometer, 974–975
Michelson-Morley experiment, 1045, 1065
microscope
   Bohr's ideal, 1161
   compound, 1029–1030
   electron, 1157
   simple, 1029
microstate, 552
Miller, D. C., 504
Millikan, Robert A., 653, 721
Millikan's oil-drop experiment, 653, 721
minimum deviation, 966
minimum-orbit earth satellite, 123–125
minor axis, 366
minority carrier, 1208, 1209
mirror
   concave, 1012–1016
   convex, 1016–1018
   plane, 1009–1011
   sign convention for, 1017
   spherical, 1012–1020
mirror equation, 1014
mirror inversion, 1011
MKS system, 58
mobility
   charge carrier, 751
   Hall, 790
mode, 591
model, 478
moderator, 262, 1198

modulates, 589
molar heat capacity, 486–488, 490, 491
   at constant pressure, 517
   at constant volume, 487
   of gases, 518
molecular beam, 504
molecular mass, 484
moment arm, 277, 774
moment of inertia, 279
   calculation of, 282–286
   of an extended body, 281–282
   radius of gyration and, 309
   of torsion pendulum, 307
momentum. *See also* linear momentum
   conservation of, 225, 1108, 1181
   energy and, 250–257
   orbital angular, 1130
   relativistic, 1073
momentum-conservation principle, 223, 226, 234, 239, 1072
momentum vector diagram, 1108
monochromatic light, 954
monolayer, 503
Moor-Hall, Chester, 1039
Morley, Edward Williams, 1045, 1065
Moseley, H. G. J., 1136
Moseley plot, 1136
Moseley's law, 1136, 1137
MOSFET, 1211, 1212, 1213
most probable speed, 497
motion
   curvilinear, 114
   oscillatory, 148–151
   projectile, 116–121
   simple harmonic, 140–148, 152–156
   uniform circular, 121–130, 142–145
motional emf, 826, 827–829
motor
   electric, 835–839
   series-wound, 849
multiple-slit interference, 982–984
   calculation of, 986–987
multiple-slit interference maxima, 983
multiplication ratio, 1085
muon, 1054, 1147, 1227
muon neutrino, 1227
muonium, 1147
mural quadrant, 366
mutual resistance, 853–854
Muybridge, Eadweard, 72
myopia, 1028

$n$-type semiconductor, 1208
natural angular frequency, 876
natural frequency, 876
near point, 1027
nearsightedness, 1028
Neddermeyer, Seth, 1235
negative charge, 636
negative energy state, 1223
negative ion, 637
negative lens, 1024
negative terminal, 730
negative vector, 86
neodymium laser, 1144
Nernst, Walther, 551
Nernst's law, 551
net work, 175–176